Lecture Notes in Mathematics Vol. 1528

ISBN 978-3-540-56251-1 © Springer-Verlag Berlin Heidelberg 2008

George Isac

Complementarity Problems

Notations: l_+ means $l\downarrow$; l_- means $l\uparrow$; l = line

Errata

1. pg. 1, l_+ 18: "the all principal" must be" "all the principal"
2. pg. 1, l_+20: "be agree", must be "to be in concordance"
3. pg. 3, l_+15: "respective studied", must be "studied respectively"
4. pg. 3, l_+ 18: "sutdy', must be "study"
5. pg. 3, l_+ 20: "researchers", must be "research"
6. pg. 7, l_- 4: "$f_2 2$", must be "f_2".
7. pg. 8, l_+ 9: "$\langle x, x\rangle > 0$", must be "$\langle x, x^*\rangle > 0$"
8. pg. 29, l_-8: "R_+", must be "R_+^n"
9. pg. 46, l_+8: "of Interregional", must be "or Interregional"
10. pg. 62, In Proposition 3.4, we must have "hemicontinuous mapping and if f is positive..."
11. pg. 69, l_+15: ";quasi-Newton", must be "quasi-Newton"
12. pg. 71, l_+1: we must have $\{x \in K \,|T(x) \in K^* \setminus \{0\}\}$
13. pg. 73, l_-14: "$\geq$", must be "$\leq$"
14. pg. 91, l_+16: "in case", must be "in this case"
15. pg. 100 :In Proposition 4.2.1, "are" must be "is"
16. pg. 107, l_-6: "transpose", must be "transposed"
17. pg. 110, l_-6: "$(1+\alpha)\, d(x, K_n) \leq (1+\alpha)\, d(x, K_n)$", must be "$(1+\alpha)\, d(x, K_m) \leq (1+\alpha)\, d(x, K_n)$"
18. pg. 115: In the case(i), we must have "$\langle x, u\rangle < r$"
19. pg. 115, l_+14: "$\langle x_r, u\rangle = 0$"
20. pg. 116: In Definition 4.3.1, we must have "$conv\,(\{x_1, x_2, ..., x_n\}) \subseteq .....$"
21. pg. 136, l_+9: we must have "is φ-asymptotically bounded with $\limsup_{r\to+\infty} \varphi(r) < +\infty$"
22. pg. 157, l_-6: we must have "for T in"
23. pg. 169, l_-10: "$f, g : K \to E^*$", must be replaced by "$f : K \to E^*$ and $g : K \to E$"
24. pg. 174: In Theorem 6.2.3,"$f(x), T_2(x)\rangle < 0$, must be "$\langle f(x), T_2(x)\rangle < 0$"

25. pg. 178: In Theorem 6.2.5, in condition 3^0), "$c_1 S(x), T_1(x)\rangle < 0$, must be $c_1 \langle S(x), T_1(x)\rangle < 0$"
26. pg. 183, l $_{-}3$: $G(x_0), x_0 - f(x_0)\rangle$" must be $\langle G(x_0), G(x_0) - f(x_0)\rangle$
27. pg. 215: In Lemma 7.4.1, "P_α" must be "p_α".
28. pg. 229, l_+4: "$x*\ D$" must be $x* \in D$"
29. pg. 237, l_+9: "$M_{n'n}(\,)b$", must be "$M_{n'n}(R)$ be"
30. pg. 251: In Theorem 8.6.3, in condition 2^0), " ss(u_0)", must be "$S(u_0)$"
31. pg. 252: In formula (8), $S(x_* \in \mathrm{K}$" must be $S(x_*) \in \mathrm{K}^*$
32. pg. 252, l_-7: "$y_n \leq Y_m$", must be "$Y_n \subseteq Y_m$"

Lecture Notes in Mathematics 1528

George Isac

Complementarity Problems

Springer-Verlag
Berlin Heidelberg New York
London Paris Tokyo
Hong Kong Barcelona
Budapest

Autor

George Isac
Département de Mathématiques
Collège Militaire Royal
St. Jean
Québec, Canada J0J 1R0

Mathematics Subject Classification (1991): 49A99, 58E35, 52A40

ISBN 3-540-56251-6 Springer-Verlag Berlin Heidelberg New York
ISBN 0-387-56251-6 Springer-Verlag New York Berlin Heidelberg

Printed in Germany

Typesetting: Camera ready by author
46/3140-543210 - Printed on acid-free paper

TABLE OF CONTENTS

INTRODUCTION

In 1984 we were invited, by the Department of Mathematics of University of Limoges, to give several lectures on a subject considered interesting in Nonlinear Analysis and Optimization.

So, we decided to present the subject "Complementarity Problems (In Infinite Dimensional Spaces)".

After this course, we became quickly conscious that a volume on all mathematical aspects of these nice problems is necessary.

The literature on this subject is already impressive and the task to write this volume was not easy.

The Complementarity Problem is considered by many mathematicians, as a large independent division of Mathematical Programming Theory, but our opinion is quite different.

The Complementarity Problem represents a very deep, very interesting and very difficult mathematical problem. This problem is a very nice research domain because it has many interesting applications and deep connections with important chapters of the Nonlinear Analysis.

Our principal aim is to present the all principal mathematical aspects about the Complementarity Problems.

To be agree with this aim we consider generally, Nonlinear Complementarity Problems in infinite dimensional spaces. But, the finite dimensional case is not neglected and several important results about the linear or the Nonlinear Complementarity Problems specific for this case are also presented.

Several problems arising in various fields (for example: Economics, Game Theory, Mathematical Programming, Mechanics, Elasticity Theory, Engineering, and, generaly, several "Equilibrium Problems") can be stated in the following unified form:

given $f: R^n_+ \to R^n$ a mapping,

$$(1): \quad \left\| \begin{array}{l} \text{find } x_o \in R^n_+ \text{ such that,} \\ f(x_o) \in R^n_+ \text{ and } \langle x_o, f(x_o)\rangle = 0, \end{array} \right.$$

(where $\langle\ ,\ \rangle$ is the inner product, $\langle x,y\rangle = \sum_{i=1}^{n} x_i y_i$).

Problem (1) is called the <u>Complementarity Problem</u> and the origin of this problem is perhaps in the <u>Kuhn-Tucker Theorem</u> for nonlinear programming (which gives the necessary conditions of optimality when certain conditions of differentiability are met), or perhaps in the old and neglected Du Val's paper. [P. Du Val, The Unloading Problem for Plane Curves, Amer. J. Math., 62 (1940), 307-311.]

Certainly, one thing is clear: in 1961 Dorn showed that, if A is a positive-definite (but not necessarily symmetric) matrix, then the minimum value of the quadratic programming problem,

$$(2): \quad \left\| \begin{array}{l} \min \langle x, Ax + q \rangle \\ x \in \mathcal{D} \\ \quad \mathcal{D} = \{x \in R^n : 0 \le x,\ 0 \le Ax + q\} \\ q \in R^n \end{array} \right.$$

is zero. [W.S. Dorn, Self-dual Quadratic Programs, SIAM J. Appl. Math., Vol.9, Nr.1, (1961), 51-54.]

Dorn's paper was the first step in treating the Complementarity Problem as an independent problem.

In 1963 Dantzig and Cottle generalized Dorn's result to the case when all the principal minors of the matrix A are positive. [G.B. Dantzig and R.W. Cottle, Positive semi-definite Programming, (Nonlinear Programming. A course. J. Abadie (ed.). North-Holland, Amsterdam, (1967), 55-73)].

The result annouced in 1963 by Dantzig and Cottle was generalized in 1964 and 1966 by Cottle to a certain class of nonlinear functions. [R.W. Cottle, a) Notes on a Fundamental Theorem in Quadratic Programming, SIAM J. of Appl. Math., Vol. 12, (1964), 663-665; b) Nonlinear Programs with Positively Bounded Jacobians, SIAM J. of Appl. Math. Vol. 14, (1966), 147-158].

Also, in 1965 Lemke proposed the Complementarity Problem as a method for solving matrix games. [C.E. Lemke, Bimatrix Equilibrium Points and Mathematical Programming, Manag. Sci., Vol. 11, Nr 7, (1965), 681-689].

Certainly, one of the first important papers on the Complementarity Problem is the Ingleton's paper, [A.W. Ingleton, A Problem in Linear Inequalities, Proc. London Math. Soc., Vol. 16, (1966), 519-536], which showed the importance of the Complementarity Problem in engineering applications.

It seems that the term "Complementarity" was proposed by Cottle, Habetler and Lemke, and the reason is the following observation.

A solution x^o of problem (1) is said to be nondegenerate if at most n compo-ponents of 2n-components vector $(x^o, f(x^o))$ equal zero.

Otherwise, it is a degenerate solution.

We denote, $N_n = \{1, 2, \dots, n\}$. If $x^o = (x_i^o)_{i=1, 2, \dots, n}$ is a nondegenerate solution of problem (1) and $y^o = f(x^o)$, then the sets,

$$A = \{i \mid x_i^o > 0\};\ B = \{i \mid y_i^o > 0\},\ \text{where } y^o = (y_i^o)_{i=1, 2, \dots, n}$$

are complementary subsets of N (that is, $A = C_{N_n} B$).

After 1970 the theory of the Complementarity Problem has known a strong and ascending development, based on several important results obtained by Cottle, Eaves, Karamardian, Mangasarian, Saigal, Gould, Garcia, Moré, Kojima, Megiddo, Kaneko and Kostreva, Pang, etc.

In Chapter 1 we present some preliminary definitions and the definitions of principal complementarity problems.

In Chapter 2 we give examples of practical problems which have as mathematical model a specific complementarity problem. Other models, especially in infinite dimensional spaces are presented in other chapters.

The important mathematical problems equivalent to complementarity problems are studied in Chapter 3.

In Chapter 4 we present the principal existence theorems and we study some properties of solution set.

The order Complementarity Problem and the Implicit Complementarity Problem are respective studied in Chapter 5 and 6.

In Chapter 7 we introduce the notion of isotone projection cone and we use this notion to study the Complementarity Problem.

The last chapter is devoted to the sutdy of several problems, about the Complementarity Problem not considered in other chapters and which are opened to new researchers.

The last time, two books on the Complementarity Problem were published: 1°) K.G. Murty: Linear complementarity, linear and nonlinear programming. Heldermann Verlag, Berlin (1988). 2°) R.W. Cottle, J.S. Pang and R.E. Stone: The linear complementarity problem. Academic Press (1992), but our volume is completely different and essentially it is a complementary book.

We hope that our notes form a satisfactory introduction to the study of the Complementarity Problem, which is a fascinating problem by its simplicity and profoundness, and by the fact that it is a cross-point of several chapters of fundamental and applied mathematics.

The part on numerical methods solving complementarity problems is not considered in this volume, since this part can be considered as a subject for another volume.

Many numerical methods for the Linear Complementarity Problem are studied in the cited books.

To select the subjects considered in this volume the author used the openness to new developments and his personal preferences, since the principal motor in mathematics is the pleasure to do mathematics.

G. Isac
St-Jean, Québec
1992

CHAPTER 1

PRELIMINARIES AND DEFINTIONS OF PRINCIPAL COMPLEMENTARITY PROBLEMS

1.1 Notations, definitions and necessary results. In this section we specify some terms and notations used systematically in this paper.

First, we suppose known the definitions and the fundamental properties of Hilbert, Banach and locally convex spaces [B8], [B9], [B16].

We denote by $(H, < , >)$ a Hilbert space, by $(E, \| \|)$ a Banach space and by $E(\tau)$ a locally convex space. In this paper we consider only real vector spaces and it is clear that every Banach or Hilbert space is a locally convex space.

If $E(\tau)$ is a locally convex space, then E^* denotes the topological dual of E.

We say that $<E, F>$ is a dual system if, E and F are vector spaces and $< , >$ is a bilinear functional on $E \times F$ such that,

1°) $<x, y> = 0$, for each $x \in E \Longrightarrow y = 0$,

2°) $<x, y> = 0$, for each $y \in F \Longrightarrow x = 0$.

If $E(\tau)$ is a locally convex space, we denote by $< E, E^* >$ the dual system defined by the bilinear functional, $<x, u> = u(x)$; for every $x \in E$ and every $u \in E^*$.

Let E be a real vector space. A subset $K \subset E$ is said to be a convex cone if the following conditions are satisfied:

$c_1 1)$ $K + K \subset K$

$c_2 2)$ $(\forall \lambda \in R_+)(\lambda K \subset K)$.

If $K \subset E$ is a convex cone, then we can define a preorder on E by:

"$x \leq y$" $\Longleftrightarrow$ $y - x \in K$.

Always the preorder defined on E by K, will be denoted by "$\leq$ ".

We can prove that, if $K \subset E$ is a pointed convex cone, that is, K is a convex cone and satisfies in addition,

$c_3)$ $K \cap (-K) = \{0\}$,

then the preorder "$\leq$" is an order, that is, it is a reflexive, transitive and antisymmetric relation.

Also, it is important to remark that, if $K \subset E$ is a pointed convex cone, then the order "$\leq$" is compatible with the linear structure, that is, the following two conditions are satisfied:

$0_1)$ $x \leq y \Longrightarrow x + z \leq y + z$; $\forall\, x, y, z \in E$,

$0_2)$ $x \leq y \Longrightarrow ax \leq ay$; $\forall\, a \in R_+$, $\forall\, x, y \in E$.

If for the vector space E is defined a pointed convex cone $K \subseteq E$, then we say that (E, K) is an ordered vector space.

Conversely, if on the vector space E is defined an order "$\leq$" satisfying 0_1) and 0_2) then the set, $K = \{x \in E | x \geq 0\}$ is a pointed convex cone.

An ordered vector space (E, K) is said to be a vector lattice if in addition, every non-empty finite subset of E has greatest lower bound.

Hence, if (E, K) is a vector lattice, then in particular, there exists sup $(x, y) \in E$ for every $x, y \in E$. In this case there exists also, inf $(x, y) \in E$ and we have, $\inf(x, y) = -\sup(-x, -y)$.

Obviously, if we consider the n-dimensional real vector space,

$R^n = \{x | x = (x_1, x_2, \ldots, x_n), x_i \in R; \forall i = 1, 2, \ldots, n\}$ then

$R^n_+ = \{(x_1, x_2, \ldots, x_n) | x_i \geq 0; \forall i = 1, 2, \ldots, n\}$ is a pointed convex cone and

" $\leq$ " is exactly the usual order considered on R^n and (R^n, R^n_+) is a vector lattice.

We observe that, R^n is a Hilbert space with respect to the inner product, $\langle x,y\rangle = \sum_{i=1}^{n} x_i y_i$, where, $x = (x_i)_{i=1,2,\ldots,n}$ and $y = (y_i)_{i=1,2,\ldots,n}$

Several interesting examples and properties of ordered vector spaces we find in [B12], [B16].

If $\langle E,F\rangle$ is a dual system and $K \subset E$ is a convex cone, then we denote,

$K^* = \{u \in F | \langle x,u\rangle \geq 0 ; \forall x \in K\}$,

$K^\circ = \{u \in F | \langle x,u\rangle \leq 0 ; \forall x \in K\}$

and we observe that K^* (resp. K°) is a convex cone.

The cone K^* (resp. K°) is called the dual (resp. the polar) cone of K. If K is closed then $K = K^{**} = (K^*)^*$.

Consider a dual system $\langle E,E^*\rangle$, where $E(\tau)$ is a locally convex space and $f:E \to R$ a functional.

The functional f is said to have a Gâteaux derivative at $x_0 \in E$ if there exists $u(x_0) \in E^*$ such that,

$$(G) : \lim_{\lambda \to 0} [f(x_0 + \lambda x) - f(x_o)]/\lambda = \langle x, u(x_o)\rangle ; \forall x \in E.$$

We denote, $\nabla f(x_0) = u(x_0)$ and $\nabla f(x_0)$ is called the Gâteaux derivative, or the gradient of f at x_0.

If for every $x_0 \in E$ holds (G), the functional f is said to be differentiable in the Gâteaux sens in E and the operator $\partial_G:E \to E^*$, which with every x_0 associates $\partial_G(x_0) = \nabla f(x_0)$ is said to be the Gâteaux differential of f in E.

Let $f:E \to R\cup\{+\infty\}$ be a convex functional and $x_0 \in E$.

The functional f is said to be <u>subdifferentiable</u> at x_0 if the set,

$$\partial f(x_0) = \{u \in E^* | f(x)-f(x_o) \geq \langle x-x_o, u \rangle ;\ \forall x \in E\}$$

is <u>non-empty</u>.

The set $\partial f(x_0)$ is said to be the <u>subgradient of f at x_0</u>.

If for every $x_0 \in E$, $\partial f(x_0) \neq \phi$, we say that f is <u>subdifferentiable in E</u> and the application, $\partial f: E \to 2^{E^*}$, which with every $x_0 \in E$ associates $\partial f(x_0)$ 2^{E^*} is called the <u>subdifferential</u> of f.

We remark that, if f is Gâteaux differentiable at x_0 then $\partial f(x_0) = \{\nabla f(x_0)\}$.

Let $K \subset E$ be a convex set. The indicator of K is the function defined by:

$$\Psi_K(x) = \begin{cases} 0 & \text{if } x \in K \\ +\infty & \text{if } x \notin K \end{cases}$$

If $K \subset E$ is a convex cone, then we have,

$$\partial \Psi_K(x_0) = \{x^* \in E^* | x^* \in K^\circ \text{ and } \langle x_o, x^* \rangle = 0\}$$

Consider again the dual system $\langle E, E^* \rangle$ and $f: E \to 2^{E^*}$ a mapping. The <u>effective domain</u> of f is, $D(f) = \{x \in E | f(x) \neq 0\}$ and its <u>graph</u> is, $Gr(f) = \{(x, f(x)) | x \in D(f)\}$.

The mapping f is said to be <u>monotone</u> if,

$$(\forall x, y\ D(f))(\forall x^* \in f(x))(\forall y^* \in f(y))(\langle x-y, x^*-y^* \rangle \geq 0).$$

and f is said to be <u>maximal monotone</u> if it is monotone and there does not exist $f: E \to 2^{E^*}$ such that f is <u>monotone</u> and $Gr(f) \underset{\neq}{\subset} Gr(f)$.

The mapping f is said to be <u>α-monotone</u> if there exists a strictly increasing function $\alpha: [0, +\infty[\longrightarrow [0, +\infty[$ such that,

α_1) $\alpha(0) = 0$,

α_2) $\lim_{t \to \infty} \alpha(t) = +\infty$,

α_3) $(\forall x, y \in D(f))(\forall x^* \in f(x))(\forall y^* \in f(y))(\langle x-y, x^*-y^* \rangle \geq \|x-y\| \alpha(\|x-y\|))$.

Also, the mapping $f: E \to 2^{E^*}$ is said to be <u>strictly monotone</u> if:

m_1) f is monotone,

m_2) $\langle x - y, x^* - y^* \rangle > 0$, if $x \neq y$, $x^* \in f(x)$ and $y^* \in f(y)$.

If f is α-monotone, where $\alpha(t) = \rho t^2$, $\rho \in R_+ \setminus \{0\}$ then we say that f is <u>strongly monotone</u>.

A mapping $f: C \to E^*$, where C is a convex subset of E is said to be <u>hemicontinuous</u> if it is continuous from the line segment of C to the weak topology of E^*.

We recall that an operator, $f: E \to F$, where $(E, \| \|)$, $(F, \| \|)$ are two Banach spaces, is <u>k-Lipschitz</u> if there exists a constant $k > 0$ such that,

$\|f(x) - f(y)\| \leq k\|x - y\|; \forall x, y \in E.$

If $0 < k < 1$ then f is said to be a <u>contraction</u>.

Finally, we denote by $M_{nxn}(R)$ the space of n x n real matrices. If for every $n \in N$ we denote, $P_n = \{1, 2, \ldots, n\}$ and $P = \{P_n\}_{n\in N}$ then a matrix $A \in M_{nxn}(R)$ is a function $A:P_n \times P_n \to R$.

We denote, $a_{ik} = A(i, k)$.

If $P \subset P_n$ and $A \in M_{nxm}(R)$, we denote by $A(P) = A_{|PxP}$ (the restriction of A to PxP).

The real number, det A(P) is called the <u>principal minor</u> of order P of A.

The following fundamental results are necessary in the development of this paper.

Let $<E, E^*>$ be a dual system where $E(\tau)$ is a locally convex space.

We say that, an element $x^* \in E^*$ is <u>normal</u> to a convex set $K \subset E$ at a point x if:

n_1) $x \in K$,

n_2) $(\forall y \in K)(<y - x, x^*> \leq 0)$

For each $x \in E$, the set of all x^* normal to K at x is called the <u>normal</u> cone to K at x.

The normal cone to K at x is weak* closed convex cone in E^*, it is empty when $x \notin K$ and it contains at least the zero element of E^* when $x \in K$.

The multivalued mapping from E to E^* which assigns to each $x \in E$ the normal cone to K at x is called the <u>normality</u> operator for K.

The normality operator for K is actually the subdifferential of the indicatrix Ψ_K of K, so that it is a maximal monotone operator with effective domain K(if K is non-empty).

The following result was proved by Rockafellar and it is a theorem of Browder-Stampacchia type.

<u>Theorem</u> [Browder-Stampacchia-Rockafellar]. [B.14]

<u>Let $(E,\| \|)$ be a Banach space and let $K \subset E$ be a non-empty closed convex subset. Let $f_1:E \to E^*$ be the normality operator for K and let $f_2 2:E \to E^*$ be a monotone operator (not multivalued and not necessarily maximal) such that, $D(f_2) \supset K$.</u>

<u>If f_2 is hemicontinuous then $f_1 + f_2$ is a maximal monotone operator.</u>

A mapping $f:E \to E^*$ is said to be <u>bounded</u>, if for every bounded set $B \subset E$, f(B) is bounded.

Theorem [MOSCO] [B10]

Let $(E, \| \ \|)$ be a reflexive Banach space and let $K \subset E$ be a closed convex cone. Suppose that $f:K \to E^*$ is a bounded, hemicontinuous strictly monotone operator and consider $\{K_r\}$ a family of non-empty closed convex subsets of K.

Then, for every r there exists a unique element $x_r \in K_r$ such that, $\langle z-x_r, f(x_r) \rangle \geq 0$; $\forall z \in K_r$.

Theorem [Rockafellar] [B14]

Let $(E, \| \ \|)$ be a reflexive Banach space and let $f:E \to E^*$ be a maximal monotone operator.

If there exists a real number $\beta > 0$ such that, $(\forall x \in D(f))(\|x\| > \beta)(\forall x^* \in f(x))(\langle x, x^* \rangle > 0)$ then there exists an element $x_o \in E$ such that, $0 \in f(x_o)$.

1.2 Complementarity problems. (Definitions and problems). Dorn [A75] considered in 1961 the following optimization problem,

$$(1.2.1): \quad \left| \begin{array}{l} \min\limits_{x \in F} f(x) \\ \text{where: } F = \{x \in R^n \mid x \geq 0 \ , \ Ax + b \geq 0\} \\ \qquad A \in M_{nxn}(R), b \in R^n \text{ and } f(x) = \langle x, Ax + b \rangle \end{array} \right.$$

and he showed that, when A is a positive definite (thought not necessarily symmetric) matrix, then the quadratic program (1.2.1) must have an optimal solution and $\min\limits_{x \in F} f(x) = 0$.

Dorn's paper was the first step in treating the complementarity problems.

In 1964 Cottle [A42] studied problem (1.2.1) under the assumption that A is a positive semidefinite matrix and he remarked that, in this case it is not true that (1.2.1) must possess an optimal solution.

However, if A is positive semidefinite and $F \neq \emptyset$ then an optimal solution for (1.2.1) exists and again, $\min\limits_{x \in F} f(x) = 0$.

After three years, Dantzig and Cottle [A71] constructively showed that, if A is a square (not necessarily symmetric) matrix, all of whose principal minors are positive, then problem (1.2.1) has an optimal solution x_* satisfying the following equation,

$$(1.2.2): \quad \langle x_*, Ax_* + b \rangle = 0.$$

This result was later generalized by Cottle [A43].

More precisely, Cottle considered the following nonlinear program associated to a continuously differentiable mapping $h:R^n \to R^n$,

(1.2.3):
$$\begin{cases} \min f(x) \\ x \in F \\ \text{where: } f(x) = \langle x, h(x)\rangle \text{ and} \\ F = \{x \in R^n \mid x \geq 0,\ h(x) \geq 0\} \end{cases}$$

and he showed that, if x_o is an optimal solution of program (1.2.3) and the Jacobian matrix $J_h(x_o)$ has positive principal minors, then x_o satisfies the following conditions:

(1.2.4):
$$\begin{cases} x_o \geq 0;\ h(x_o) \geq 0 \quad \text{and} \\ \langle x_o, h(x_o)\rangle = 0 \end{cases}$$

Thus, in relation with program (1.2.3) Cottle obtained in 1966 the first nice result on the Nonlinear Complementarity Problem. This result is the following.

Consider a differentiable mapping $h: R^n \to R^n$.

We say that h has a positively bounded Jacobian matrix $J_h(x)$, if there exists a real number $0 < \delta < 1$ such that for every $x \in R^n$, each principal minor of $J_h(x)$ is an element of the interval $[\delta, \delta^{-1}]$.

A solution (y, x) of the equation $y - h(x) = 0$ is nondegenerate if at most n of its 2n components are zero.

Theorem 1.2.1 [Cottle] [A43]

If $h: R^n \to R^n$ is a continuous differentiable mapping such that the solutions of $y - h(x) = 0$ are non degenerate and if h has a positively bounded Jacobian matrix $J_h(x)$, then there exists an element $x_o \in R^n_+$ such that, $h(x_o) \geq 0$ and $\langle x_o, h(x_o)\rangle = 0$.

We note that another important result which contributed to the development of the Complementarity Theory is the Lemke's paper [A 173]. In this paper Lemke proposed the Complementarity Theory as method for solving matrix games.

Certainly, the development of the Complementarity Theory was imposed by a large variety of applications in fields as: Optimization, Economics, Games Theory, Mechanics, Variational Calculus, Stochastic Optimal Control Theory etc.

The Complementarity Theory is closely linked with two other problems, the solution of variational inequalities and the determination of the fixed points for a given mapping.

Thus, the existence theorems and the methods used in the study of the last two problems are widely used in the Complementarity Theory and conversely, the ideas and methods developed specially for complementarity problems are used to solve variational inequalities or to solve fixed point problems.

In this section we present the principal complementarity problems studied till now.

Some of these problems has been much studied till now, but other ones are very little known.

Concerning the Complementarity Problem we distinguish two entirely distinct class of problems: the Topological Complementarity Problem (T.C.P.) and the Order Complementarity Problem (O.C.P.).

A. Topological Complementarity Problems.

In this class we have the following problems.

A.1-The generalized complementarity problem.

Let $<E, F>$ be a dual system of locally convex spaces.

For a given closed convex cone $K \subset E$ and a mapping $f:K \to F$, the Generalized Complementarity Problem (associated to K and f) is,

$$\text{(G.C.P.)}: \left\| \begin{array}{l} \text{find } x_0 \in K \text{ such that,} \\ f(x_0) \in K^* \text{ and } <x_o, f(x_o)> = 0. \end{array} \right.$$

Remarks

(1.2.1) If $f(x) = L(x) + b$, where $L:E \to F$ is a linear mapping and b an element of F then we have the Linear Complementarity Problem (L.C.P.).

(1.2.2) If $f:K \to F$ is a nonlinear mapping, then we have the Nonlinear Complementarity Problem (N.C.P.).

(1.2.3) If $E = F = R^n$, $K = R^n_+$, $<x, y> = \sum_{i=1}^n x_i y_i$, where $x = (x_i)$, $y = (y_i) \in R^n$, $A \in M_{nxm}(R)$ and $b \in R^n$, then we obtain the classical Linear Complementarity Problem,

$$\text{(L.C.P.)}: \left\| \begin{array}{l} \text{find } x_0 \geq 0 \text{ such that} \\ Ax_o + b \geq 0 \text{ and } <x_o, Ax_o + b> = 0 \end{array} \right.$$

(In this case, $K = K^*$).

(1.2.4) The following Special Linear Complementarity Problem is used in the study of some structural engineering problems [A185], [A152].

We consider, $E = F = R^n$, $K = R^n_+$, $<x, y> = \sum_{i=1}^n x_i y_i$, (where $x = (x_i)$, $y = (y_i)$), $M, N, P \in M_{nxm}(R)$, $q, r \in R^n$ and supposing that $f(x, v) = q + Mv + Nx$ and $y(v)=r-Pv$, we are interested to study the following complementarity problem:

$$\text{(S.L.C.P.)}: \left\| \begin{array}{l} \text{find } x_0, v_o \in R^n_+ \text{ such that,} \\ f(x_0, v_o) \in R^n_+, \; y(v_o) \in R^n_+, \\ <v_o, f(x_o, v_o)> = 0 \text{ and } <x_o, y(v_o)> = 0 \end{array} \right.$$

We can show that problem (S.L.C.P.) is equivalent to an ordinary linear complementarity problem.

Indeed, if we set,

$$Z = \begin{bmatrix} v \\ x \end{bmatrix}; \; q_o = \begin{bmatrix} q \\ r \end{bmatrix}; \; L = \begin{bmatrix} M & N \\ -P & 0 \end{bmatrix}; \; F(Z) = LZ + q_o$$

and if we consider, $E = F = R^{2n}$, $K = R^{2n}_+$, $X, Y = \sum_{i=1}^{2n} X_i Y_i$; $X = (X_i)$; $Y = (Y_i)$, then problem (S.L.C.P.) is equivalent to the following linear complementarity problem:

$$(1.2.5): \quad \left| \begin{array}{l} \text{find } Z_o \in R_+^{2n} \text{ such that,} \\ \\ \text{find } F(Z_o) \in R_+^{2n} \text{ and } \langle Z_o, F(Z_o) \rangle = 0 \end{array} \right.$$

A.2-<u>The Generalized Multivalued Complementarity Problem</u>.

This problem is important in the study of some practical problems, as for example some problems defined in Economics.

If $\langle E, F \rangle$ is a dual system of locally convex spaces, $K \subset E$ a closed convex cone and $f: K \to F$ a multivalued mapping (that is $f: K \to 2^F$), then <u>the Generalized Multivalued Complementarity Problem</u>, associated to f and K is:

$$(\text{G.M.C.P.}): \left| \begin{array}{l} \text{find } x_o \in K \text{ and } y_o \in F \text{ such that,} \\ y_o \in f(x_o) \cap K^* \text{ and } \langle x_o, y_o \rangle = 0. \end{array} \right.$$

A.3-<u>Parametric Complementarity Problems</u>.

Supposing defined a dual system of locally convex spaces $\langle E, F \rangle$, a closed convex cone $K \subset E$, a topological space T and a mapping $f: K \times T \to F$, the <u>Generalized Parametric Complementarity Problem</u> is:

$$(\text{G.P.C.P.}): \left| \begin{array}{l} \text{find the point-to-set mapping} \\ X_o: T \to E, \text{ such that, for every } X_o(t) \neq \phi, \\ \text{if } x_o(t) \in X_o(t) \text{ then, } x_o(t) \in K, \ f(x_o(t), t) \in K^* \\ \text{and } \langle x_o(t), f(x_o(t), t) \rangle = 0 \end{array} \right.$$

<u>Remarks</u>

<u>(1.2.6)</u> If $E = F = R^n$, $T = R_+$, $\langle x, y \rangle = \sum_{i=1}^{n} x_i y_i$; where $x = (x_i)$, $y = (y_i)$, $K = R_+^n$, $A \in \mathcal{M}_{nxm}(R)$, $q, p \in R^n$ and $f(x, t) = Ax + q + tp$, then we obtain the Parametric Complementarity Problem defined in the <u>Elastoplastic Structures Theory</u> [A185], [A186].

<u>(1.2.7)</u> The Parametric Complementarity Problem has close relations with the Sensitivity Theory of nonlinear programming problems and in these cases is interesting to study the existence of continuously or differentiable selections of the solution mapping X_o.

In finite dimensional spaces we distinguish another interesting parametric complementarity problem. This problem was considered by Meister [A212].

Let $\mathcal{D} \subset R^n$ be a set of the form, $\mathcal{D} = R_+^k \times R_+^k \times Q$, where $\mathcal{D}$ takes the form $R_+^k \times R_+^k$ for $m = 2k$ and $\mathcal{D}$ is an arbitrary interval in R^m for $k = 0$.

We remark that $\mathcal{D}$ is supposed to be with non-empty interior.

Given a mapping $f: \mathcal{D} \to R^n$ and supposing, $<x, y> = \sum_{i=1}^{n} x_i y_i$, we consider the <u>Special Parametric Complementarity Problem</u>,

$$(S.P.C.P.): \left\| \begin{array}{l} \text{find } (x_0, y_0, z_o) \in \mathcal{D} \text{ such that,} \\ f(x_0, y_0, z_o) = 0 \text{ and } <x_o, y_o> = 0 \end{array} \right.$$

Certainly, we can define this problem in infinite dimensional spaces but its study is more complicated.

A.4-<u>Implicit Complementarity Problems</u>.

The origin of the <u>Implicit Complementarity Problems</u> is the dynamic programming approach of stochastic impulse and of continuous optimal control.

It is not without interest to know that there exist deep and interesting relations between the Implicit Complementarity Problems and the Quasivariational Inequalities Theory [A18], [A19], [A20], [A21], [A22], [A37], [A222].

We consider a locally convex space $E(\tau)$, a closed convex cone $K \subset E$, an element $b \in E$ and two mappings $A, M: E \to E$.

Given a bilinear functional $< , >$ on $E \times E$, the Implicit Complementarity Problem is,

$$(I.C.P.): \left\| \begin{array}{l} \text{find } x_0 \in E \text{ such that,} \\ M(x_0) - x_0 \in K,\ b - A(x_0 \in K \text{ and} \\ <A(x_o) - b,\ x_o - M(x_o > = 0. \end{array} \right.$$

If $<E, F>$ is a dual system of locally convex spaces and $K \subset E$ is a closed convex cone, then given $M: E \to E$ and $A: E \to E$ two mappings and $b \in F$ an arbitrary element, the <u>Generalized Implicit Complementarity Problem</u> is,

$$(G.I.C.P.): \left\| \begin{array}{l} \text{find } x_0 \in E \text{ such that,} \\ M(x_0) - x_0 \in K,\ b - A(x_0) \in K^* \\ \text{and } < A(x_0) - b,\ x_0 - M(x_o)> = 0 \end{array} \right.$$

More general, we can consider the following multivalued implicit complementarity problem.

Let $<E, F>$ be a dual system of locally convex spaces and consider:

$M: E \to E$, a point-to-point mapping,

$f: E \to F$, a point-to-set mapping and

$L: E \to E$, a cone-valued mapping, that is for every $x \in E$, $L(x) \subset E$ is a closed convex cone.

The <u>Multivalued Implicit Complementarity Problem</u> is,

$$(M.I.C.P.): \left\| \begin{array}{l} \text{find } x_0 \in M(x_0) + L(x_0) \text{ and } y \in F \\ \text{such that, } y \in f(x_0) \cap L(x_0)^* \text{ and} \\ <y,\ x_o - M(x_o)> = 0. \end{array} \right.$$

Remark.

(1.2.8) If $E = F = R^n$, $K = R^n_+$, $\langle x, y\rangle = \sum_{i=1}^{n} x_i y_i$, $L(x) = R^n_+$ for every $x \in R^n$ and f is an affine single-valued mapping, then (M.I.C.P.) reduces to the classical Implicit Complementarity Problem.

It is important to remark that a special consideration must give to system of implicit complementarity equations.

Precisely, let $\langle E, F\rangle$ be a dual system of locally convex spaces and let $K \subset E$ be a closed convex cone.

Given $f:E \to F$ and $g:E \to E$ two mappings and $y \in F$ an element, consider the following System of Implicit Complementarity Equations,

$$\text{(S.I.C.E.)}: \left\| \begin{array}{l} \text{find } u \in (g(u) - K) \cap K,\ v \in (f(u) - y + K^*) \cap K^* \\ \text{such that, } \langle u, v\rangle = 0 \text{ and} \\ \langle g(u) - u, v - f(u) + y\rangle = 0. \end{array} \right.$$

This problem is equivalent to a quasivariational inequality [A294].

A.5 ε-Complementarity Problems.

The origin of the definition of ε-complementarity problem is the McLinden's paper [A204].

Let $\langle E, F\rangle$ be a dual system of locally convex spaces and let $K \subset E$ be a closed convex cone.

Given a mapping $f:K \to F$ the ε-Complementarity Problem (associated to f and K) is:

$$(\varepsilon\text{-C.P.}): \left\| \begin{array}{l} \text{for a given } \varepsilon > 0 \text{ find } x_0 \in K \\ \text{such that, } f(x_0) \in K^* \text{ and } \langle x_0, f(x_0)\rangle \leqslant \varepsilon \end{array} \right.$$

This problem is used in the sensitivity analysis of optimization problems and it was studied by McLinden [A204] and Borwein [A31].

B. Order Complementarity Problems.

In this section we consider a vector lattice E and we denote, $K = \{x \in E \mid x \geq 0\}$ and the lattice operations sup (resp. inf) by $\vee$ (resp. $\wedge$).

Given a mapping $f:K \to E$ the Order Complementarity Problem is:

$$\text{(O.C.P.)}: \left\| \begin{array}{l} \text{find } x_0 \in K \text{ such that,} \\ x_0 \wedge f(x_0) = 0 \end{array} \right.$$

This problem was considered by Isac [A130], [A131], Borwein [A32] and Borwein and Dempster [A33].

Remarks

Since $x_0 \wedge f(x_0) = 0$ then it is clear that $f(x_0) \in K$.

(1.2.10) Problem (O.C.P.) has interesting applications in Economics.

In this case, $E = C(\Omega, R)$(the space of real continuous functions defined on a topological space Ω). On $C(\Omega, R)$ we consider the vector lattice structure defined by the pointwise order, that is by the convex cone, $K = \{x \in C(\Omega, | x(t) \geq 0; \forall t \in \Omega\}$.

We say that an ordered Hilbert space $(H, \langle , \rangle, K)$ is a Hilbert lattice if and only if:

h_1) H is a vector lattice,

h_2) $(\forall x, y \in H)(|x| \leq |y|) \Longrightarrow \|x\| \leq \|y\|$

(with, $|x| = x^+ + x^-$, $x^+ = 0 \vee x$ and $x^- = 0 \vee (-x)$.

Examples: If H is a Hilbert lattice then $K = K^*$

1°) If $H = R^n$, $K = R^n_+$ and the inner-product is $\langle x, y\rangle = \sum_{i=1}^{n} x_i y_i$; $x = (x_i)$, $y = (y_i)$ then $(H, \langle , \rangle, K)$ is a Hilbert lattice.

2°) If $H = L_2(\Omega, \mu)$, for an appropriate Radon measure and the order on $L_2(\Omega, \mu)$ is the pointwise order, then H is a Hilbert lattice. We have the following result.

Proposition 1.2.1

If $(H, \langle , \rangle, K)$ is a Hilbert lattice and $x, y \in K$, then $x \wedge y = 0$ if and only if $\langle x, y\rangle = 0$. //

By this proposition we obtain that, for Hilbert lattices, problems (O.C.P.) and (G.C.P.) are equivalent.

We note that there exists a multivalued analogous to problem (O.C.P.) [A101].

Indeed, consider a vector lattice E and set $K = \{x \in E | x \geq 0\}$.

For a point-to-set mapping $f:K \to E$ (that is, $f:K \to 2^E$) we have the following Multivalued Order Complementarity Problem,

$$\text{(M.O.C.P.)}: \left\| \begin{array}{l} \text{find } x_0 \in K \text{ and } y_0 \in E \\ \text{such that, } y_0 \in f(x_0) \text{ and } x_0 \wedge y_0 = 0 \end{array} \right.$$

If E is a vector lattice and $K = \{x \in E | x \geq 0\}$, then we say that a bilinear form $\langle , \rangle$ on $E \times E$ is K-local iff, for every $x, y \in K$ such that, $x \wedge y = 0$ we have, $\langle x, y \rangle = 0$.

In this case, we have that, for a given mapping, $f:K \to E$, a solution of problem (O.C.P.) is a solution of complementarity problem,

$$\left\| \begin{array}{l} \text{find } x_0 \in K \text{ such that,} \\ f(x_0) \in K \text{ and } \langle x_o, f(x_o) \rangle = 0. \end{array} \right.$$

It is very interesting to study both problems (T.C.P.) and (O.C.P.) since we obtain two different theories with interesting and deep connections.

Finally, given a complementarity problem, we are interested to study the following problems:

1°) existence of solutions,

2°) uniqueness or the topological study of the set of solutions, if we have no uniqueness,

3°) approximation of solutions,

4°) the study of dependence of solutions with respect to parameters, if we have a parametric complementarity problem,

5°) the study of stability of solutions, if we have a perturbed complementarity problem.

CHAPTER 2

MODELS AND APPLICATIONS

The complementarity theory derives its importance from the fact that it unifies problems in fields such as: mathematical programming, game theory, the theory of equilibrium in a competitive economy, equilibrium of traffic flows, mechanics, engineering, lubricant evaporation in the cavity of a cylindrical bearing, elasticity theory, fluid flow through a semiimpermeable membrane, maximizing oil production, computation of fixed point etc.

In this sense, let us consider in this chapter some applied examples which lead to complementarity problems.

2.1 Mathematical Programming

In this section we consider several examples and models in finite dimensional spaces.

Hence, we consider the topological vector space R^n ordered by the pointed closed convex cone R^n_+ and we denote by $< , >$ the inner product, $< x, y > = \sum_{i=1}^{n} x_i y_i$; $x = (x_i)$; $y = (y_i)$.

2.1.1 Linear Programming

Let $c = (c_i) \in R^n$, $b = (b_i) \in R^m$ be two vectors and let $A = (a_{ij}) \in \mathcal{M}_{mxn}(R)$ be a a matrix.

Consider the primal linear program,

$$(\text{p.L.P.}): \left\| \begin{array}{l} \text{minimize } <c, x> \\ x \in \mathcal{F}_1 \\ \text{where, } \mathcal{F}_1 = \{x \in R^n | x \in R^n_+ \text{ and } Ax - b \in R^m_+\} \end{array} \right.$$

and its dual,

$$(\text{d.L.P.}): \left\| \begin{array}{l} \text{maximize } <y, b> \\ y \in \mathcal{F}_2 \\ \text{where, } \mathcal{F}_2 = \{y \in R^m | y \in R^m_+ \text{ and } A^t y - c \in - R^n_+\} \end{array} \right.$$

A fundamental result of linear programming is the following,

Theorem 2.1

If there exist $x_0 \in F_1$ and $y_0 \in F_2$ such that $<c, x_o> = <y_o, b>$ then x_0 is a solution of problem (p.L.P.) and y_0 is a solution of problem (d.L.P.).

Using this result we can associate to the problems, (p.L.P.) and (d.L.P.) a complementarity problem.

Indeed, adding slack variables $u \in R^n$ and $v \in R^m$ such that, $Ax - v = b$ and $A^t y + u = c$ and denoting, $z = \begin{bmatrix} x \\ y \end{bmatrix}$; $w = \begin{bmatrix} u \\ v \end{bmatrix}$; $q = \begin{bmatrix} c \\ -b \end{bmatrix}$; $M = \begin{bmatrix} 0 & -A^t \\ A & 0 \end{bmatrix}$ we obtain the following complementarity problem,

$$\text{(L.C.P.)}: \quad \left\| \begin{array}{l} \text{find } z \in R^{n+m} \text{ such that,} \\ \\ z \in R_+^{n+m}, \ w = Mz + q \in R_+^{n+m} \text{ and } <z, w> = 0. \end{array} \right.$$

We observe that this linear complementarity problem is equivalent to the couple primal-dual of linear programs (p.L.P.) - (d.L.P.).

Remarks.

i) Condition, $<z, w> = 0$ in the definition of problem (L.C.P.) was obtained observing that this condition expresses exactly the fact that $<c, x> = <y, b>$.

ii) The principal contribution of complementarity problem to the linear programming is that it transforms an optimization problem in an equation.

2.1.2 Quadratic Programming

Consider the quadratic programming problem,

$$(1): \quad \left\| \begin{array}{l} \text{minimize } f(x) \\ \quad x \in F \\ \text{where, } F = \{x \in R^n \mid x \in R_+^n, \ b - Ax \in R_+^m\} \\ \qquad f(x) = \frac{1}{2} <x, Qx> + <c, x>, \\ \qquad c \in R^n, \ Q \in M_{nxn}(R), \ (Q \text{ symmetric}), \\ \qquad A \in M_{mxn}(R) \text{ and } b \in R^m \end{array} \right.$$

Denoting the Lagrangian multiplier vectors of the constraints $Ax \le b$ and $x \ge 0$ by $\lambda \in R^m$ and $u \in R^n$ respectively and denoting the vector of slack variables by $v \in R^m$, the Kuhn-Tucker necessary optimality conditions could be written as:

$$(2): \quad \left\| \begin{array}{l} c + A^t\lambda + Qx - u = 0 \\ Ax + v = b \\ \\ u \in R_+^n, \ v \in R_+^m, \ x \in R_+^n, \ \lambda \in R_+^m \\ <u, x> = 0 \text{ and } <v, \lambda> = 0. \end{array} \right.$$

Now, observe that conditions (2) can be written also as:

$$(3): \quad \left\| \begin{array}{l} \begin{bmatrix} u \\ v \end{bmatrix} = \begin{bmatrix} c \\ b \end{bmatrix} + \begin{bmatrix} Q & A^t \\ -A & 0 \end{bmatrix} \begin{bmatrix} x \\ \lambda \end{bmatrix} \\ \\ \begin{bmatrix} u \\ v \end{bmatrix} \in R_+^{n+m}; \ \begin{bmatrix} x \\ \lambda \end{bmatrix} \in R_+^{n+m}; \ < \begin{bmatrix} u \\ v \end{bmatrix}, \begin{bmatrix} x \\ \lambda \end{bmatrix} > = 0 \end{array} \right.$$

If we denote, $z = \begin{bmatrix} x \\ \lambda \end{bmatrix}$; $q = \begin{bmatrix} c \\ b \end{bmatrix}$; $M = \begin{bmatrix} Q & A^t \\ -A & 0 \end{bmatrix}$ and $f(z) = \begin{bmatrix} u \\ v \end{bmatrix}$

we obtain that the Kuhn-Tucker conditions (2) are equivalent to the following linear complementarity problem,

$$(4): \quad \left\| \begin{array}{l} \text{find } z \in R^{n+m} \text{ such that,} \\ \\ z \in R_+^{n+m}; \ f(z) \in R_+^{n+m} \text{ and } < z, f(z) > = 0. \end{array} \right.$$

It is remarquable to note that there exists another connection between linear programming, quadratic programming and the linear complementarity problem.

Consider the linear programming problem,

$$(5): \quad \left\| \begin{array}{l} \text{minimize} < p, x > \\ x \in \mathcal{F} \\ \text{where: } \mathcal{F} = \{x \in R^n | Ax - b \in R_+^m\} \\ \qquad\qquad p \in R^n, \ b \in R^m \text{ and } A \in \mathcal{M}_{mxn}(R) \end{array} \right.$$

Suppose that every row of A is different from zero and consider a quadratic perturbation on $\mathcal{F}$ of problem (5) of the form,

$$(6): \quad \left\| \begin{array}{l} \text{minimize } [< p, x> + \frac{\varepsilon}{2} < x, x>] \\ x \in \mathcal{F} \end{array} \right.$$

In 1979 Mangasarian and Meyer proved the following result. [Mangasarian O.L. and Meyer R.R.: Nonlinear perturbation of linear programs. Siams J. Control and Optimization Vol. 17 Nr. 6 (1979), 745-752].

Theorem 2.2.

If program (5) has an optimal solution then program (6) has a unique solution x_o for every $\varepsilon \in [0, \alpha]$ and some $\alpha > 0$.

Moreover, the solution x_o is independent of ε and it is also a solution of program (5).

Consider now a more general case, precisely the quadratic program,

$$(7): \quad \left\| \begin{array}{l} \text{minimize } [\frac{1}{2} < x, Qx> + <p, x>] \\ \quad x \in \mathcal{F} \\ \text{where; } \mathcal{F} = \{x \in R^n | Ax - b \in R_+^m\}; \\ \qquad p \in R^n; \ b \in R^m; \ Q \in \mathcal{M}_{nxn}(R) (Q \text{ symmetric, positive definite}) \\ \qquad \text{and } A \in \mathcal{M}_{mxn}(R) \end{array} \right.$$

The dual program of (7) is,

$$(8):\quad \left\| \begin{array}{l} \text{maximize } [-\frac{1}{2} <x, Qx> + <b, u>] \\ (x, u) \in F_1 \\ \text{where: } F_1 = \{(x, u) | x \in R^n, u \in R^m_+ \text{ and } Qx - A^t u + p = 0\} \end{array} \right.$$

which under the positive definite assumption on Q is, [upon eliminating x, since from (8) $x = Q^{-1}(A^t u - p)$], equivalent to,

$$(9):\quad \left\| \begin{array}{l} \text{minimize } [\frac{1}{2} <u, AQ^{-1}A^t u> - <b + AQ^{-1}p, u>] \\ u \in F_2 \\ \text{where: } F_2 = \{u \in R^m | u \in R^m_+\} \end{array} \right.$$

Since $AQ^{-1}A^t$ is positive semidefinite, (9) is equivalent to the following <u>symmetric linear complementarity problem</u>.

$$(S.L.C.P.):\quad \left\| \begin{array}{l} \text{find } u \in R^m \text{ such that,} \\ u \in R^m_+;\ v = AQA^t u - (b + AQ^{-1}p) \in R^m_+ \\ \text{and } <u, v> = 0 \end{array} \right.$$

We find more details on this subject in [A194].

<u>2.1.3</u> <u>Nonlinear Programming</u>

Consider the convex program,

$$(1):\quad \left\| \begin{array}{l} \text{minimize } f(x) \\ x \in F \\ \text{where: } F = \{x \in R^n | x \geq 0 \text{ and } g_i(x) \leq 0;\ i = 1,2,\ldots,m\} \end{array} \right.$$

In this programming problem suppose all the functions convex and differentiable. The Lagrangian function $L(x, u)$ for (1) is given by,

$$L(x, u) = f(x) + \sum_{i=1}^{m} u_i g_i(x).$$

Hence, $u = (u_i) \in R^m$ and the Kuhn-Tucker necessary conditions for optimality can be written as:

$$(2):\quad \left\| \begin{array}{l} \frac{\partial L(x, u)}{\partial x_j} = h_j(x, u) \geq 0;\ j = 1, 2, \ldots, n \\ -\frac{\partial L(x, u)}{\partial u_i} = h_{n+i}(x, u) \geq 0;\ i = 1, 2, \ldots, m \\ x \geq 0,\ u \geq 0 \\ \sum_{j=1}^{n} x_j h_j(x, u) = 0 \text{ and } \sum_{i=1}^{m} u_i h_{n+i}(x, u) = 0 \end{array} \right.$$

If we denote, $z = \begin{bmatrix} x \\ u \end{bmatrix}$ and $h(z) = \begin{bmatrix} h_1(z) \\ \vdots \\ h_n(z) \\ \vdots \\ h_{n+m}(z) \end{bmatrix}$

then the Kuhn-Tucker conditions (2) may be stated as the following complementarity problem,

$$
\text{(C.P.)}: \left\| \begin{array}{l} \text{find } z \in R^{n+m} \text{ such that,} \\ z \in R_+^{n+m},\ h(z) \in R_+^{n+m} \text{ and} \\ \langle z, h(z) \rangle = 0 \end{array} \right.
$$

Remark

We have a similar construction for a nonlinear program (not necessary convex), where f and $g_i (i = 1 \ldots m)$, are C^1-functions on an open set U, such that $U \supset R_+^n$.

2.1.4 The Saddle Point Theory

Let $\mathcal{D} \subset R^n \times R^m$ be an open subset such that, $R_+^n \times R_+^m \subset \mathcal{D}$.

Given a differentiable function $f: \mathcal{D} \to R$, consider the following problem,

$$
(1): \left\| \begin{array}{l} \text{find } z_* \in R_+^n,\ y_* \in R_+^m \text{ such that,} \\ f(z, y_*) \le f(z_*, y_*) \le f(z_*, y);\ \forall\ (z, y) \in R_+^n \times R_+^m \end{array} \right.
$$

Every solution (z_*, y_*) of problem (1) is called a saddle point or a max-min point for f on $R_+^n \times R_+^m$.

Now, consider the function $F = (F_1, \ldots, F_n, \ldots, F_{n+m})$ defined by,

$$F_i(z, y) = -\frac{\partial f}{\partial z_i}(z, y);\ i = 1, 2, \ldots, n,$$

$$F_{n+j}(z,y) = \frac{\partial f}{\partial y_j}(z, y);\quad j = 1, 2, \ldots, m.$$

If we denote, $x = \begin{bmatrix} x \\ y \end{bmatrix}$ then we obtain that, if (z_*, y_*) is a positive saddle point for (1), then it is also a solution of the following complementarity problem,

$$
\text{(N.C.P.)}: \left\| \begin{array}{l} \text{find } x \in R_+^n \times R_+^m \text{ such that,} \\ F(x) \in R_+^n \times R_+^m \text{ and } \langle x, F(x) \rangle = 0 \end{array} \right.
$$

2.1.5 Mathematical Programming and Complementarity Problem with Restrictions

In this section we define a family of extensions of the linear complementarity problem.

If M, N, P, Q, R and S are precisely defined matrices we distinguish the following problems:

i) Second linear complementarity problem

(S.L.C.P.):

find $x \in R^n$ and $u \in R^m$ such that,

$y = q + Mx + Nu$

$0 = p + Rx + Su$

$x \geq 0,\ y \geq 0,\ \langle x, y \rangle = 0$

where, $y \in R^n,\ q \in R^n,\ p \in R^m$

ii) Minimum linear complementarity problem

(M.L.C.P.):

$\underset{(x,\, y,\, u)}{\text{minimize}} \{\langle p, x \rangle + \langle q, y \rangle + \langle r, u \rangle\}$

subject to: $Px + Qy + Ru = b,$

$x \geq 0,\ y \geq 0,\ \langle x, y \rangle = 0$

$x, y \in R^n,\ b \in R^m,\ u \in R^\ell$

$p, q \in R^n,\ r \in R^\ell$

iii) Second minimum linear complementarity problem

(S.M.L.C.P.):

$\underset{(x,\, y,\, u)}{\text{minimize}} \{\langle p, x \rangle + \langle q, y \rangle + \langle r, u \rangle\}$

Subject to: $Px + Qy + Ru = b$

$x \geq 0,\ x, y \in R^n,\ u \in R^\ell$

$b \in R^m,\ \langle x, y \rangle = 0$

$p, q \in R^n,\ r \in R^\ell$

These problems have interesting applications in mathematical programming.

Examples

a) Consider the general linear programming problem,

(1):

$\underset{(x,\, y) \in F_p}{\text{minimize}} \{\langle c, x \rangle + \langle d, y \rangle\}$

where: $F_p = \{(x,y) \mid Ax + By = b;\ Ex + Fy \geq g,\ x \in R^n,\ y \in R^m\}$

$c \in R^n,\ d \in R^m,\ b \in R^r,\ g \in R^s,$

$A \in M_{r\times n}(R),\ B \in M_{r\times m}(R),\ E \in M_{s\times n}(R),\ F \in M_{s\times m}(R)$

The dual program of (1) is,

$$(2):\quad \left\| \begin{array}{l} \text{minimize } \{\langle b, u\rangle + \langle g, v\rangle\} \\ (u, v) \in F_d \\ \text{where: } F_d = \{(u,v) \mid A^t u + E^t v \le c;\ B^t u + F^t v = d;\ u \in R^r;\ v \in R^s_+\} \end{array} \right.$$

Suppose that (x_*, y_*) is an optimal solution of the linear program (1).

The fundamental duality theorem of linear programming, implies that the dual program (2) also has an optimal solution (u_*, v_*).

The complementarity slackness property concerning the nonnegative variables z_* and w_* corresponding to the inequalities in the dual and primal program holds.

We obtain that $(z_*, w_*, x_*, v_*, y_*, u_*)$ is a solution of the following second linear complementarity problem,

$$(3):\quad \left\| \begin{array}{l} \begin{bmatrix} z \\ w \end{bmatrix} = \begin{bmatrix} c \\ -g \end{bmatrix} + \begin{bmatrix} 0 & -E^t \\ E & 0 \end{bmatrix}\begin{bmatrix} x \\ v \end{bmatrix} + \begin{bmatrix} 0 & -A^t \\ F & 0 \end{bmatrix}\begin{bmatrix} y \\ u \end{bmatrix} \\ \begin{bmatrix} 0 \\ 0 \end{bmatrix} = \begin{bmatrix} d \\ -b \end{bmatrix} + \begin{bmatrix} 0 & -F^t \\ A & 0 \end{bmatrix}\begin{bmatrix} x \\ v \end{bmatrix} + \begin{bmatrix} 0 & -B^t \\ B & 0 \end{bmatrix}\begin{bmatrix} y \\ u \end{bmatrix} \\ z \in R^n_+;\ x \in R^n_+;\ y \in R^m;\ w \in R^s_+;\ v \in R^s_+;\ u \in R^r \\ \text{and } \langle z, x\rangle + \langle w, v\rangle = 0. \end{array} \right.$$

Moreover, since we can prove that $\langle c, x_*\rangle + \langle d, y_*\rangle = \langle b, u_*\rangle + \langle g, v_*\rangle$, we obtain that <u>there is a one-to-one correspondence between the solutions of the linear program (1) and the (S.L.C.P.)(3)</u>.

b) Consider the following general quadratic programming problem.

Suppose A, B, E, F, P, Q, R, S well defined matrices and x, y, c, d, b, g vectors such that the next operations are well defined.

Denoting, $z = \begin{bmatrix} x \\ y \end{bmatrix}$; $q = \begin{bmatrix} c \\ d \end{bmatrix}$; $D = \begin{bmatrix} P & R \\ S & Q \end{bmatrix}$ and supposing that P and Q are symmetric matrices and $R = S^t$, consider the following general quadratic program,

$$(4):\quad \left\| \begin{array}{l} \text{minimize } \{\langle q, z\rangle + \frac{1}{2}\langle z, Dz\rangle\} \\ z \in F \\ \text{where: } F = \{(x, y) \mid Ax + By = b,\ Ex + Fy \ge g,\ x \ge 0;\ -\infty < y < +\infty\} \end{array} \right.$$

<u>Proposition 2.1</u>

If $z_* = \begin{bmatrix} x_* \\ y_* \end{bmatrix}$ is an optimal solution of program (4) then z_* is also an optimal solution of the linear program,

$$(5):\quad \left\| \begin{array}{l} \text{minimize } \langle q + Dz_*, z\rangle \\ z \in F \end{array} \right.$$

Proof

See: [K. Murty: Linear and combinatorial programming. Wiley (1976) p. 491]. □

Since the dual of program (5) is,

$$\left\| \begin{array}{l} \text{maximize } \{<b, u> + <g, v>\} \\ (u, v) \in F_d \\ \text{where: } F_d = \left\{ (u,v) \,\middle|\, \begin{array}{l} A^t u + E^+ v \le c + Px_* + Ry_*, \\ B^t u + F^t v = d + Sx_* + Qy_*, \\ v \ge 0, \; -\infty < u < +\infty \end{array} \right\} \end{array} \right.$$

then by example (a) and above proposition we obtain the following result.

Proposition 2.2

If (x_*, y_*) is an optimal solution of the quadratic program (4) then there exist vectors u_*, v_*, w_* and t_* such that $(x_*, y_*, u_*, v_*, w_*, t_*)$ is a solution of the following second linear complementarity problem (S.L.C.P.),

$$(7): \left\| \begin{array}{l} \begin{bmatrix} w \\ t \end{bmatrix} = \begin{bmatrix} c \\ -g \end{bmatrix} + \begin{bmatrix} P & -E^t \\ E & 0 \end{bmatrix} \begin{bmatrix} x \\ v \end{bmatrix} + \begin{bmatrix} R & -A^t \\ F & 0 \end{bmatrix} \begin{bmatrix} y \\ u \end{bmatrix} \\ \begin{bmatrix} 0 \\ 0 \end{bmatrix} = \begin{bmatrix} d \\ -b \end{bmatrix} + \begin{bmatrix} S & -F^t \\ A & 0 \end{bmatrix} \begin{bmatrix} x \\ v \end{bmatrix} + \begin{bmatrix} Q & -B^t \\ B & 0 \end{bmatrix} \begin{bmatrix} y \\ u \end{bmatrix} \\ w \ge 0;\; t \ge 0;\; x \ge 0;\; v \ge 0;\; -\infty < y < +\infty;\; -\infty < u < +\infty \\ \text{and } <w, x> + <t, v> = 0 \end{array} \right.$$

c) Consider the case of zero-one integer programming problem which is a very important optimization problem, since many nonlinear optimization problems and combinatorial optimization problems, may be restated as a zero-one (mixed) integer programming problem.

A zero-one (mixed) integer programming problem may be stated as,

$$(8): \left\| \begin{array}{l} \text{minimize } \{<c, x> + <d, u>\} \\ (x, u) \in F \\ \text{where: } F = \left\{ (x, u) \,\middle|\, \begin{array}{l} Ax + Bu \ge \alpha,\; u \ge 0, \\ x_i = 0 \text{ or } 1 \text{ for all } i = 1, 2, \ldots, n \end{array} \right\} \\ \qquad x \in R^n,\; u \in R^m;\; A \in M_{rxn}(R), \\ \qquad B \in M_{rxm}(R);\; \alpha \in R^r;\; c \in R^n;\; d \in R^m \end{array} \right.$$

By introducing a slack variable y_i $(i = 1, 2, \ldots, n)$ for each i, the 0-1 condition can be written, $x_i + y_i = 1$; $x_i \ge 0$; $y_i \ge 0$; $x_i y_i = 0$; for $i = 1,2, \ldots,n$.

Denote by e the vector $(1, 1, \ldots, 1)$.

Thus, this problem is equivalent to the following (M.L.C.P.),

$$(9): \quad \left\| \begin{array}{l} \underset{(x,\,u)}{\text{minimize}} \ \{<c, x> + <d, u>\} \\ \text{subject to:} \quad \begin{bmatrix} y \\ z \end{bmatrix} = \begin{bmatrix} e \\ -\alpha \end{bmatrix} + \begin{bmatrix} -1 & 0 \\ A & B \end{bmatrix} \begin{bmatrix} x \\ u \end{bmatrix} \\ \qquad\qquad\quad x \geq 0;\ y \geq 0;\ z \geq 0;\ u \geq 0 \text{ and} \\ \qquad\qquad\quad <x, y> = 0 \end{array} \right.$$

d) Consider now the case of variable separable programming.

Precisely, a variable separable program may be stated in the following form,

$$(10): \quad \left\| \begin{array}{l} \underset{x \in F}{\text{minimize}} \ \sum_{i=1}^{n} {}_i(x_i) \\ \text{where: } F = \left\{ x \in R^n \,\middle|\, \begin{array}{l} \sum_{i=1}^{n} \Psi_{ji}(x_i) \geq b_j;\ j = 1,2,\ldots,m \\ x_i \in [c_i, d_i];\ i = 1, 2, \ldots, n \end{array} \right\} \end{array} \right.$$

It is shown in [141] that this problem, in some situations, is equivalent to a (S.M.L.C.P.).

2.2 Game Theory

Game theory is the mathematical analysis of conflict and strategy and reflects deep conflict that arise in social, economic, military or political situations.

2.2.1 Bimatrix Games

First, consider the two-person game or bimatrix game.

Presicely, there are two players, where the first player solves the program,

$$(G_1): \quad \left\| \begin{array}{l} \text{maximize} <x^1, A_1x^2> \\ x \in F_1 \\ \text{where: } F_1 = \{x^1 \in R^n_+ \mid \sum_{j=1}^{n} x^1_j = 1\} \\ \text{given } x^2 \end{array} \right.$$

and the second player solves the program,

$$(G_2): \quad \left\| \begin{array}{l} \text{maximize} <x^1, A_2x^2> \\ x^2 \in F_2 \\ \text{where: } F_2 = \{x^2 \in R^m_+ \mid \sum_{j=1}^{m} x^2_j = 1\} \\ \text{given } x^1. \end{array} \right.$$

where A_1, A_2 are two suitable matrices.

This game will be denoted by (G_1, G_2).

<u>Remark</u>

We have as examples of two-person games, chess, checkers, backgammon etc.

A pair of vectors (x_*^1, x_*^2) is a Nash equilibrium point for (G_1, G_2) if and only if,

$n_1)$ x_*^1 solves (G_1) given x_*^2 and

$n_2)$ x_*^2 solves (G_2) given x_*^1.

In addition we suppose that each element of A_1 and A_2 is negative, that is we suppose $A_i < 0$; $i = 1, 2$.

If not, we substract a scalar from each element and clearly the solution of (G_1, G_2) is unchanged.

Denote by e the unit vector (each component of e is equal to 1) of appropriate dimension.

If we suppose that (x_*^1, x_*^2) is a Nash equilibrium point for (G_1, G_2) then for every x^i $(i = 1, 2)$ such that,

(1): $\langle e, x^i \rangle = 1;\ x^i \geq 0;\ i = 1, 2$

we have,

$$(2):\quad \left\| \begin{array}{l} \langle x_*^1, A_1 x_*^2 \rangle \geq \langle x^1, A_1 x_*^2 \rangle , \\ \langle x_*^1, A_2 x_*^2 \rangle \geq \langle x_*^1, A_2 x_2 \rangle . \end{array} \right.$$

We can show that, $x_* = (x_*^1, x_*^2)$ is a Nash equilibrium point for (G_1, G_2), if and only if, (2) holds for all x satisfying (1).

Next we show that if (2) holds for all x satisfying (1) then we have,

$$(3):\quad \left\| \begin{array}{l} \langle x_*^1, A_1 x_*^2 \rangle e \geq A_1 x_*^2, \\ \langle x_*^1, A_2 x_*^2 \rangle e \geq (A_2)^t x_*^1 \end{array} \right.$$

and conversely, if (3) holds then (2) holds for every x satisfying (1).

Indeed, consider (2) holds for every x satisfying (1).

Let $x^i = e^j$; $j = 1, 2, \ldots, n$ for $i = 1$ and $j = 1, 2, \ldots, m$ for $i = 2$, where e^j is the vector with all zeros except for a 1 in the jth position.

From (2) we have,

$$(4):\quad \left\| \begin{array}{l} \langle x_*^1, A_1 x_*^2 \rangle \geq \langle e^j, A_1 x_*^2 \rangle ;\ \forall\, j = 1, 2, \ldots, n \\ \langle x_*^1, A_2 x_*^2 \rangle \geq \langle x_*^1, A_2 e^j \rangle ;\ \forall\, j = 1, 2, \ldots, m \end{array} \right.$$

and we observe that (4) implies (3).

Conversely, if (3) holds and x^i $(i = 1, 2)$ are vectors satisfying (1) then we deduce,

$$(5): \quad \left\| \begin{array}{l} \langle x_*^1, A_1 x_*^2 \rangle \langle x^1, e \rangle \geq \langle x^1, A_1 x_*^2 \rangle , \\ \langle x_*^1, A_2 x_*^2 \rangle \langle e, x^2 \rangle \geq \langle x_*^1, A_2 x^2 \rangle \end{array} \right.$$

and since $\langle x^i, e \rangle = 1$, for $i = 1, 2$ we observe that (2) holds.

Now, we prove that the bimatrix game (G_1, G_2) is equivalent to the following <u>linear complementarity problem</u>,

$$(\text{L.C.P.}): \quad \left\| \begin{array}{l} \text{find } X \in R_+^{n+m} \text{ such that,} \\ \\ Y = q + MX \in R_+^{n+m} \text{ and } \langle X, Y \rangle = 0 \\ \\ \text{where: } X = \begin{bmatrix} x^1 \\ x^2 \end{bmatrix}; \; q = \begin{bmatrix} -e \\ -e \end{bmatrix}; \; Y = \begin{bmatrix} y^1 \\ y^2 \end{bmatrix}; \; M = \begin{bmatrix} 0 & -A_1 \\ -(A_2)^t & 0 \end{bmatrix} \end{array} \right.$$

Indeed, let (x^1, x^2) be a solution of problem (L.C.P.).

Then, $x^1 \in R_+^n$, $x^2 \in R_+^m$ and $x^i \neq 0$; $i = 1, 2$, which imply, $\langle e, x^i \rangle > 0$; $i = 1, 2$.

Denoting,

$$(6): \quad x_*^i = \frac{x^i}{\langle e, x^i \rangle} \; ; \; i = 1. \; 2,$$

we remark that x_*^i; $(i = 1, 2)$ satisfy (3).

Really, since (x^1, x^2) is a solution of problem (L.C.P.) we have,

$$(7): \quad A_1 x^2 \leq -e; \; (A_2)^t x^1 \leq -e,$$

$$(8): \quad \langle x^1, A_1 x^2 \rangle = -\langle e, x^1 \rangle; \; \langle x^1, A_2 x^2 \rangle = - \langle e, x^2 \rangle.$$

Using (6) from (7) and (8) we deduce,

$$(9): \quad A_1 x_*^2 \leq \frac{-e}{\langle e, x^2 \rangle} \; ; \; (A_2)^t x_*^1 \leq \frac{-e}{\langle e, x^1 \rangle} ,$$

$$(10): \quad \langle x_*^1, A_1 x_*^2 \rangle = - \frac{1}{\langle e, x^2 \rangle}; \; \langle x_*^1, A_2 x_*^2 \rangle = - \frac{1}{\langle e, x^1 \rangle}$$

which imply,

$$A_1 x_*^2 \leq \langle x_*^1, A_1 x_*^2 \rangle e \text{ and } (A_2)^t x_*^1 \leq \langle x_*^1, A_2 x_*^2 \rangle e$$

that is (3) holds.

Conversely, if we suppose that x_*^1, x_*^2 satisfy (3) then (2) holds for every x satisfying (1) and by an elementary calculus we can show that,

$$x^1 = - \frac{x_*^1}{< x_*^1, A_2 x_*^2 >} \; ; \; x^2 = - \frac{x_*^2}{< x_*^1, A_1 x_*^2>}$$

is a solution of problem (L.C.P.).

2.2.2 Polymatrix Games

Consider now a polymatrix game associated to r^2 matrices $A_{ij} \in M_{n_i x n_j}(R)$.

In this case there are $r \geq 2$ players such that the player i has n_i number of strategies.

Suppose $A_{ii} = 0$ (i = 1, 2, ..., r) and $A_{ij} > 0$ for all $i \neq j$.

Denote by e the unit vector of appropriate dimension.

Let $X^i \geq 0$ and $eX^i = 1$ (i = 1, 2, ..., r) be vectors of n_i components defining mixed stategies for the r players.

Let X be the set of all mixed strategies, $X = [X^1, X^2, \ldots, X^r]$.

For this polymatrix game the expected payoff for the player i is defined as,

$$E_i(X) = (X^i)^t \sum_{j=1}^{r} A_{ij} X^j$$

and an equilibrium point $X^* = [X^{*1}, X^{*2}, \ldots, X^{*r}]$ is such that for all mixed strategies X the following inequality holds,

$$E_i(X^*) = (X^{*i})^t \sum_{j=1}^{r} A_{ij} X^{*j} \leq (X^i)^t \sum_{j=1}^{r} A_{ij} X^{*j}$$

For every $v_i = E_i(X)$ we introduce a complementarity artificial variable u_i such that, $u_i = eX^i - 1$.

In the theory of polymatrix games is shown that the equilibrium point satisfies the following equations,

$$(1): \quad \left\| \begin{array}{l} Y^i = \sum_{j=1}^{r} A_{ij} X^j - v_i e \\ u_i = eX^i - 1 \\ X^i, Y^i \geq 0; \; u_i, v_i \geq 0 \\ <X^i, Y^i> = 0; \; u_i v_i = 0, \end{array} \right.$$

where Y^i is a vector of n_i components and u_i, v_i are scalars.

We observe that (1) is exactly a complementarity problem in R^m with respect to R^m_+ where $m = r + \sum_{i=1}^{r} n_i$.

2.3 Variational Inequalities and Complementarity

In the last twenty years variational inequalities have gained a great importance, both from the theoretical and the practical points of view.

Variational inequalities are used in the study of calculus of variations and generally in optimization problems.

Precisely, let $f:R^n \to R$ be a C^1-function and let $K \subset R^n$ be a closed convex set.

Consider the following classical result.

If there exists $x_o \in K$ such that,

$$(1): \quad \left\| \; f(x_o) = \min_{x \in K} f(x), \right.$$

then x_o satisfies,

$$(2): \quad \left\| \begin{array}{l} x_o \in K \text{ and} \\ \langle f'(x_o), x - x_o \rangle \geq 0; \; \forall \, x \in K \end{array} \right.$$

Generally, a solution of problem (2) is not always a solution of problem (1), but if f is a convex C^{-1}-function then problem (1) is equivalent to problem (2).

Certainly, the following problem is a natural generalization of problem (2).

Given a continuous mapping $f:R^n \to R^n$ we are interested to solve the problem,

$$(3): \quad \left\| \begin{array}{l} \text{find } x_o \in R^n \text{ such that,} \\ x_o \in K \text{ and } \langle f(x_o), x - x_o \rangle \geq 0; \; \forall \, x \in K \end{array} \right.$$

Problem (3) is a variational inequality.

In the next chapter we will prove that if K is a convex cone then problem (3) is equivalent to the following complementarity problem,

$$(4): \quad \left\| \begin{array}{l} \text{find } x_o \in K \text{ such that,} \\ f(x_o) \in K^* \text{ and } \langle x_o, f(x_o) \rangle = 0 \end{array} \right.$$

It is well known that variational inequalities theory is the result of researches of Stampacchia, Browder, Brezis, Lions, Rockafellar, Mosco, Baiocchi, Kinderlehrer etc and it has interesting applications in the study of: obstacles probems, confined plasmas, filtration phenomena, free-boundary problems plasticity and viscoplasticity phenomena, elasticity problems, stochastic optimal control problems etc.

Another source of complementarity problems defined through variational inequalities are differential equations of evolution.

Consider a differential equation of evolution of the form,

$$(5): \quad \left\| \begin{array}{l} \dfrac{dx(t)}{dt} = Ax(t); \; t \geq 0 \\ x(0) = x_o; \; x_o \in K \end{array} \right.$$

where $K \subset R^n$ is a closed convex set and $A:R^n \longrightarrow R^n$ a mapping.

In some practical problems as for example the study of economical system we are interested to find x(t) such that $x(t) \in K$ and we are interested to study the problem,

$$(6): \quad \left\| \begin{array}{l} \text{find } x \text{ such that,} \\ x(t) \in K, \ \forall\, t \geq 0, \ x(0) = x_0 \text{ and} \\ \langle \frac{dx(t)}{dt} - Ax(t),\ y - x(t)\rangle \geq 0;\ \forall\, t \geq 0;\ \forall\, y \in K. \end{array} \right.$$

If K is a convex cone then problem (6) is exactly a complementarity problem.

Finally, we note that, every time when we have a variational inequality on a convex cone, we have a complementarity problem.

2.4 Mechanics and Complementarity

2.4.1 The Contact Problem

The relation between the contact problem and complementarity problems is based on the following idea of Fridman and Chernina. [V.M. Fridman and V.S. Chernina: An iteration process for the solution of the finite-dimensional contact problem. Zh. Vychisl. Mat. Fiz. 7 (1967), 160-163].

Consider two elastic bodies in contact at a finite number of points. Denote:

(1°) x_i = the contact stress at the i-th point,

(2°) a_{ij} = the effect of the j-th stress on the relative deflection at the i-th point of the surface of the bodies in contact.

(3°) A = $[a_{ij}]$ the matrix of components a_{ij}

(4°) q_i = the distance that would exist between corresponding i-th points of the surfaces if free penetration were permitted.

In this model, the clearance condition is expressed by the linear unilateral contraints.

The contact stress x_i are nonnegative and the important contact condition says that the contact stress x_i can be positve only if the bodies are really in contact at the i-th point i.e., $(q + Ax)_i = 0$. Thus we obtain the following complementarity problem, which express the equilibrium,

$$(\text{L.C.P.}): \quad \left\| \begin{array}{l} \text{find } x \in R^n \text{ such that,} \\ x \in R^n_+;\ q + Ax \in R_+ \text{ and} \\ \langle x,\ q + Ax \rangle = 0 \end{array} \right.$$

This connection between the contact problem and complementarity theory implies interesting applications in engineering.

2.4.2 Structural Engineering Applications

A fundamental problem in structural engineering is to determine the behaviour of the structure subjected to a set of loads.

The developments in these applications are largely due to Maier [A185]

Maier considered the elasto-plastic analysis of discrete structures subjected to external loads.

In these models many technical considerations are involved including the nature of the applied loads and the behavior characteristics of the material.

To describe the set of loads which leave a structural element in the elastic range the yield function is used. Suppose that this set is taken to be convex and polyhedral.

The yield conditions are given by linear inequalities.

To be determined are the plastic multipliers (and possibly displacements) and other quantities of interest.

The complementarity condition is introduced through the condition that a plastic multiplier cannot be positive in the elastic range.

To understand the model defined by Maier we shall consider a simplified model.

Consider a discrete structure subjected to external loads.

Suppose that the behaviour of the entire structure can be represented by the values of stresses and strains at n "critical points" in the structure.

Suppose also that the stress and strain at each of the critical points are one-dimensional and consider an important class of structures called piecewise linear elastic-plastic.

In particular, reinforced concrete frames belong to this class.

Under appropriate conditions, the set of mechanical principles which governs the behaviour of the elastic-plastic structure can be represented by the following (nonlinear) complementarity problem:

$$(1): \quad \left\| \begin{array}{l} \text{find } x \in R^n \text{ such that} \\ x \in R^n_+, \\ \\ q - Mu + h(x) - MAMx \in R_+ \text{ and} \\ \langle x, q - Mu + h(x) - MAMx \rangle = 0, \end{array} \right.$$

where: x represents the plastic activity (x is the vector of plastic multipliers). $q \in R^n_+$, h is a function such that, $h(x) = (h_1(x_1), \ldots, h_n(x_n))$, where $h_j(x_j)$ is a concave, strictly monotone increasing real-valued function of x_j with, $h_j(0) = 0$; $j = 1, 2, \ldots, n$, $M \in \mathcal{M}_{n\times n}(R)$; $M = (m_{ij})$ where, $m_{ii} = +1$ or -1, and $m_{ij} = 0 \;\; \forall\, i \neq j$, $u \in R^n$ such that $Mu > 0$, $A \in \mathcal{M}_{n\times n}(R)$ such that, $A = -BSB^t$, where $S \in \mathcal{M}_{m\times m}(R)$ and $B \in \mathcal{M}_{n\times m}(R)$.

In (1) all data can be determined by mechanical properties of the structure under consideration and the loads applied to it.

Note that m is the number of critical points with a certain property called redundant and S denotes the stiffness matrix with respect to the redundancies. We have, $m < n$ and S is positive definite.

If x is a solution of problem (1) then the vector of <u>stress</u> is,

(2): $$s = u + AMx$$

and the vector of <u>strains</u> is,

(3): $$v = Cs + Mx,$$

where $C \in M_{nxn}(R)$ is called the <u>flexibility matrix</u> (with respect to all critical points).

Now, we give very briefly some mechanical principles which give rise to problem (1).

Denoting, $q' = q + h(x)$ and using the expression (2) we may consider the inequality, $q - Mu + h(x) - MAMx \geq 0$ of problem (1) as requiring the "stress point" s lie in the convex polyhedron,

(4): $$E = \{s \in R^n | q' - Ms \geq 0\}.$$

The set E, called the <u>elastic domain</u> defines the region such that the "plastic activity" takes place only if the stress point reaches the boundary of E.

The vector x of plastic multipliers affects the structural behaviour in two ways.

First, it contributes to the values of stresses and strains directly through the relations, $x \in R^n_+$ and $<x, q - Mu + h(x) - MAMx> = 0$ of problem (1).

The value of x also affects the value of q, hence the shape of the elastic domain, through h.

The map h is called the (work-) hardening rule.

Hence the <u>plastic flôw rules</u> specify conditions, x must satisfy,

(5): $$x \geq 0$$

(6): $$<x, q' - Ms> = 0.$$

The boundary hyperplanes of E represent <u>yield limits</u>.

Thus, the rules (5) and (6) together with $s \in E$ state that the i-th plastic multiplier x_i can take on a positive value only if the stress point reaches the i-th yield limit.

It is important to remark that this model implies also a <u>parametric complementarity problem</u>.

In this sense we remark that in representation of problem (1) the effect of the applied loads is accounted for in the form of the vector u of <u>linear-elastic responses</u> to the loads.

For many practical problem it is important to determine the complete evolution of stresses and strains during a loading process where the vector of loads of the form λg is applied for $\lambda \in [0, \bar{\lambda}]$ with $\bar{\lambda} > 0$.

This situation implies the following <u>parametric complementarity problem</u>,

(7): $$\left|\left| \begin{array}{l} \text{find } x \in R^n_+ \text{ such that,} \\ q - \lambda Mu + h(x) - MAMx \in R_+ \text{ and} \\ < x, q - \lambda Mu + h(x) - MAMx > = 0. \end{array} \right.\right.$$

Certainly, if we denote the solution of problem (7) by $x(\lambda)$ it is important to know if $x(\lambda)$ is a monotone increasing function (with respect to λ).

In the same area of practical problems we consider now another model proposed by Strang. [G. Strang: Discrete plasticity and the complementarity problem. Proceedings fo the U.S.-Germany symposium on finite elements, M.I.T. Press, Cambridge, Massachusetts (1977)]. The construction of this model is based on Maier's model.

The importance of this model is that it is a second linear complementarity problem.

Consider the deformation theory, the static case and then we have,

(1°) $\varepsilon = Bu$ (compatiblity of the strains ε and displacements u),

(2°) $B^t\sigma = f$ (equilibrium of stresses σ with external forces f),

(3°) $\varepsilon = S^{-1}\sigma + p$ (a splitting of strains into elastic and plastic parts),

(4°) $p = N\lambda$; $\lambda \geq 0$ (plastic strains related to normals to piecewise linear yield surfaces),

(5°) $\emptyset = N^t\sigma - H\lambda - k \leq 0$ (the yield condition with hardening matrix H)

(6°) $\langle\emptyset, \lambda\rangle = 0$ (the complementarity condition).

Hence we obtain the model,

$$(8):\quad \left\| \begin{array}{l} -\emptyset = (H + N^tSN)\lambda - N^tSBu + k \geq 0 \\ -B^tSN\lambda + B^tSBu + f = 0 \\ \lambda \geq 0,\ u \text{ (free) and} \\ \langle\emptyset, \lambda\rangle = 0 \end{array} \right.$$

Here, the stiffness matrix B^tSB is symmetric and positive semi-definite.

If we seek λ and u then problem (8) is exactly a second linear complementarity problem (S.L.C.P.).

In the positive definite case, u can be eliminated and a standard (L.C.P.) is obtained.

2.4.3 Free Boundary Problems

Many steady-state free boundary problems can be formulated as variational inequalities of the form:

$$(1):\quad \left\| \begin{array}{l} \text{find } x \in D \text{ satisfying,} \\ a(x, y - x) + \langle q, y - x\rangle \geq 0;\ \forall y \in D. \end{array} \right.$$

where D is a closed convex set in an infinite Hilbert space E, $q \in E\ (= E^*)$ and $a: E \times E \to R$ is a continuous bilinear operator.

If $(E, \langle\ ,\ \rangle, \leq)$ is an ordered Hilbert space and $D = K = \{x \in E \mid x \geq 0\}$ we consider the dual cone K^* of K and let $M: E \to E^*$ (= E) the operator defined by, $a(x, y) = \langle y, Mx\rangle$.

In this case problem (1) is equivalent to the infinite-dimensional linear complementarity problem,

$$\text{(L.C.P.)}: \quad \left\| \begin{array}{l} \text{find } x \in E \text{ satisfying,} \\ x \in K,\ Mx + q \in K^* \text{ and} \\ \langle x, Mx + q \rangle = 0 \end{array} \right.$$

When (1) or (L.C.P.) is approximated using finite differences or finite elements one obtains a finite-dimensional (L.C.P.) of the form:

$$(2): \quad \left\| \begin{array}{l} \text{find } x \in R^n_+ \text{ such that,} \\ Ax - b \in R^n_+ \text{ and} \\ \langle x, Ax - b \rangle = 0 \end{array} \right.$$

where b is a known n-vector and A is a n x n matrix.

In practice, a large number of free-boundary problems, as for example, porous flow through dams, journal bearing lubrication and elastic-plastic torsion can be formulated as (L.C.P.) as follows.

Given a domain $\mathcal{D} \subseteq R^n$ with boundary $\partial\mathcal{D}$ and given functions f and g find v defined in $\mathcal{D}$ such that,

$$(3): \quad \left\| \begin{array}{l} -Lv(x) + f(x) \geq 0;\ x \in \mathcal{D} \\ v(x) \geq 0;\ x \in \mathcal{D} \\ v(x)[-Lv(x) + f(x)] = 0;\ x \in \mathcal{D} \\ v(x) = g(x);\ x \in \partial\mathcal{D} \end{array} \right.$$

where L is a given second-order elliptic operator.

If $\mathcal{D}$ is approximated by a regular grid, then the grid can be divided into $n = |G|$ interior points G and $|\partial G|$ boundary points ∂G.

When (3) is approximated using finite differences in G one obtain a finite-dimensional (L.C.P.),

$$(4): \quad \left\| \begin{array}{l} -L\,u(x) + f(x) \geq 0;\ x \in G, \\ u(x) \geq 0;\ x \in G, \\ u(x)[-L\,u(x) + f(x)] = 0;\ x \in G, \\ u(x) = g(x);\ x \in \partial G. \end{array} \right.$$

where u(x) is an approximation of v(x) at the grid points x and where L_h is a difference operator which approximates L.

By eliminating the known values of u(x) in ∂G, the (L.C.P.) (4) may be written in matrix form,

$$(5): \quad \left\| \begin{array}{l} \text{find } u \in R^n_+ \text{ such that,} \\ Au + b \in R^n_+ \text{ and} \\ \langle u, Au + b \rangle = 0 \end{array} \right.$$

where A is a n x n matrix symmetric and positive definite and u is the n-vector of values of u(x) in G.

Problem (5) has a unique solution and it has certain special features:

a) A is a large matrix, for example a $10^4 \times 10^4$ matrix,

b) A is a sparse matrix. Typically, each row of A will have no more than five non zero elements. However, A^{-1} is a full matrix.

We consider now two typical examples.

A. Porous flow through a dam [A48][A184].

We consider the well-known free-boundary problem of the flow of water through a porous dam. The geometry is shown in fig. 1.

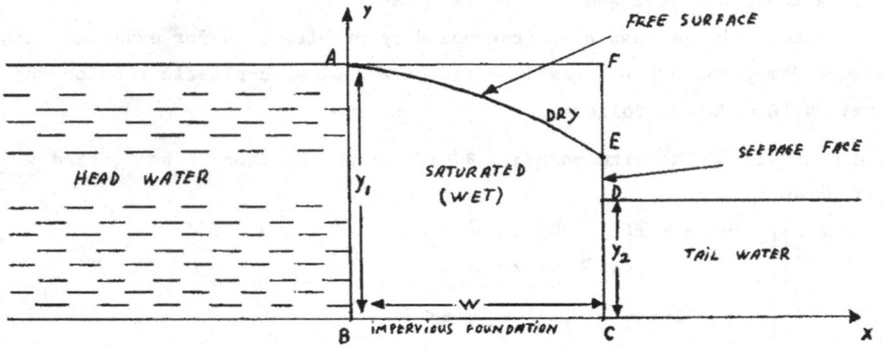

Figure 1

Water seeps from a reservoir of height y_1 through a rectangular dam of width w to a reservoir of height y_2.

Part of the dam is saturated and the remainder of the dam is dry. The wet and dry regions are separated by an unknown free boundary which must be found as part of the solution.

As shown by Baiocchi [C. Baiocchi: Sur un problème à frontière libre traduisant le filtrage de liquides à travers des milieux poreaux. C.R. Acad. Sci. Paris. A273 (1971), 1215-1217] the problem can be formulated as follows:

$$
(6): \quad \left\| \begin{array}{l}
\text{find } u \text{ on the rectangle } \mathcal{R} = ABCF \text{ sucht that,} \\
-\nabla^2 u + 1 \geq 0; \ \text{in } \mathcal{R} \\
\quad u \geq 0; \qquad \text{in } \mathcal{R} \\
u\,(-\nabla^2 u + 1) = 0; \ \text{in } \mathcal{R} \\
u = g = \begin{cases} (y_1 - y)^2/2; & \text{on AB} \\ (y_2 - y)^2/2; & \text{on CD} \\ \left[y_1^2(w - x) + y_2^2\,(x)\right]/2w; & \text{on BC} \\ 0 & ; \text{on DFA.} \end{cases}
\end{array} \right.
$$

where ∇^2 is the Laplace operator.

B. The journal bearing problem [A68][A48][A172].

We describe now the free-boundary problem of the finite-length journal bearing.

A journal bearing consists of a rotating shaft (the journal) separated from a surface (the bearing) by a thin film of lubricating fluid.

Journal bearing are among the most used basic engineering components; their annual production is in the billions.

A journal bearing of general engineering interest is shown in fig. 2.

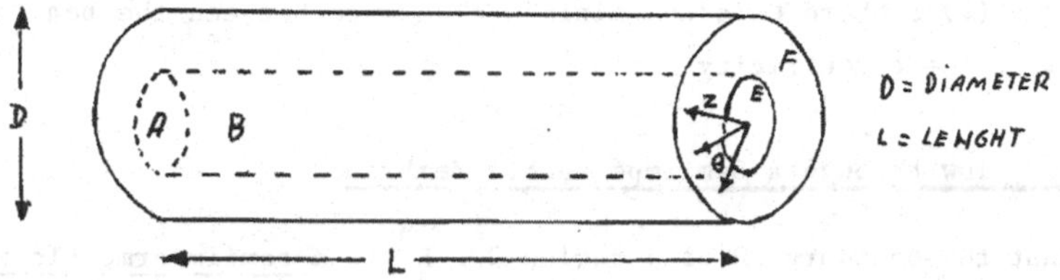

Figure 2

One wishes to find the presure distribution of the thin lubricating film. Since the gap between the journal and the bearing is very narrow, the pressure may be assumed not to vary across the gap; therefore, the problem becomes a two-dimensional problem in the rectangular domain R = ABEF in the θZ-plane (fig. 3).

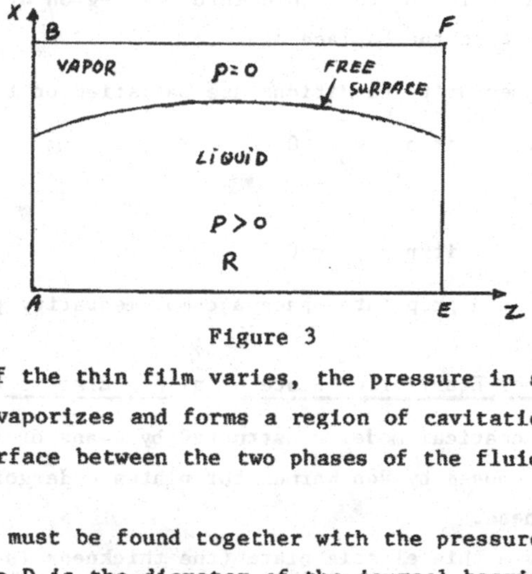

Figure 3

As the thickness of the thin film varies, the pressure in some region may become so low that the fluid vaporizes and forms a region of cavitation.

The resulting interface between the two phases of the fluid is called the free boundary.

This free boundary must be found together with the pressures distribution.

Let $x = D\theta/2$, where D is the diameter of the journal bearing.

Let $k(x,z)$ be the film thickness, U be the surface velocity of the journal, ρ and μ the density and viscosity of the lubricant.

The problem can be formulated as [G. Cimmati: On a problem of the theory of lubrication governed by a variational inequality. Appl. Math. Optim. Nr 3 (1977) 227-243]:

$$(7): \left\| \begin{array}{l} \text{find } p \text{ on the rectangle } \mathcal{R} = ABEF, \text{ such that,} \\ -Lp + f \geq 0; \text{ in } \mathcal{R}, \\ \quad p \geq 0; \quad \text{in } \mathcal{R}, \\ p(-Lp + f) = 0; \quad \text{in } \mathcal{R} \\ p = 0; \text{ on } \partial\mathcal{R} \end{array} \right.$$

where, $L = \frac{\partial}{\partial x}\left(\frac{\partial}{\mu} k^3 \frac{\partial}{\partial x}\right) + \frac{\partial}{\partial z}\left(\frac{\rho}{\mu} k^3 \frac{\partial}{\partial z}\right)$ and $f = 6U \frac{\partial(\rho\, k)}{\partial x}$

The density ρ and viscosity μ are usually assumed to be constants.

The film thickness k depends only on the θ coordinate and can be approximated by $k = C(1 + \varepsilon \cos(\theta))$; where C is the minimum clearance between the bearing and the journal and ε is the eccentricity ratio.

2.4.4 Fluid Flow through a Semiimpermeable Membrane

We say that the boundary $\partial\Omega$ of a region $\Omega \subsetneq R^3$ is a semiimpermeable membrane if a fluid may only flow into the region Ω.

Consider a region Ω with a semiimpermeable boundary. In this case, if p is the fluid pressure of a fluid flowing into the region Ω, we have, $\frac{\partial p}{\partial \nu} \geq 0$; where ν is the outer normal to the boundary $\partial\Omega$.

The fluid outside the region Ω creates a pressure $p_o(x)$; $x \in \partial\Omega$.

It is known that the inflow of the fluid into the region Ω is described by the equation, $\frac{\partial p}{\partial t} = \Delta p$, where Δ is the Laplace operator.

The following complementarity conditions are satisfied on the boundary $\partial\Omega$:

$$\left\| \begin{array}{ll} \frac{\partial p}{\partial \nu} = 0, & \text{if } p - p_o > 0 \\ \text{and} & \\ \frac{\partial p}{\partial \nu} > 0, & \text{if } p - p_o = 0 \end{array} \right.$$

Hence, we obtain in an appropriate space a complementarity problem.

2.4.5 The Post-Critical Equilibrium State of a Thin Elastic Plate

We consider the mathematical model constructed by means of the classical nonlinear description proposed by Von Karman for plates undergoing large delfections relative to their thickness.

We suppose that Ω is a thin elastic plate (the thickness is supposed to be constant) resting without friction on a flat rigid support. The material is supposed to be homogeneous and isotropic.

Mathematically speaking, Ω is identified to a bounded open connected subset of R^2.

The plate Ω is assumed to be clamped on $\gamma_1 \subset \gamma$ and simply supported on $\gamma_2 = \gamma \backslash \gamma_1$, where γ is the boundary of Ω which is supposed to be sufficiently regular.

The middle of the plate is referred to an Euclidean coordinate system $(0, x_1, x_2, z)$ and the z-axis represents the vertical displacement of the plate.

We suppose that a lateral variable load (λL_α) with λ positive and increasing is applied to the boundary of Ω.

Because Ω is a thin elastic plate, we observe that if λ exceeds a critical value termed the critical load (specific for each plate Ω), then the plate deflects out of its plane and we say that it buckles.

We consider the Sobolev space,

$$H^2(\Omega) = \{u \in L^2(\Omega) \,|\, \frac{\partial u}{\partial x_i}, \frac{\partial^2 u}{\partial x_i \partial x_j} \in L^2(\Omega), \forall\, i, j = 1, 2\}$$

equipped with the usual norm $\|\cdot\|_{H^2(\Omega)}$.

If $\vec{n}$ stands for the normal to γ exterior to Ω and if $\frac{\partial \cdot}{\partial n}$ denotes the normal exterior derivative, the physical problem leads us to consider the closed subspace E of $H^2(\Omega)$ defined by,

$$E = \{z \in H^2(\Omega) \,|\, z_{|\gamma} = 0 \text{ and } \frac{\partial z}{\partial n}_{|\gamma_1} = 0 \text{ a.e.}\}$$

As described in[A123] we may define a continuous bilinear form $a(v, w)$ on $E \times E$ which is coercive whenever γ_1 is for instance nonempty.

Therefore, we consider E equipped with the scalar product $<,>$ defined by this form (the associated norm $\|\cdot\|$ is equivalent to the initial norm $\|\cdot\|_{H^2(\Omega)}$).

For a fixed λ the equilibrium post-critic of the plate is governed by the following complementarity problem,

$$\text{(N.C.P.)}: \left\| \begin{array}{l} \text{find } z \in K \text{ such that,} \\ z - \lambda L(z) + C(z) \in K^* \text{ and} \\ <z, z - \lambda L(z) + C(z)> = 0 \end{array} \right.$$

where: $K = \{z \in E \,|\, z \geq 0 \text{ a.e. on } \Omega\}$ is a closed convex cone and, represents the vertical admissible displacements, $\lambda \geq 0$ is exactly the intensity of the lateral load (λL_α),

L is a self-adjoint linear compact operator defined by the nature of the load applied on the boundary of Ω,

C is a nonlinear continuous compact operator connected with the expansive properties of the plate. It is positively homogeneous of order $p = 3$.

We note that C is the Fréchet derivative of $z \to \frac{1}{4} <z, C(z)>$ and satisfies also the following properties:

a) $<z, C(z)> \geq 0$; for each $z \in E$,

b) $z \in E$ and $<z, C(z)> = 0$ imply $z = 0$.

2.5 Maximizing oil production

The following model was studied by Meerov, Bershchanski and Litvak [A206], [A26] as mathematical model of maximizing oil production.

The idea of this model is the following.

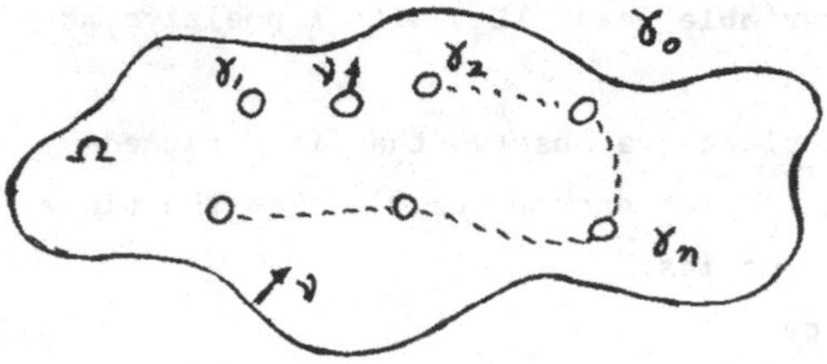

Consider a two-dimensional region Ω bounded by an outer boundary γ_o and n inner boundaries $\gamma_1, \gamma_2, \ldots, \gamma_n$ representing the wells.

The pressure distribution $p = p(x)$ in this region is described by the Laplace equation $\Delta p = 0$ with the boundary conditions, $p = p_o > 0$ on γ_o and $p = p_i \geq 0$ on $\gamma_i (i = 1, 2, \ldots, n)$.

Also, on $\gamma_i (i = 1, 2, \ldots, n)$ we have $Q_i \geq 0$, where Q_i is the output of the i-th well given by,

$$Q_i = - \int_{\gamma_i} \left(\frac{\partial p}{\partial \nu}\right) d\gamma,$$

where ν is the outer normal to the boundary of Ω.

These conditions imply that a fixed constant pressure is maintained on the outer boundary.

The pressure cannot fall below a certain level specified by engineering and physical factors.

In these conditions the fluid which is a mixture of oil and water flows out of Ω through the boundaries γ_i.

We remark that the proportion k_i of oil in the fluid pumped from well i is known.

The problem is to find the pressure $p_1, p_2, \ldots, p_n$ such that the output of pure oil,

$$Q = \sum_{i=1}^{n} k_i Q_i$$

is maximized.

This is a linear programming problem having the property that there is an implicit linear relationship between oil outputs and pressures.

It is shown [A206] [A26] that in the optimal solution for each i we have either $p_i = 0$ or $Q_i = 0$.

In this model the complementarity conditions ($p_i = 0$ or $Q_i = 0$) do not give the optimal solution of the dual problem coincides with the unique solution of the corresponding linear complementarity problem.

2.6 Complementarity Problems in Economics

The complementarity theory is much used in the study of equilibrium of diverse economic phenomena.

It is known that the equilibrium problem of an economy is traditionally stated in terms of excess demand functions determined by the endowments of the economy, the preferences of its members and its technology.

We consider in this section several economic situations which imply diverse complementarity problems.

2.6.1 Equilibrium in a Competitive Economy

We consider a system with n different commodities and m traders buying and selling these commodities.

Certainly, each trader maximizes his preference function $u^i(x)$; $(i=1,2, \ldots, m)$, subject to $<p, x> \leq <p, w^i>$, where p is the price vector $(p \geq 0)$, w^i is the vector of commodities available to the i-th trader and x is the vector of commodities that he wishes to obtain.

If the solution of this maximization problem is denoted by $x^i(p)$, we consider the function, $f(p) = \sum_{i=1}^{m} (x^i(p) - w^i)$, which is called the excees demand function.

In this case, the Walras law asserts that, for the equilibrium price p we have, $f(p) \geq 0$ and $<p, f(p)> = 0$.

That is, we obtain that to find the economic equilibrium in this case is necessary to solve the following complementarity problem:

$$(\text{C.P.}): \left\| \begin{array}{l} \text{find } p \in R^n_+ \text{ such that,} \\ \\ f(p) \in R^n_+ \text{ and } <p, f(p)> = 0. \end{array} \right.$$

2.6.2 Equilibrium of an Economy or a Sector with Production

Consider an economy or a sector with production and to simplify, we will restrict ourselves to an economy with competitive behaviour throughout with no price distortions.

Suppose that our economy has m commodities and n activities with constant returns to scale production.

Our model is based on the assumption that, production is characterized by a linear technology matrix with fixed input-output coefficients.

For $i = 1, 2, \ldots m$ and $j = 1, 2, \ldots, n$ we denote:

$p = (p_i)$ the vector of prices,

$b = (b_i)$ the vector of endowments,

$d(p) = (di(p))$ the market demand functions, which we assume to be point-to-point and continuously differentiable but not necessarily linear,

$y = (y_j)$ the vector of activity levels,

$c = (c_j)$ the vector of unit costs of operating the activities,

$A = (a_{ij})$ the technology matrix of input-output coefficients consistent with unit production, where $a_{ij} > 0$ ($a_{ij} < 0$) denotes an output (input).

It is known that there exist several ways to characterize an equilibrium, but we use here the Scarf's definition.

[See: Scarf H.E. and Hansen T.: Computation of economic equilibria. Yale University Press, New Haven, Conn. (1983)].

A price vector p_* and a vector of activity levels y_* constitute a competitive equilibrium if:

(1°): no activity earns a positive profit,

$$c - A^t p_* \geq 0,$$

(2°): no commodity is in excess demand,

$$b + Ay_* - d(p_*) \geq 0,$$

(3°): no prices or activity levels are negative

$$p_* \geq 0,\ y_* \geq 0,$$

(4°): an activity earning a deficit is not used and an operated activity has no loss,

$$(c - A^t p_*)^t y_* = 0$$

(5°): a commodity in excess supply has zero price and a positive price implies market clearance,

$$p_*^t (b + Ay_* - d(p_*)) = 0.$$

We note that the vector c of operating costs represents factors of production that are exogenous to the economy or sector under consideration.

If (1°) - (5°) describes a general equilibrium problem of a closed economy, then the cost vector $c = 0$, because all prices will be determined simultaneously and no single price will be exogeneously given.

In this case, demands $d_i(p)$ for $i = 1, 2, \ldots, m$ are functions of all prices in the economy, i.e. both product and factor prices.

Furthermore, these demand functions will usually be specified in a manner consistent with individual household utility maximization, that is,

$d_i(p) = \sum_h x_i^h$, where x_i^h is the h-th household's utility maximizing demand of commodity i.

Housefhold's excess demands are given by $d(p) - b$.

We remark that if the demands satisfy each individual household's budget and there is nonsatiation, then $p^t d(p) = p^t b$, and the demand functions $d(p)$ are homogeneous of degree 0 in all prices.

We observe that when $c = 0$ conditions (1°) - (5°) determine only relative prices, that is, if the vector p_* represents equilibrium prices, so does λp_* for any scalar $\lambda > 0$.

Hence, we are free to normalize the prices.

Now, we observe that we can associate to this model the following complementarity problem,

$$(\text{C.P.}): \left\| \begin{array}{l} \text{find } z \in R^r \text{ such that,} \\ z \geq 0,\ F(z) \geq 0 \text{ and} \\ \langle z, F(z) \rangle = 0, \end{array} \right.$$

where, $z = \begin{bmatrix} y \\ p \end{bmatrix}$; $F(z) = \begin{bmatrix} c - A^t p \\ b + Ay - d(p) \end{bmatrix}$

and the order "$\leq$" is defined by R^r_+.

<u>Remark</u>

The problem (C.P.) is a nonlinear complementarity problem since $d(p)$ is not necessary linear.

Also in more complicated situation A is possible to be a nonlinear mapping, as for example, if we consider the dynamical model of two countries in the presence of unemployment.

<u>2.6.3 Equilibrium of an Economy with Institutional Constraints on Prices</u>

In the theory of perfect competition, it is supposed that there are no institutional restrictions upon prices.

The presence of such constraints implies, for example, that the market price and the marginal productivity (shadow price) of the factors of production will not necessarily coincide.

Unless such constraints are introduced, models cannot explain the simultaneous existence of excess supply of an item and yet a positive market price.

If there is a gap between market and shadow prices then this implies a serious problem.

By what set of prices are the economic agents, actions guided.

In this model, considered in this section, we assume that one sector of the economy (which may be interpreted as the private sector) is guided by market prices. The other (the public sector) is guided by shadow prices.

We remark that with conventional optimization techniques, it is awkward and sometimes impossible to handle this type of problem.

Firstly, we consider a "small" economy that can sell unlimited amounts of its outputs upon world markets.

Consider such an economy with n productive processes.

The matrix $A = (a_{ij})$ describes the technology available to the economy.

The coefficients $a_{ij} \geq 0$ ($i = 1, 2, \ldots, m$, $j = 1, 2, \ldots, n$) denote the amount of item i required to operate activity j at unit level.

Consider the vector $B = \begin{pmatrix} b_1 \\ b_2 \\ \vdots \\ b_m \end{pmatrix}$. If $B \geq 0$ then it denotes the resources and if a component were negative this would denote a delivery requirement rather than a resource available.

We assume that the pay off from each activity is determined by world market prices and that these are independent of the activity levels in the economy. Let $C = \begin{pmatrix} c_1 \\ \vdots \\ c_n \end{pmatrix}$, where c_j denotes the value of output of activity j when it is operated at the unit level.

Finally, let $X = \begin{pmatrix} x_1 \\ x_2 \\ \vdots \\ x_n \end{pmatrix}$ denote the vector of activity levels and let $Z = \begin{pmatrix} z_1 \\ z_2 \\ \vdots \\ z_m \end{pmatrix}$ denote the vector of item prices.

<u>A competitive equilibrium is characterized by a nonnegative vector of activity levels X_* and a nonnegative vector of prices Z_* such that:</u>

(1°): <u>$W_* = B - AX_* \geq 0$, (the production plan is feasible)</u>

(2°): <u>$U_* = -C + A^t Z_* \geq 0$, (no activity makes a positive profit)</u>,

(3°): <u>$\langle Z_*, W_* \rangle = 0$, (an item in excess supply has a zero price)</u>,

(4°): <u>$\langle X_*, U_* \rangle = 0$, (no activity that makes a negative profit is operated at a positive level).</u>

Clearly, the problem of computing a competitive equilibrium is equivalent to solving the following complementarity problem:

$$\text{(C.P.)}: \quad \left\| \begin{array}{l} \text{find } \tilde{X} \geq 0 \text{ such that,} \\ \tilde{W} = D - A\tilde{X} \geq 0 \text{ and} \\ \langle \tilde{X}, \tilde{W} \rangle = 0 \end{array} \right.$$

where: $\tilde{X} = \begin{bmatrix} X \\ Z \end{bmatrix}$; $A = \begin{bmatrix} 0 & -A^t \\ A & 0 \end{bmatrix}$; $\tilde{W} = \begin{bmatrix} U \\ W \end{bmatrix}$ and $D = \begin{bmatrix} -C \\ B \end{bmatrix}$

Obviously the problem (C.P.) is considered in R^{n+m} ordered by R_+^{n+m}.

We note that in this case the problem (C.P.) is equivalent to solving the linear programming problem,

$$\text{(L.P.)}: \quad \left\| \begin{array}{l} \text{maximize} \langle C, X \rangle \\ \text{subject to: } AX \leq B \text{ and } X \geq 0 \end{array} \right.$$

It is important to remark that in the economy described above, there were no institutional constraints upon prices.

In the next example we shall introduce two types of constraints:

α) lower bounds upon individual prices,

β) upper bounds upon individual prices.

The presence of such constraints implies for example, that the factors of production are not necessarily paid according to their marginal productivity.

In the case of a minimum price, marginal productivity will coincide with the market price only if the marginal productivity of that item exceeds the minimum price. The converse holds for an upper bound upon a price.

<u>Suppose now, that there exists a divergence between market prices and shadow prices.</u>

In this case we assume that one sector of the economy (which may be interpreted as the private sector) is guided by market prices. The other (the public sector) is guided by shadow prices.

We suppose that the first <u>n_1 activities</u> refer to the private sector and $n_2 = n - n_1$ activities are publicly controlled.

Thus we partition X and C so that,

$X = \begin{bmatrix} X_1 \\ X_2 \end{bmatrix}$; $C = \begin{bmatrix} C_1 \\ C_2 \end{bmatrix}$ and where X_1 and C_1 refer to private activities whereas X_2 and C_2 refer to public activities.

We suppose also, that there is a maximum price constraint on the first m_1 items, whereas the remaining m_2 items have a minimum price constraint.

The latter also include those items where the only price constraint is nonnegativity. Partition the matrix A,

$$A = \begin{bmatrix} A_{11} & A_{12} \\ A_{21} & A_{22} \end{bmatrix}$$

where, $A_{11} \in \mathcal{M}_{m_1 x n_1}(R)$; $A_{12} \in \mathcal{M}_{m_1 x n_2}(R)$; $A_{21} \in \mathcal{M}_{m_2 x n_1}(R)$ and $A_{22} \in \mathcal{M}_{m_2 x n_2}$.

The vector $Z = \begin{bmatrix} z_1 \\ \vdots \\ z_m \end{bmatrix}$ denotes the shadow prices and $Y = \begin{bmatrix} y_1 \\ \vdots \\ y_m \end{bmatrix}$ denotes the market prices.

Partition Z and Y, $Z = \begin{bmatrix} Z_1 \\ Z_2 \end{bmatrix}$, $Y = \begin{bmatrix} Y_1 \\ Y_2 \end{bmatrix}$

such that Z_1 and Y_1 refer to factors with a maximum price constraint and Z_2 and Y_2 refer to items with a minimum price constraint.

We denote by Q_1 and Q_2 maximum and minimum prices respectively, that is $Y_1 \le Q_1$ and $Y_2 \ge Q_2$.

An equilibrium (if it exists), is characterized by a vector of activity levels X^o, market prices Y^o and shadow prices Z^o such that:

(1°): $W^o = B - AX^o \ge 0$ (i_1)(the production plan is feasible)

(2°): $U_1^o = [A_{11}^t \; A_{21}^t]\, Y^o - C_1 \ge 0$ (i_2),

$U_2^o = [A_{12}^t \; A_{22}^t]\, Z^o - C_2 \ge 0$ (i_3),

(no activity makes a positive profit),

(3°) $V_1^o = Q_1 - Y_1^o \ge 0$

$V_2^o = Y_2^o - Q_2 \ge 0$

(4°) $T_1^o = Z_1^o - Y_1^o \ge 0$ (i_4),

$T_2^o = Y_2^o - Z_2^o \ge 0$ (i_5)

(there is a nonnegative wedge between market and shadow prices),

(5°) $< Z^o, W^o > = 0$,

(if a factor is in excess supply, it has a zero shadow price),

(6°) $< X^o, U^o > = 0$ (if an activity makes a negative profit, it is operated at a zero level),

(7°) $< T^o, V^o > = 0$ (the shadow price equals the market price if the institutional constraint is not binding).

If we substitute (3°) into (i_2), (i_4) and (i_5) we obtain,

$$U_1^o = - C_1 + A_{11}^t Q_1 + A_{21}^t Q_2 - A_{11}^t V_1^o + A_{21}^t V_2^o,$$

$$T_1^o = Z^o + V_1^o - Q_1; \; T_2^o = V_2^o + Q_2 - Z_2^o$$

We obtain that, the problem of computing an equilibrium is equivalent to solving the following special complementarity problem:

(S.C.P.): Find vectors $(X^o, Z^o, Y^o, U^o, W^o, T^o)$ such that

$$M_1\tilde{W} = D - M_2\tilde{X},$$

$$\tilde{W} \geq 0,\ \tilde{X} \geq 0 \text{ and}$$

$$\langle X^o, U^o\rangle + \langle Z^o, W^o\rangle + \langle V^o, T^o\rangle = 0$$

where: $\tilde{X} = \begin{bmatrix} X^o \\ Z^o \end{bmatrix}$; $\tilde{W} = \begin{bmatrix} V^o \\ U^o \\ W^o \\ T^o \end{bmatrix}$; $X^o = \begin{bmatrix} X_1^o \\ X_2^o \end{bmatrix}$; $Z^o = \begin{bmatrix} Z_1^o \\ Z_2^o \end{bmatrix}$;

$V_o = \begin{bmatrix} V_1^o \\ V_2^o \end{bmatrix}$; $U^o = \begin{bmatrix} U_1^o \\ U_2^o \end{bmatrix}$; $W^o = \begin{bmatrix} W_1^o \\ W_2^o \end{bmatrix}$; $T^o = \begin{bmatrix} T_1^o \\ T_2^o \end{bmatrix}$;

$$M_1 = \begin{bmatrix} A_{11}^t & -A_{21}^t & I_{n_1} & 0 & 0 & 0 & 0 & 0 \\ 0 & 0 & 0 & I_{m_2} & 0 & 0 & 0 & 0 \\ 0 & 0 & 0 & 0 & I_{m_1} & 0 & 0 & 0 \\ 0 & 0 & 0 & 0 & 0 & I_{m_2} & 0 & 0 \\ -I_{m_1} & 0 & 0 & 0 & 0 & 0 & I_{m_1} & 0 \\ 0 & -I_{m_2} & 0 & 0 & 0 & 0 & 0 & I_{m_2} \end{bmatrix}$$

$$M_2 = \begin{bmatrix} 0 & 0 & 0 & 0 \\ 0 & 0 & -A_{12}^t & -A_{22}^t \\ A_{11} & A_{12} & 0 & 0 \\ A_{21} & A_{22} & 0 & 0 \\ 0 & 0 & -I_{m_1} & 0 \\ 0 & 0 & 0 & I_{m_2} \end{bmatrix}; \quad D = \begin{bmatrix} -C_1 + A_{11}^t Q_1 + A_{21}^t Q_2 \\ -C_2 \\ B_1 \\ B_2 \\ -Q_1 \\ Q_2 \end{bmatrix}$$

I_s $(s = n_1;\ n_2;\ m_1;\ m_2)$ is an identity matrix of order $s \times s$

<u>Remark</u>

The complementarity problem (S.C.P.) does not correspond to a linear programming problem and it is not sufficiently studied.

The general model of this problem is the following special implicit complementarity problem:

$$\text{(S.I.C.P.)}: \quad \left\| \begin{array}{l} \text{find } X, Z, U, V, W, T \text{ such that,} \\ X, Z, U, V, W, T \geq 0, \\ F(X, Z, U, W, V, T) = 0 \text{ and} \\ \langle X, U \rangle + \langle Z, W \rangle + \langle V, T \rangle = 0 \end{array} \right.$$

This problem is of the form of the problem studied by Meister[A122].

2.6.4 Equilibrium of International of Interregional Trade in a Single Commodity.

Consider now another model used in economics. This model was considered by Glassey [A110] and it concerns the problem of equilibrium of international or interregional trade in a single commodity.

We suppose that in each region there is a market characterized by classical supply and demand curves.

In this model the equilibrium price and quantity produced and consumed will be determined, in the absence of imports or exports, by the intersection of these curves.

If imports are introduced into this local market consumption will exceed production but at a lower equilibrium price.

The first simplification in this model is to observe that, from global view, the internal supply and demand of each region are irrelevant; it is only the net import quantity and the local market equilibrium price that matter.

Suppose that there is a linear relation between price and net imports of the form,

$$(1): \quad p_\alpha = a_\alpha - b_\alpha y_\alpha.$$

for every region α.

We denote:

n = the total number of regions implied in market,

p_α = the equilibrium price in the α-th region,

y_α = the net import of the α-th region

a_α = the equilibrium price in the absence of imports (and exports) ($a_\alpha \geq 0$),

$x_{\alpha\beta}$ = exports from region α to region β ($x_{\alpha\beta} \geq 0$)(is a flow variable),

b_β = is related to elasticity of supply and demand ($b_\alpha \geq 0$),

$c_{\alpha\beta}$ = cost per unit shipped from α to β.

Here $\alpha, \beta = 1, 2, \ldots, n$.

We note that if p_α exceeds a_α, then supply locally exceeds demand and the difference is available for export.

Thus in (1) y_α is not restricted to be nonnegative; negative values of y_α are simply interpreted as exports.

The price p_α has been determined from global equilibrium considerations, the local supply and demand quantities are uniquely determined.

We suppose also, that all shipments take place over the least cost route and therefore the $c_{\alpha\beta}$ will obey the triangle inequality:

$$c_{\alpha\beta} \le c_{\alpha\gamma} + c_{\gamma\beta};\ \forall\ \alpha, \beta, \gamma \in \{1, 2, \dots, n\}.$$

In this case to find the trade equilibrium is to solve the following problem:

$$(2):\quad \left\|\begin{array}{l} \text{find } p_\alpha;\ y_\alpha;\ x_{\alpha\beta};\ \forall\ \alpha, \beta = 1, 2, \dots n \text{ such that} \\ p_\alpha = a_\alpha - b_\alpha y_\alpha; \qquad (\theta_1) \\ y_\alpha = \sum_{\beta=1}^{n} x_{\beta\alpha} - \sum_{\beta=1}^{n} x_{\alpha\beta}; \qquad (\theta_2) \\ p_\alpha + c_{\alpha\beta} - p_\beta \ge 0 \\ x_{\alpha\beta} \ge 0 \text{ and} \\ x_{\alpha\beta}\,(p_\alpha + c_{\alpha\beta} - p_\beta) = 0 \end{array}\right.$$

We denote, $I = \{(\alpha, \beta) | \alpha \ne \beta;\ \alpha, \beta = 1, 2, \dots, n\}$ and we have, $|I| = n\,(n-1)$.

If $k = (\alpha, \beta) \in I$ we denote:

$z_k = x_{\alpha\beta};\ z = (z_k)_{k\in I};\ q_k = a_\alpha - a_\beta + c_{\alpha\beta};\ q = (q_k)_{k\in I}$

and we consider the matrix $M = (m_{k\ell})_{k,\ell\in I}$ defined by:

$$m_{k\ell} = \begin{cases} b_\alpha + b_\beta; & \text{if } k = \ell = (\alpha, \beta) \\ -(b_\alpha + b_\beta); & \text{if } k = (\alpha, \beta),\ \ell = (\beta, \alpha) \\ b_\beta; & \text{if } k = (\alpha, \beta),\ \ell = (\gamma, \beta) \text{ and } \alpha \ne \gamma \\ b_\alpha; & \text{if } k = (\alpha, \beta),\ \ell = (\alpha, \gamma) \text{ and } \beta \ne \gamma \\ -b_\beta; & \text{if } k = (\alpha, \beta)\ \ell = (\beta, \gamma) \text{ and } \alpha \ne \gamma \\ -b_\alpha; & \text{if } k = (\alpha, \beta),\ \ell = (\gamma, \alpha) \text{ and } \beta \ne \gamma \\ 0 & \text{: for the all non considered cases.} \end{cases}$$

We remark now, that problem (2) is equivalent to the following complementarity problem:

$$(\text{C.P.}):\quad \left\|\begin{array}{l} \text{find } z \in R^{n(n-1)} \text{ such that,} \\ z \in R_+^{n(n-1)};\ q + Mz \in R_+^{n(n-1)} \text{ and} \\ \langle z, q + Mz \rangle = 0 \end{array}\right.$$

__Remarks__

1) To obtain the problem (C.P.) we eliminate p and y_α in problem (2) using relations (θ_1) and (θ_2).

2) If p_α is not supposed to be linear, for every α then we obtain a nonlinear complementarity problem.

2.7 Equilibrium of Traffic Flows

It is known that network equilibrium models arise in applied contexts as varied as urban transportation, energy distribution, electrical networks and water resource planning.

We give now here one of these applications, namely for predicting traffic flow on a congested transportation network, using the model known as Wardrop's model of traffic equilibrium.

The following construction was proposed by Aashtiani and Magnanti [A1].

The equilibrium model is defined on a transportation network [N, A] with nodes N, directed arcs A and with a given set I of origin-destination (O, D) node pairs.

Nodes represent concentrations of population, business districts, street intersections etc and arcs model streets and arteries or might be introduced to model connections (and wait time) between legs of a trip, between modes or between streets at an intersection.

We denote:

I = the set of (O, D) pairs,

P_i = the set of "available" paths for flow for (O, D) - pair i (which might, but need not, be all path joining the (O, D) - pair),

h_p = the flow on path p,

h = the vector of $\{h_p\}$ with dimension $n_1 = \sum_{i\in I} P_i$ equal to the total number of (O, D) - pairs and path combinations,

u_i = an accessibility variable, shortest travel time, (or generalized cost) for (O,D) pair i,

u = the vector $\{u_i\}$ with dimension $n_2 = |I|$

$D_i(u)$ = the demand function for (O, D) - pair i, $D_i: R_+^{n_2} \to R_+$,

$T_p(h)$ = the delay time, or general disutility, function for path p;

$$T_p: R_+^{n_1} \to R_+.$$

The problem to find the equilibrium on this transportation network is to solve the following problem:

$$(I): \left\|\begin{array}{lr} \text{find } h \text{ and } u \text{ such that,} & \\ (T_p(h) - u_i)\, h_p = 0; \ \forall\, p \in P_i \text{ and } \forall\, i \in I & (a) \\ T_p(h) - u_i \geq 0; \ \ \forall\, p \in P_i \text{ and } \forall\, i \in I & (b) \\ \sum_{p\in P_i} h_p - D_i(u) = 0; \ \forall\, i \in I & (c) \\ h \geq 0 & (d) \\ u \geq 0 & (e) \end{array}\right.$$

In this model, $P = \bigcup_{i \in I} P_i$ denotes the set of all "available" paths in the network and we assume that the network is strongly connected; i.e., for any (O, D) - pair $i \in I$ there is at least one path joining the origin to the destination (that is, $|P_i| \geq 1$).

The equations (a) and (b) in (1) require that for any (O, D) - pair i, the travel time (generalized travel time for all paths $p \in P_i$ with positive flow $h_p > 0$, is the same and equal to u_i, which is less than or equal to the tavel time for any path with zero flow.

Equation (c) requires that the total flow among different paths between any (O, D) - pair i equal the total demand $D_i(u)$, which in turn depends upon the congestion in the network through the shortest path variable u.

Finally, conditions (d) and (e) in (1) state that both flow on paths and minimum travel times should be nonnegative.

If $x = (h, u) \in R^n$, where $n = n_1 + n_2$ we denote,

$$f_p(x) = T_p(h) - u_i; \quad \forall\, p \in P_i \text{ and } \forall\, i \in I,$$

$$g_i(x) = \sum_{p \in P_i} h_p - D_i(u); \quad \forall\, i \in I,$$

and we consider the function $F: R^n \to R^n$ defined by,

$$F(x) = (f_p(x);\ \forall\, p \in P_i \text{ and } \forall\, i \in I;\ g_i(x);\ \forall\, i \in I)$$

We consider now, the following complementarity problem:

$$(2)(\text{C.P.}): \left\| \begin{array}{l} \text{find } x \in R^n \text{ such that,} \\ x \geq R^n_+,\ g_i(x) \geq 0;\ \forall\, i \in I; \\ f_p(x) \geq 0;\ \forall\, p \in P_i \text{ and } \forall\, i \in I; \\ f_p(x)\, h_p = 0;\ \forall\, p \in P_i \text{ and } \forall\, i \in I \text{ and} \\ g_i(x)\, u_i = 0;\ \forall\, i \in I. \end{array} \right.$$

We observe that, since any solution $x_* = (h_*, u_*)$ to the traffic equilibrium; problem satisfies $g_i(x_*) = 0$ for all $i \in I$, the solution x_* solves the nonlinear complementarity problem (2) (C.P.) independent of the nature of the delay functions $T_p(h)$ and the demand function $D_i(u)$.

In [A1] is proved the following result.

<u>Proposition 2.3</u>

<u>If, for all $p \in P$ the function $T_p: R^{n_1}_+ \longrightarrow R_+$ is a positive function (that is, $h \in R^{n_1}_+ \setminus \{0\}$ implies $T_p(h) > 0$), then problem (1) is equivalent to the nonlinear complementarity problem (2).</u>

Proof

Since every solution of problem (1) is a solution of problem (2) it is sufficient to show that any solution to (2) is a solution to (1).

Suppose the contrary, that is, suppose that there is an $x = (h, u)$ satisfying (2) but that $g_i(x) = \sum_{p \in P_i} h_p - D_i(u) > 0$, for some $i \in I$.

In this case, $g_i(x)\, u_i = 0$ implies $u_i = 0$, and from the definition of D_i we have, $\sum_{p \in P_i} h_p > D_i(u) \geq 0$, which implies that $h_p > 0$, for some $p \in P_i$.

Then, for this particular p the relation, $f_p(x)\, h_p = 0$ implies that,

$$f_p(x) = T_p(h) - u_i = 0 \text{ or } T_p(h) = u_i$$

Because $u_i = 0$ we obtain, $T_p(h) = 0$ which contradicts the assumption $T_p(h) > 0$.

□

2.8 The Linear Complementarity Problem and Circuit Simulation

Recent researches prove that the Linear Complementarity Problem can be used with success in the simulation of electronic circuits.

The Linear Complementarity Problem used in this case is,

$$\left\| \begin{array}{l} \text{find } x_o \in R^n_+ \text{ such that,} \\ Ax_o + b \in R^n_+ \text{ and } \langle x_o, Ax_o + b \rangle = 0 \end{array} \right.$$

where the matrix A has the following properties,

i) A is a large sparse matrix,

ii) A is generally not restricted to a certain class of matrices used currently in the Complementarity Theory as for example: Positive definite matrices, P-matrices, Strictly copositive matrices, Strictly semimonotone matrices etc.,

iii) problem (1) has one or more solutions.

These characteristics imposed for problem (1) a special study and the necessity to find special numerical methods.

The reader find more details in: [J.T.J. Van Eijndhoven: Solving the linear complementarity problem in circuit simulation. SIAM J. Control and Optim. Vol. 24 Nr. 5 (1986), 1050-1062].

2.9 Complementarity and Fixed Point Theory

As a final interesting application of complementarity theory, we remark only in this section, the possibility to use the complementarity theory in the study and approximation of fixed points for a mapping $f: K \to K$, where K is a closed convex cone in a Hilbert space. We shall give some details on this problem in the next chapter.

Comments

Models 2.1.1, 2.2.1, 2.1.3 and 2.1.4 are the first applications of the Complementarity Problem. These models were studied by, Cottle [A 42 - A 52], Dorn [A 75], Cottle and Dantzig [A 53 - A 54], Lemke [A 173 - A 178], Eaves [A 80 - A 87], Habetler and Price [A 114 - A 115], Gould and Tolle [A 111], Karamardian [A 155 - A 158], Kaneko [A 142 - A 153], Kojima [A 161 - A 162], Kostreva [A 168 - A 171], Mangasarian [A 191 - A 200], Moré [A 219 - A 220], Murty [A 224 - A 230], Garcia [A 102 - A 105], Tamir [A 280-283], Mohan [A 215 - A 218] etc.

The Complementarity Problem with restrictions (models 2.1.5) was studied by Ibaraki [A 124] and systematically by Judice and Mitra [A 141].

The connection between the Game Theory and the Complementarity Theory (models 2.2.1 and 2.2.2) was proved by Lemke [A 173] and studied by Lemke and Howson [A 179], Mitra [A 213] etc.

We find the equivalence between the Variational Inequality on a convex cone and the Complementarity Problem in [A 157], [A 48] and [A 259].

The first applications of the Complementarity Problems in Mechanics and Engineering were established by Du Val [A 79], Ingleton [A 125] and Maier [A 185 - A 186]. Interesting applications of the Complementarity Problem in Engineering we find in [A 187 - A 188], [A 143], [A 147], [A 151 - A 153].

Interesting applications of the Complementarity Theory to tree boundary problems (particularly to the Lubrication Theory) we find in [A 172], [A 65 - A 66], [A 68 - A 69].

The Complementarity Theory was used in the study of the post-critical equilibrium state of a thin elastic plate by Isac [A 128] and Isac and Théra [A 132 - A 133].

The Complementarity Problem was used as mathematical model of maximizing oil production by Meerov, Berschanski and Litvak [A 206], [A 26 - A 27].

The Complementarity Theory is intensively used in Economics [A 29 - A 30], [A 70], [A 110], [A 99], [A 120], [A 131], [A 136], [A 160], [A 202], [A 247], [A 249], [A 250], [A 255], [A 274], [A 303] and in study of equilibrium of traffic flow [A 1], [A 190].

The recent paper [J.T.J. Van Eijundhaven: solving the linear complementarity problem in circuit simulation. SIAM J. Control and Optim. Vol. 24, Nr. 5 (1986), 1050-1062] proves that the Complementarity Problem can be used in the simulation of electronic circuits.

CHAPTER 3

EQUIVALENCES

This chapter is important since we will prove that the general complementarity problem is equivalent (if some assumptions are satsified) to different problems, as for example, to a variational inequality, to the least element problem, to an unilateral minimization problem etc.

So, by these connections we obtain both, new methods to study complementarity problems and the possibility to use the complementarity theory in the study of these problems.

Moreover, by these equivalences we may use the complementarity problem to interpret by a different way results obtained in the variational inequalities theory, in optimization theory or in the study of particular phenomena modeled by variational inequalities etc.

Let $(E, \|\ \|)$ be a Banach space and let $\langle E, E^*\rangle$ be the duality defined by, $\langle x, u\rangle = u(x)$, for every $x \in E$ and every $u \in E^*$, where E^* is the topological dual of E.

If $K \subset E$ is a closed convex cone and K^* is the dual of K, that is,

$$K^* = \{u \in E^* \mid \langle x, u\rangle \geq 0;\ \forall\, x \in K\},$$

we denote by "$\leq$" the preorder relation defined on E by, $(x \leq y) \iff (y - x \in K)$ and respective on E^* by, $(u \leq v) \iff (v - u \in K^*)$.

If E with respect to "$\leq$" is a vector lattice then E^* with respect to the order "$\leq$" defined by K^* is also a vector lattice [J.L. Kelley and I. Namioka: Linear topological spaces. Springer-Verlag, New York (1976). Appendix p. 224-231].

Given a closed convex cone $K \subset E$ and two mappings, $f: E \to E^*$ and $\phi: E \to R$, we denote by $\mathcal{F}$ the feasible set of f with respect to K, that is, $\mathcal{F} = \{x \in E \mid x \in K$ and $f(x) \in K^*\}$ and we consider the following problems:

(I): for a given element $u \in E^*$ find $x_* \in \mathcal{F}$ such that,

$$u(x_*) = \min_{x \in \mathcal{F}} u(x).$$

[Nonlinear Program]

(II): find $x_* \in \mathcal{F}$ such that,

$x_* \leq x$, for every $x \in \mathcal{F}$

[The least element problem]

(III): $\left\| \begin{array}{l} \text{find } x_* \in K \text{ such that,} \\ \Phi(x_*) = \min\limits_{x \in K} \Phi(x) \end{array} \right.$ [Unilateral minimization]

(IV): $\left\| \begin{array}{l} \text{find } x_* \in K \text{ such that,} \\ \langle x - x_*, f(x_*)\rangle \geq 0; \ \forall\, x \in K \end{array} \right.$ [Variational inequality]

(V): $\left\| \begin{array}{l} \text{find } x_* \in \mathcal{F} \text{ such that,} \\ \langle x_*, f(x_*)\rangle = 0 \end{array} \right.$ [Complementarity problem]

If E is a Hilbert space and $f:K \to E$ has the particular form, $f(x) = x - g(x)$; where $g:K \to K$, then we consider the problem:

(VI): $\left\| \begin{array}{l} \text{find } x_* \in K \text{ such that,} \\ g(x_*) = x_* \end{array} \right.$ [Fixed point problem].

Definition 3.1

A mapping $f:E \to E^*$ is said to be:

1°) positive at infinity, if for every $x \in E$ there exists a real number $\rho(x) \geq 0$, such that, $\langle y - x, f(y)\rangle > 0$, for every $y \in E$ such that $\|y\| \geq \rho(x)$,

2°) coercive, if there exists a function $c:R_+ \to R_+$ such that,

i) $c(t) > 0$, for every $t > 0$,

ii) $\lim\limits_{t \to \infty} c(t) = +\infty$,

iii) $\langle x, f(x)\rangle \geq c(\|x\|)\|x\|$,

3°) bounded, if there exists a mapping $b:R_+ \to R_+$ such that, $\|f(x)\| \leq b(\|x\|)$, for every $x \in E$. □

Remark

If f is coercive, bounded and $\lim\limits_{t \to \infty} \dfrac{b(t)}{tc(t)} = 0$, then using the inequality,

$$\langle y - x, f(y)\rangle \geq \langle y, f(y)\rangle - \|f(y)\| \cdot \|x\|$$

we can prove that f is positive at infinity.

If $(E, \|\ \|)$ and $(F, \|\ \|)$ are Banach spaces, we consider on $L(E, F)$ the structure of Banach space defined by the norm,

$$\|u\| = \sup\{\|u(x)\| \mid x \in E, \|x\| \leq 1\}; \text{ for every } u \in L(E, F).$$

We recall that a mapping $f:E \to F$ is Gâteaux differentiable at the point $x_o \in E$, if there exists an element $u \in L(E, F)$ such that, for every $x \in E$ we have,

(3.1): $$\|f(x_o + tx) - f(x_o) - u(tx)\|/t \xrightarrow{t \to 0} 0.$$

We denote in this case, $[(d/dt)f(x_o + tx)]\big|_{t=0} = u(x)$.

The linear mapping u associated to x_o in formula (3.1) is unique and we denote, $u = f'(x_o)$.

The mapping $f':E \to L(E, F)$ defined by $x_o \longrightarrow f'(x_o)$ (if it exists) is called the Gâteaux derivative of f.

If there exists an element $u \in L(E, F)$ such that,

$$\lim_{\|h\| \to 0} \frac{\|f(x_o+h) - f(x_o) - u(h)\|}{\|h\|} = 0$$

then we say that u is the Fréchet derivative of f at $x_o \in E$.

If for f there exists the Fréchet derivative at $x_o \in E$, then f has the Gâteaux derivative at x_o.

We recall also the following classical result.

Proposition 3.1

If for a mapping $f:E \to E^*$ the following assumptions are satisfied:

i) f is a C^1 - function,

ii) for every $x_o \in E$ the mapping, $(y, z) \to \langle f'(x_o) y, z \rangle$; $\forall y, z \in E$, is a symmetrical bilinear form (that is, $\langle f'(x_o)y, z \rangle = \langle f'(x_o)z, y \rangle$; $\forall y, z \in E$), then there exists a function $\varphi:E \to R$ such that, $f = \varphi'$ the Gâteaux derivative of φ). □

Definition 3.2

Let $(E(\tau), \leq)$ be an ordered locally convex space which is supposed to be a vector lattice.

We denote, $K = \{x \in E \mid x \geq 0\}$.

A mapping $f:E \to E^*$ is said to be a Z-mapping with respect to K, if and only if, for every $x, x_o, x_1 \in K$ such that, $\inf(x, x_1 - x_o) = 0$ we have, $\langle x, f(x_1)-f(x_o)\rangle \leq 0$.

□

Remark

A linear mapping $f:E \to E^*$ is a Z-mapping with respect to K, if and only if, for every $x, y \in K$ such that $\inf(y, x) = 0$, we have, $\langle y, f(x)\rangle \leq 0$.

Examples

(3.1) If $E = R^n$, $K = R^n_+$ and $f(x) = Ax$, where $A = (a_{ij}) \in M_{nxn}(R)$, then f is a Z-mapping, if and only if, A is a Z-matrix, that is, $a_{ij} \leq 0$; $\forall i \neq j$; $i, j = 1, 2, \ldots, n$.

(3.2) We suppose again, $E = R^n$ and $K = R^n_+$.

Given $f:R^n \to R^n$, where $f = (f_1, f_2, \ldots f_n)$, we denote, $e^j = (e^j_k) \in R^n$ the vector defined by,

$$e_k^j = \begin{cases} 0 \text{ if } k \neq j \\ 1 \text{ if } k = j. \end{cases}$$

For every i, j = 1, 2, ..., n and $x \in R_+^n$, we consider the mapping, $f_{ij}: R_+ \to R$ defined by, $f_{ij}(t) = f_i(x + te^j)$; for every $t \in R_+$.

We say that f is a Z-mapping in the Rheinboldt's sense on R_+^n if for every $x \in R_+^n$ and i, j = 1, 2, ..., n, $i \neq j$ the mappings f_{ij} are monotone decreasing (antitone).

[W.C. Rheinboldt: On M-functions and their applications to nonlinear Gauss-Seidel iterations and to network flows. J. of Math. Anal. and Appl. 32 (1970) p. 274-304].

If a mapping is a Z-mapping in this sense, then it is a Z-mapping.

Z-matrices and Z-mapping in the Rheinboldt's sense are currently used in Economics.

(3.3) Let $E = H_0^1(\Omega)$, where Ω is a bounded domain (open connected set) in R^n and $H_0^1(\Omega)$ is the Sobolev space of once-differentiable functions vanishing on $\partial\Omega$. [R.A. Adams: Sobolev spaces. Academic Press (1975)].

We consider in this case,

$$K = \{x \in E \mid x(t) \geq 0, \text{ a.e.}\}$$

and it is known that $E^* = H^{-1}(\Omega)$.

$E = H_0^1(\Omega)$ is a vector lattice, since the functions,

$$\sup (x, y)(t) = \sup (x(t), y(t)); \ \forall\, t \in \Omega$$

$$\inf (x, y)(t) = \inf (x(t), y(t); \ \forall\, t \in \Omega,$$

for every $x, y \in E$ are representations of elements in $H_0^1(\Omega)$. [H. Lewy and G. Stampacchia: On the regularity of the solution of a variational inequality. Comm. Pure Appl. Math. 22 (1969) p. 153-188].

We consider the linear self-adjoint operator,

$$(Mu)(t) = -\sum_{i,j=1}^{n} \frac{\partial}{\partial t_j}\left(a_{ij}(t)\frac{\partial u}{\partial t_i}(t)\right); \ t \in \Omega,$$

where $a_{ij}(t)$ are continuously differentiable functions.

We assume that -M is uniformly elliptic, so that, there exists α 0 satisfying,

$$\sum_{i,j=1}^{n} a_{ij}(t)\,\xi_i\xi_j \geq \alpha\,|\xi|^2; \ \forall\, t \in \Omega,$$

for all $\xi = (\xi_i) \in R^n$, where $|\xi| = [\sum_{i=1}^{n} \xi_i^2]^{\frac{1}{2}}$.

As defined, the operator M can only be applied to functions u, twice differentiable.

The standard theory of elliptic operators [J.L. Lions and E. Magenes: Non-homogeneous boundary value problems and applications-Ì. Springer-Verlag (1972)] gives a method to extend the domain of definition of M such that we can consider M as a mapping from $E = H_0^1(\Omega)$ to its dual $E^* = H^{-1}(\Omega)$.

In [A68] is proved that M is a Z-mapping with respect to K.

Proposition 3.2

A Gâteaux continuous differentiable mapping $f: E \longrightarrow E^*$ is a Z-mapping with respect to a closed convex cone $K \subset E$, if and only if, for every $x_o \in K$, $f'(x_o)$ is a linear Z-mapping.

Proof

We suppose that f is a Z-mapping and $x_o \in K$ an arbitrary element.

For $x, y \in K$ such that, $\inf(x, y) = 0$, we denote, $x_t = x_o + tx$; $\forall\, t \in R_+$.

Since, $\inf(tx, y) = 0$, for every $t \in [0, 1]$, we have, $\inf(x_t - x_o, y) = 0$ and because, f is a Z-mapping we get,

$$\langle y, f(x_t) - f(x_o)\rangle \le 0;\ \forall\, t \in [0, 1],$$

which implies,

$$\langle y, f'(x_o)x\rangle = \lim_{\substack{t\to 0\\ t>0}} \langle y, f(x_t) - f(x_o)\rangle / t \le 0,$$

that is, $f'(x_o)$ is a Z-mapping.

Conversely, we suppose that $f'(x_o)$ is a Z-mapping for every $x_o \in K$.

For x_1, x_2 and y elements of K such that, $\inf(x_1 - x_2, y) = 0$, we denote, $x = x_1 - x_2$; $x_t = x_2 + tx$; $\forall\, t \in R_+$ and integrating the formula, $\frac{d}{dt} f(x_t) = f'(x_t)(x)$ on [0, 1] we obtain,

$$\text{(3.2)} \qquad \langle y, f(x_1) - f(x_2)\rangle = \langle y, \int_0^1 f'(x_t)(x)dt\rangle = \int_0^1 \langle y, f'(x_t)(x)\rangle dt.$$

Now, since $f'(x_t)$ is a Z-mapping we deduce from (3.2) that, $\langle y, f(x_1)-f(x_2)\rangle \le 0$ and the proof is finished.□

The following results on variational inequalities are necessary to prove the principal results of this chapter.

We note that the next theorem is a generalization to locally convex spaces of the Hartman-Stampacchia theorem.

Theorem 3.1 [Hartman-Stampacchia].

Let $\mathcal{C}$ be a compact convex subset of a locally convex space E and let $f: \mathcal{C} \to E^*$ be continuous (with respect to the strong topology).

Then there exists $x_* \in \mathcal{C}$ such that,

$$\text{(3.3)} \qquad \langle x - x_*, f(x_*)\rangle \ge 0;\ \text{for every } x \in \mathcal{C}.$$

Proof

We find an elegant proof of this result base on Fan-Kakutani fixed point theorem in [R.B. Holms: Geometric functional analysis and its applications. Springer-Verlag (1975)].

Lemma 1

Let $f:E \longrightarrow E^*$ be a monotone hemicontinuous operator, where E is a reflexive Banach space.

If $K \subset E$ is a closed convex cone, then an element $u_o \in K$ satisfies,

(3.4) $\quad < x - u_o, f(u_o) > \geq 0$; for every $x \in K$, if and only if,

(3.5) $\quad < x - x_o, f(x) > \geq 0$; for every $x \in K$.

Proof

(3.5) ==>(3.4). If $x \in K$ is an arbitrary element, we denote, $x_t = (1-t)\, u_o + tx$; $0 < t < 1$ and substituting in (3.5) x by x_t we get,

(3.6) $\quad < x_t - u_o, f(x_t) > \geq 0$, that is, $< t(x - u_o), f(x_t) > \geq 0$ or,

(3.7) $\quad < x - u_o, f(x_t) > \geq 0$.

But, since f is hemicontinuous, supposing $t \to 0$ we obtain that $f(x_t)$ is weakly convergent to $f(u_o)$ and from (3.7) we deduce,

$$< x - u_o, f(u_o) > \geq 0; \text{ for every } x \in K,$$

that is (3.4) is true.

(3.4) ==> (3.5). Indeed, supposing (3.4) true, since f is monotone we obtain,

$< x - u_o, f(x) > \geq \; < x - u_o, f(u_o) > \geq 0$; for every $x \in K$. □

Lemma 2

Let $K \subset R^n$ be a closed convex cone and let $f:K \longrightarrow R^n$ be a continuous mapping. For every $r > 0$ there exists $u_o \in \{x \in K \,|\, \|x\| \leq r\}$ such that,

$$< x - u_o, f(u_o) > \geq 0; \text{ for every } x \in K, \text{ such that, } \|x\| \leq r.$$

Proof

We apply Theorem 3.1 [Hartman-Stampacchia], where $C = \{x \in K \,|\, \|x\| \leq r\}$. □

Lemma 3

Consider a Banach space E, $r > 0$ a real number and $K \subset E$ a closed convex cone. Suppose also $f:E \to E^*$ to be a mapping such that,

(3.8) $\quad < x, f(x) > \geq 0$; for every $x \in K$, such that, $\|x\| = r$.

If $u_o \in \{x \in K \,|\, \|x\| \leq r\}$ satisfies (3.4) for every $x \in \{x \in K \,|\, \|x\| \leq r\}$ then,

i) $\|u_o\| < r$,

ii) u_o satisfies (3.4).

Proof

Consider, $C = \{x \in K \,|\, \|x\| \leq r\}$. If we suppose,

$$< x - u_o, f(u_o) > \geq 0; \text{ for every } x \in C,$$

then for $x = 0$ we obtain, $< u_o, f(u_o) > \leq 0$ and since, $u_o \in C$ from (3.8) we get, $\|u_o\| < r$.

To prove ii) we consider an arbitrary element $x \in K$ and we denote, $x_t = (1-t)u_o + tx;\ 0 < t < 1$.

We have, $x_t \in K$, for every $0 < t < 1$. and for t sufficiently small we observe that $x_t \in C$.

Now, our assumption on u_o implies, $< x_t - u_o, f(u_o) > \geq 0$ and finally,

$$< x - u_o, f(u_o) > \geq 0; \text{ for every } x \in K. \quad \square$$

Theorem 3.2 [Browder-Hartman-Stampacchia]

Consider a reflexive Banach space E, $f: E \to E^*$ a monotone hemicontinuous operator and $K \subset E$ a closed convex cone.

If there exists $r > 0$ such that, $< x, f(x) > > 0$, for every $x \in K$ such that, $\|x\| = r$, then there exists an element $x_o \in K$, such that, $\|u_o\| \leq r$ and $< x - x_o, f(x_o) > \geq 0$; for every $x \in K$.

Moreover, if f is strictly monotone then x_o is unique.

Proof

Let $\mathcal{F}$ be the directed family of finite dimensional subspaces F of E such that, $F \cap K \neq \phi$, orderd by $F_\alpha \leq F_\beta \iff F_\alpha \subseteq F_\beta$.

If $F \in \mathcal{F}$ we denote by i_F the mapping, $i_F: F \to E$ defined by, $i_F(x) = x$ and by i_F^* its adjoint and we consider the mapping, $f_F = i_F^* o f o i_F$.

The operator f_F is monotone and continuous (since it is hemicontinuous and dim F $+ \infty$).

If $K_F = F \cap K$, then it is clear that $< x, f_F(x) > > 0$, for every $x \in K_F$, such that, $\|x\| = r$.

From Lemma 2 there exists an element $x_F \in K_F$ satisfying, $\|x_F\| \leq r$ and such that,

$$< x_F - x\ ,_F f(x\ ) > \geq 0; \text{ for every } x_F \in K\ , \text{ satisfying, } \|x\| \leq r.$$

But Lemma 3 implies,

(3.9) $\qquad < x_F - x\ ,_F f(x\ ) > \geq 0;$ for every $x_F \in K$.

Consider now the net $\{x_F\}_{F \in \mathcal{F}}$, which is weakly compact since, $\{x_F\}_{F \in \mathcal{F}} \subset \{x \in E \mid \|x\| \leq r\}$, and we know that there exists a weakly convergent subnet $\{x_{F_i}\}_{i \in I}$ of $\{x_F\}_{F \in \mathcal{F}}$. We denote $x_o = (w) - \lim\limits_{i \in I} x_{F_i}$.

Since I is a cofinal subset of $\mathcal{F}$, if $x \in K$ is an arbitrary element, there exists $i_o \in I$ such that, $x \in F_{i_o}$. For every x_{F_i} $(i \in I)$ such that $x_{F_i} \in F_i \supset F_{i_o}$, we have, $K_{F_i} = K \cap F_i \supset K \cap F_{i_o} = K_{F_{i_o}}$, and hence $x \in K_{F_i}$, which implies,

$$< x - x_{F_i}, f(x_{F_i}) > \geq 0; \ x_{F_i} \in F_i \supset F_{i_o};\ i \in I.$$

Using the monotony of f we get,

$$<x - x_{F_i}, f(x)> \geq \; < x - x_{F_i}, f(x_{F_i})> \; \geq 0; \text{ for every } i \in I \text{ such that, }$$

$F_i \supset F_{i_o}$.

Hence, $<x - x_{F_i}, f(x) > \geq 0$; for every $i \in I$, such that, $F_i \supset F_{i_o}$ and computing the weak limit we obtain, $< x - x_o, f(x) > \geq 0$, or $< x - x_o, f(x) > \geq 0$; for every $x \in K$, which implies (using Lemma 1) that,

$$<x - x_o, f(x_o) > \geq 0; \text{ for every } x \in K.$$

If we suppose now that f is strictly monotone and x_o, x_1 satisfy,

$$<x_1 - x_o, f(x_o) > \geq 0,$$
$$<x_o - x_1, f(x_1) > \geq 0,$$

we obtain, $<x_1 - x_o, f(x_1) - f(x_o) > \leq 0$, which implies finally, $x_o = x_1$. □

Considering now the definitions of problems (I) - (VI) we prove several fundamental equivalence theorems.

Theorem 3.3

If an operator $f: E \to E^*$ is the Gâteaux derivative of a function $\Phi:E \dashrightarrow R$, then every solution x_* of problem (III) is a solution of problem (IV).

Proof

Indeed, we suppose, $\Phi(x_*) = \min_{x \in K} \Phi(x)$ and if $x \in K$ is an arbitrary element, then the set, $A = \{x_* + t(x - x_*) | 0 \leq t \leq 1\}$ is a subset of K and we have,

$$\Phi (x_*) = \min_{x \in A} \Phi(x),$$

which implies that x_* is a solution of problem (IV) since,

$$<x - x_*, f(x_*)> = [\frac{d}{dt} \Phi (x_* + t (x - x_o))]\Big|_{t=0} \geq 0. \qquad □$$

Theorem 3.4

If $f:E \to E^*$ is a monotone operator which is also the Gâteaux derivative of a function $\Phi:E \dashrightarrow R$, then every solution x_* of problem (IV) is a solution of problem (III).

Proof

This fact is a direct consequence of the following classical equivalences, well known in convex analysis:

i) Φ is a convex mapping,

ii) $\Phi(x) - \Phi(x_*) \geq <x - x_*, \Phi'(x_*) >; \forall x, x_* \in E$,

iii) Φ' is a monotone operator,

where Φ' is the Gâteaux derivative of Φ. □

Theorem 3.5

Suppose $E(\tau)$ to be a locally convex space, $K \subset E$ a closed convex cone and $f:E \to E^*$ a mapping.

An element $x_* \in K$ is a solution of problem (IV) if and only if x_* is a solution of problem (V).

Proof

If $x_* \in K$ is a solution of problem (IV) then we have,

(3.10) $\quad \langle x - x_*, f(x_*) \rangle \geq 0; \ \forall\, x \in K$

and if $y \in K$ is an arbitrary element and $x = y + x_*$, then from (3.10) we get,

$$\langle y, f(x_*) \rangle \geq 0; \ \forall\, y \in K$$

which implies that $x_* \in \mathcal{F}$.

Now, if we consider $x = 2x_*$ in (IV) we obtain, $\langle x_*, f(x_*) \rangle \geq 0$ and if we put $x = 0$ also in (IV) we deduce, $\langle x_*, f(x_*) \rangle = 0$, that is, we have, that x_* is a solution of problem (V).

Conversely, if we suppose that x_* is a solution of problem (V) we have, $\langle x_*, f(x_*) \rangle = 0$ and $\langle x, f(x_*) \rangle \geq 0$, for every $x \in K$, which clearly imply,

$$\langle x - x_*, f(x_*) \rangle \geq 0; \ \forall\, x \in K,$$

that is, x_* is a solution of problem (IV). □

Theorem 3.6

Suppose $E(\tau)$ to be locally convex space, which is also a vector lattice, $K = \{x \in E \mid x \geq 0\}$ and consider $f:E \to E^*$ to be a Z-mapping strictly monotone.

If x_* is a solution of problem (IV), then x_* is a solution of problem (II) too.

Proof

Suppose that $x_* \in K$ is a solution of problem (IV), that is,

(3.11) $\quad \langle x - x_*, f(x_*) \rangle \geq 0; \ \forall\, x \in K.$

From Theorem 3.5 we have that x_* is a solution of problem (V) and hence, $x_* \in \mathcal{F}$.

We prove now that, $x_* \leq x$; for every $x \in \mathcal{F}$.

To prove this fact, we will prove that, $x_* = x_0 = \inf(x, x_*)$.

Indeed, $x_0 \geq 0$ and from (2) we obtain, $\langle x_0 - x_*, f(x_*) \rangle \geq 0$, which implies,

(3.12) $\quad \langle x_* - x_0, f(x_*) - f(x_0) \rangle = \langle x_0 - x_*, f(x_0) - f(x_*) \rangle =$

$$= \langle x_0 - x_*, f(x_0) \rangle - \langle x_0 - x_*, f(x_*) \rangle \leq \langle x_0 - x_*, f(x_0) \rangle .$$

Since, $f(x) \geq 0$ and $x_0 - x_* \leq 0$, we have,

(3.13) $\quad \langle x_0 - x_*, f(x) \rangle \geq 0.$

If we denote, $y = x_* - x_0$, then we observe that,

$$\inf(x - x_0, y) = \inf(x - x_0, x_* - x_0) = \inf(x, x_*) - x_0 = 0$$

and because f is a Z-mapping, we deduce,

(3.14) $\quad \langle y, f(x) - f(x_o) \rangle \leq 0.$

By addition, from (3.14) and (3.13) we obtain, $< x_o - x_*, f(x_o) > \leq 0$ and using (3.12) we get,

$$< x_* - x_o, f(x_*) - f(x_o) > \leqslant 0$$

which implies (since f is strictly monotone), $x_* = x_o$ and the proof is finished.□

Theorem 3.7

Let $E(\tau)$ be a locally convex space ordered by a closed convex cone $K \subset E$ and let $f: E \to E^*$ be a mapping.

For an arbitrary element $u \in K^*$, if x_* is a solution of problem (II) then x_* is a solution of problem (I).

Proof

Indeed, if x_* is a solution of problem (II) then we have, $x_* \leq x$; for every $x \in F$, which implies, $u(x - x_*) \geq 0$, that is $u(x_*) \leq u(x)$.□

The following result is a distinguished property of Z-mappings.

Proposition 3.3

Let $E(\tau)$ be a vector lattice, supposed locally convex. Let $f: E \to E^*$ be a strictly monotone Z-mapping with respect to $K = \{x \in E | x \geq 0\}$.

If f satisfies the property,

(P): $(\forall z \in K)(\exists v \in K)(\forall w \in K)(< w - v, f(z + v) > \geq 0)$

then the set F has the property,

(Q): $x_1, x_2 \in F \Longrightarrow \inf(x_1, x_2) \in F$.

Proof

If $x_1, x_2 \in F$ then we have, $x_1, x_2 \in K$ and $f(x_1), f(x_2) \in K^*$.

Since $z = \inf(x_1, x_2)$ is an element of K, it is sufficient to prove that $f(z) \in K^*$.

Using the property (P) we obtain an element $x_o \in K$ such that, $y = z + x_o$ satisfies,

(3.15) $\quad < w - x_o, f(y) > \geq 0; \ \forall w \in K.$

Particularly, if we denote, $w = x + x_o$, for every $x \in K$, then from (3.16) we obtain $< x, f(y) > \geq 0$; for every $x \in K$, that is, $f(y) \in K^*$.

Now, we prove that $z = y$. Indeed, we denote, $u = \inf(x_1, y)$ and since $y, x_1 \geq z$, we have that $u \geq z$ and hence, $w = u - z \in K$ can be used in (3.15) and we obtain,

(3.16) $\quad < u - y, f(y) > \geq 0.$

Because, $\inf(x_1 - u, y - u) = \inf(x_1, y) - u = 0$

and f is a Z-mapping we get,

(3.17) $\langle y - u, f(x_1) - f(u) \rangle \le 0.$

But, since $f(x_1) \in K^*$ and $u - y \in -K$ we have also,

(3.18) $\langle u - y, f(x_1) \rangle \le 0.$

By addition, from (3.17) and (3.18) we deduce,

(3.19) $\langle u - y, f(u) \rangle \le 0$

and from (3.19) and (3.16) we obtain,

(3.20) $\langle u - y, f(u) - f(y) \rangle \le 0.$

Now, from (3.20) since f is strictly monotone, we have, $u = y$ and because $u = \inf(x_1, y)$, we get, $y \le x_1$.

By a similar calculus we obtain, $y \le x_2$, which implies, $y \le \inf(x_1, x_2) = z$ and because, $y = z + x_0 \ge z$, we have $y = z$ and the proof is finished. □

The next result proves that the property (P) is satisfied in an important practical case.

Proposition 3.4

If E is a reflexive Banach space and $f: E \to E^*$ a monotone hemicontinuous mapping.

If f is positive at infinity, then it satisfies the property (P).

Moreover, if f is strictly monotone, then for every $z \in K$ the element v satisfying the property (P) is unique.

Proof

For an arbitrary element $z \in E$ we consider the hemicontinuous monotone mapping, $f_z(x) = f(z + x)$.

Since f is positive at infinity we consider, $\rho_1 = \|z\| + \rho(z)$ (where $\rho(z)$ is the number used in definition 3.1) and we denote, $u = z + x$.

If $z \in K$, then for every $x \in K$ such that, $\|x\| \ge \rho_1$, we have, $u \ge 0$, $\|u\| \ge \rho(z)$ and hence, $\langle x, f_z(x) \rangle = \langle u - z, f(u) \rangle > 0$; for every $x \in K$ such that $\|x\| \ge \rho_1$.

From Theorem 3.2 (Browder-Hartman-Stampacchia, we obtain an element $v \in K$ such that, $\langle w - v, f_z(v) \rangle \ge 0$; for every $w \in K$, that is, f satisfies the property (P).

If f is strongly monotone, then a similar calculus as in the proof of Theorem 3.2 implies that v is unique. □

Theorem 3.8

Let $E(\tau)$ be a locally convex space and we denote, $K = \{x \in E \mid x \ge 0\}$. Suppose E to be a vector lattice.

Let $f: E \longrightarrow E^*$ be a mapping such that f satisfies the property (Q) and consider an element $u \in K^*$ such that, $u(x) > 0$; for every $x \in K \setminus \{0\}$.

If problem (I) associated to u has a unique solution, then a solution x_* of problem (I) is also a solution of problem (II).

Proof

Let x_* be a solution of problem (I) associated to u and let $x \in F$ be an arbitrary element.

Since $\inf(x_*, x) \in F$ and u is positive we have,

$$u(x_*) \le u(\inf(x_*, x)) \le u(x_*) = \min_{y \in F} u(y),$$

which implies that $\inf(x_*, x)$ is a solution of problem (I).

Clearly, the uniqueness implies that $x_* = \inf(x_*, x)$, that is, $x_* \le x$, for every $x \in F$. □

Proposition 3.5

Let E be an ordered reflexive Banach space. Suppose that E is a vector lattice with respect to order defined by $K = \{x \in E \mid x \ge 0\}$.

If $f: E \to E^*$ is a Z-mapping strictly monotone hemicontinuous and positive at infinity and $u \in K^*$ is a strictly positive element $\{x \in K \setminus \{0\} \Rightarrow u(x) > 0\}$, then problem (I) associated to u has a unique solution and F satisfies the property (Q).

Proof

If we consider $z = 0$ in Proposition 3.4, we obtain that problem (IV) has a solution, which implies, using Theorem 3.6 that problem (II) has a solution and finally from Theorem 3.7 we deduce that problem (I) has a solution.

Since we can apply Proposition 3.3 we obtain that F satisfies the property (Q).

Now it is sufficient to prove that problem (I) has a unique solution.

Indeed, we consider two solutions $x_1, x_2 \in F$ of problem (I).

Because F satisfies the property (Q), we have, $\inf(x_1, x_2) \in F$ and u being strictly positive we obtain, $u(\inf(x_1, x_2)) \le u(x_1)$, with a strict inequality if $\inf(x_1, x_2) \le x_2$ and $\inf(x_1, x_2) \ne x_1$.

But this inequality is impossible since x_1 is a solution of problem (I) and hence, $x_1 = \inf(x_1, x_2)$.

By a similar calculus we obtain, $x_2 = \inf(x_1, x_2)$ and because $\inf(x_1, x_2)$ is unique, we have $x_1 = x_2$. □

Finally, we obtain the following important result.

Theorem 3.9

Let E be a reflexive Banach space which is a vector lattice. Denote, $K = \{x \in E \mid x \ge 0\}$ and consider $f: E \to E^*$ to be a Z-mapping strictly monotone, hemicontinuous and positive at infinity.

If $u \in K^*$ is a strictly positive element, then there exists $x_* \in F$ which is a solution of problems (I), (II), (V) and (IV).

Moreover, the solution x_* is unique and if $f = \Phi'$, the Gâteaux derivative of $\Phi: E \longrightarrow R$, then x_* is also a unique solution of problem (III). □

We study now the equivalence between problems (V) and (VI), that is, we are interested to know when the Complementarity Problem is equivalent to a Fixed Point Problem.

We consider this equivalence in a Hilbert space.

Theorem 3.10

Let $(H, <,>)$ be a Hilbert space and let $K \subset H$ be a convex cone. If $f:K \longrightarrow H$ has the form, $f(x) = x - g(x)$, where $g:K \longrightarrow K$, then x_* is a solution of problem (V), if and only if, x_* is a solution of problem (VI).

Proof

If x_* is a solution of problem (VI) then x_* is a fixed point for g and $f(x_*)=0$, which implies that x_* is a solution of problem (V).

Conversely, if we suppose that x_* is a solution of problem (V) then by Theorem 3.5 we have that,

$$x_* \in K \text{ and } < x - x_*, f(x_*) > \geq 0, \text{ for every } x \in K.$$

But since $f(x_*) = x_* - g(x_*)$ and $g(x_*) \in K$ we deduce that

$$< x_* - g(x_*), g(x_*) - x_* > \geq 0,$$

which implies $0 \leq \; < x_* - g(x_*), x_* - g(x_*) > \leq 0$, that is, $g(x_*) = x_*$. □

The case when f does not have the special form, $f(x) = x - g(x)$, where $g:K \longrightarrow K$, is more complicated.

To study this case we consider the projection operator onto a closed convex cone in a Hilbert space.

Proposition 3.6

Let $(H, <,>)$ be a Hilbert space and let $K \subset H$ be a closed convex cone.

For every $x \in H$ there exists a unique element $P_K(x) \in K$ such that,

$$\| x - P_K(x) \| \leq \| x - y \|; \text{ for every } y \in K.$$

Proof

If $x \in H$ is chosen we consider the continuous function $f:K \rightarrow R$ defined by,

$$(3.21) \qquad f(y) = \frac{1}{2} \| x - y \|^2; \text{ for every } y \in K$$

and if we denote, $\alpha = \inf_{y \in K} f(y)$ we remark that, $\alpha > -\infty$.

Our proposition will be proved if we prove that there exists a unique element $y_o \in K$ such that $f(y_o) = \alpha$.

The definition of greatest lower bound implies that,

$$(3.22) \qquad (\forall n \in N)(y_n \in K)\left(f(y_n) \leq \alpha + \frac{1}{n}\right)$$

and we observe that the sequence $\{y_n\}$ is bounded.

Since H is a Hilbert space, from the sequence $\{y_n\}_{n\in N}$ we can extract a weakly convergent subsequence.

Let $\{y_{n_k}\}_{k\in N}$ be the subsequence extracted and y_o its (weak) limit.

Since K is weakly closed, $y_o \in K$ and because $\{y_{n_k}\}_{n\in N}$ is weakly convergent to y_o we have that, $\underline{\lim} \|y_{n_k}\| \geq \|y_o\|$.

Now, applying the operator $\underline{\lim}$ to

(3.23) $\qquad \alpha + \frac{1}{n_k} \geq f(y_{n_k})$

we get, $\alpha \geq f(y_o) \geq \alpha$.

Finally, y_o is unique since f is strictly convex and denoting $P_K(x) = y_o$ the proof is finished.□

Proposition 3.7

Let (H, < , >) be a Hilbert space and let $K \subset H$ be a closed convex cone. If $x \in H$ is an arbitrary element, then the following statements are equivalent:

i): $\|x - P_K(x)\| \leq \|x - y\|$; for every $y \in K$,

ii): $\langle x - P_K(x), P_K(x) - y \rangle \geq 0$; for every $y \in K$.

Proof

(ii) ==> (i). This implication is a consequence of the following inequality:

$$\|x - P_K(x)\|^2 - \|x - y\|^2 = \|x - P_K(x)\|^2 - \|(x - P_K(x)) + (P_K(x) - y)\|^2 =$$

$$= -2 \langle x - P_K(x), P_K(x) - y \rangle - \|P_K(x) - y\|^2 \leq 0.$$

(i) ==> (ii). Indeed, we have (for $0 < t \leq 1$),

$$\|x - P_K(x)\|^2 - \|x - [ty + (1 - t) P_K(x)]\|^2 =$$

$$= -2t \langle x - P_K(x), P_K(x) - y \rangle - t^2 \|P_K(x) - y\|^2,$$

and if (i) is satisfied we obtain,

$$0 \geq -2t \langle x - P_K(x), P_K(x) - y \rangle - t^2 \|P_K(x) - y\|^2.$$

Dividing by t and compution $\lim_{t\to 0}$, we obtain formula (ii).□

Proposition 3.8

If (H, < , >) is a Hilbert space and $K \subset H$ a closed convex cone, then for an arbitrary element $x \in H$ the projection $P_K(x)$ is characterized by the following relations:

(iii): $\langle P_K(x) - x, y \rangle \geq 0$; for every $y \in K$

(iv): $\langle P_K(x) - x, P_K(x) \rangle = 0$.

Proof

(iii) and (iv) $\Longrightarrow$ (ii)(Prop. 3.7). Indeed, we have for every $y \in K$,

$$< P_K(x) - x, P_K(x) - y > \; = \; < P_K(x) - x, P_K(x) > - < P_K(x) - x, y > \; \leq 0.$$

(ii)(Prop. 3.7) $\Longrightarrow$ (iii) and (iv). If we put in formula (ii), $y = P_K(x) + v$, where $v \in K$ we get, $<x - P_K(x), P_K(x) - P_K(x) - v> \; \geq 0$, that is formula (iii).

If we consider now $y = 0$ in formula (ii) we have, $< P_K(x) - x, P_K(x) > \; \leq 0$ and putting $y = P_K(x)$ in formula (iii) (which is now true) we obtain, $P_K(x) - x, P_K(x) > \; \geq 0$, that is we have formula (iv). □

Using Propositions 3.7 and 3.8 we obtain the following result.

Let $(H, < , >)$ be a Hilbert space and let $K \subset H$ be a closed convex cone.

If $f:K \to H$ is an arbitrary mapping we consider the complementarity problem,

$$\text{C.P.}(f,K): \quad \left\| \begin{array}{l} \text{find } x_* \in K \text{ such that,} \\ f(x_*) \in K^* \text{ and } < x_*, f(x_*) > = 0 \end{array} \right.$$

Also, for an arbitrary $\tau \in R_+ \setminus \{0\}$ we consider the mapping $T:H \longrightarrow K$ defined by, $T(x) = P_K(x - \tau f(x))$.

Theorem 3.11

The problem C.P.(f,K) is equivalent to the following fixed point problem,

$$\text{(F.P.)}: \quad \left\| \begin{array}{l} \text{find } x_* \in K \text{ such that} \\ T(x_*) = x_*. \end{array} \right.$$

Proof

By Propositions 3.7 and 3.8 we deduce that $x_* \in K$ is a fixed point for T if and only if, $x_* - [x_* - \tau f(x_*)] \in K^*$ and $< x_*, x_* - [x_* - \tau f(x_*)] > = 0$, that is, if and only if, $f(x_*) \in K^*$ and $< x_*, f(x_*) > = 0$. □

We recall that the polar cone of a convex cone $K \subset H$ is, $K^o = \{x \in H | <x,y> \leq 0; \forall y \in K\}$ and it is well known that, if K is closed then $K = (K^o)^o$.

If K and Q are two closed convex cones in H, then we say that K and Q are mutually polar if $K = Q^o$ (which implies, $K^o = Q$).

The following result has many interesting applications in the theory of ordered Hilbert spaces and also in mechanics.

Theorem 3.12 [Moreau]

If K and Q are two mutually polar convex cones in H and $x, y, z \in H$, then the following statements are equivalent:

(a): $z = x + y$, $x \in K$, $y \in Q$ and $< x, y > = 0$

(b): $x = P_K(z)$ and $y = P_Q(z)$.

Proof

(a) ⟹ (b). We suppose that $x, y, z \in H$ satisfy assumption (a). Using characterization (ii) of Proposition 3.7 of the projection P_K and the following inequality, $\langle z - x, u - x \rangle = \langle y, u - x \rangle = \langle y, u \rangle \leq 0$; for every $u \in K$, we obtain $x = P_K(z)$.

By a similar calculus we obtain that, $y = P_Q(z)$.

(b) ⟹ (a). Indeed, if $z \in H$ is an arbitrary element, we put, $x = P_K(z)$ and $y' = z - x$.

For every $u \in K$ we have,

(3.24): $\langle z - x, u - x \rangle \leq 0$

(using (ii) of Proposition 3.7) and if $u = \lambda x$; $\lambda \geq 0$ from (3.24) we deduce,

$(\lambda - 1) \langle y', x \rangle \leq 0$.

Since $\lambda - 1$ can be positive or negative we obtain, $\langle y', x \rangle = 0$ and from (3.24) we deduce, $\langle y', u \rangle \leq 0$; for every $u \in K$, that is, $y' \in Q$.

Hence, x, y', z satisfy property (a) and by a similar calculus as in the proof of implication (a) ⟹ (b), we obtain that $y' = P_Q(z)$. □

Theorem 3.12

Let $(H, \langle , \rangle)$ be a Hilbert space and let $K \subset H$ be a closed convex cone.

The Complementarity Problem C.P.(f, K) has a solution, if and only if, the mapping,

$$\Phi(x) = P_K(x) - f(P_K(x)); \text{ for every } x \in H,$$

has a fixed point in H. If x_o is a fixed point of Φ then $x_* = P_K(x_o)$ is a solution of problem C.P. (f, K).

Proof

We suppose that Φ has a fixed point, for example, $x_o = \Phi(x_o)$, that is, $x_o = P_K(x_o) - f(P_K(x_o))$.

We denote, $x_* = P_K(x_o)$, which implies, $x_* \in K$ and $x_o = x_* - f(x_*)$, or $x_* - x_o = f(x_*)$.

From Proposition 3.8 we obtain, $\langle f(x_*), y \rangle \geq 0$; for every $y \in K$, that is, $f(x_*) \in K^*$.

Using again Proposition 3.8 we get, $\langle f(x_*), x_* \rangle = 0$ and hence x_* is a solution of problem C.P. (f, K).

Conversely, we suppose that $x_* \in K$ is a solution of problem C.P.(f, K).

We denote, $x_o = x_* - f(x_*)$ and from Moreau's Theorem we deduce (since x_* is a solution of problem C.P. (f, K)) that $P_K(x_o) = x_*$ and finally,

$$\Phi(x_o) = P_K(x_o) - f(P_K(x_o)) = x_* - f(x_*) = x_o,$$

that is x_o is a fixed point of Φ. □

An interesting equivalence of the Complementarity Problem to a nonlinear system was obtained by Mangasarian [A 191]. This equivalence is about the Complementarity Problem in finite dimensional vector spaces.

We consider the space R^n ordered by the pointed closed convex cone $K = R^n_+$ and we denote by $<,>$ the inner product, $< x, y > = \sum_{i=1}^{n} x_i y_i$, where $x = (x_i)_{i=1,\ldots,n}$, $y = (y_i)_{i=1,\ldots,n} \in R^n$.

If $f:R^n \longrightarrow R^n$ is an arbitrary mapping we denote by $f_i (i = 1, 2, \ldots, n)$ the components of f.

We are interested to solve the following complementarity problem,

$$(3.25): \left\| \begin{array}{l} \text{find } x^* \in R^n \text{ such that,} \\ x^* \in R^n_+,\ f(x^*) \in R^n_+ \text{ and } <x^*, f(x^*)> = 0 \end{array} \right.$$

Definition 3.3

We say that a function $\Phi:R \longrightarrow R$ is strictly increasing if $\Phi(x) < \Phi(y)$ is equivalent to $x < y$.

The following result was proved by Mangasarian in [A 191].

Theorem 3.13

Let $\Phi:R \longrightarrow R$ be a strictly increasing function such that $\Phi(0) = 0$.

Then, x^* solves the complementarity problem (3.21) if and only if, x^* solves the following nonlinear system,

$$(3.26): \left\| \begin{array}{l} \Phi(|f_i(x) - x_i|) - \Phi(f_i(x)) - \Phi(x_i) = 0 \\ i = 1, 2, \ldots, n. \end{array} \right.$$

Proof

Sufficiency. Let x^* be a solution of system (3.26). To prove that $f(x^*) \geq 0$ we suppose the contrary, that is, $f_i(x^*) < 0$, for some $i = 1, 2, \ldots, n$. In this case we have,

$$0 \leq \Phi(|f_i(x^*) - x_i^*|) = \Phi(f_i(x^*)) + \Phi(x_i^*) < \Phi(x_i^*)$$

which implies that,

$$0 < x_i^* \text{ and } x_i^* - f_i(x^*) = |x_i^* - f_i(x^*)| < x_i^*.$$

that is, $f_i(x^*) > 0$, which is impossible.

Similarly as above, if we interchange the roles of x_i^* and $f_i(x^*)$ we obtain that $x^* \geq 0$.

We prove now that $<x^*, f(x^*)> = 0$.

Indeed, if we suppose the contrary, that is, $x_i^* > 0$ and $f_i(x^*) > 0$, for some $i = 1, 2, \ldots, n$, then supposing $f_i(x^*) \geq x_i^*$ (the proof is similar if $x_i^* \geq f_i(x^*)$) we obtain,

$$\Phi(|f_i(x^*) - x_i^*|) = \Phi(f_i(x^*) - x_i^*) < \Phi(f_i(x^*)) < \Phi(f_i(x^*)) + \Phi(x_i^*),$$

which contradicts, $\Phi(|f_i(x^*) - x_i^*|) - \Phi(f_i(x^*)) - \Phi(x_i^*) = 0$.

<u>Necessity</u>. If x^* is a solution of problem (3.25) then for each $i = 1, 2, \ldots, n$ either $x_i^* = 0$ or $f_i(x^*) = 0$. If $x_i^* = 0$ then we have,

$$\Phi(|f_i(x^*) - x_i^*|) - \Phi(f_i(x^*)) - \Phi(x_i^*) =$$
$$= \Phi(f_i(x^*)) - \Phi(f_i(x^*)) - 0 = 0.$$

Similarly, if $f_i(x^*) = 0$ we deduce,

$$\Phi(|f_i(x^*) - x_i^*|) - \Phi(f_i(x^*)) - \Phi(x_i^*) =$$

$= \Phi(x_i^*) - 0 - \Phi(x_i^*) = 0$, and hence x^* is a solution of system (3.26). □

<u>Remark</u>

The method which is largely used to solve the nonlinear system (3.26) is the Newton (or the ;quasi-Newton) method.

But to use the Newton method it is important to have that the Jacobian of the mapping used in system (3.26) is nonsingular at the solution x^* of problem (3.25) (which is also a solution of system (3.26)).

<u>Let x^* be a solution of problem (3.25) satisfying the nondegeneracy condition, $x^* + f(x^*) > 0$.</u>

<u>If the principal minors of the Jacobian of f at x^* are nonsingular and $\Phi: R \longrightarrow R$ is a differentiable strictly increasing function such that, $\Phi'(0) + \Phi'(r) > 0$ for all $r > 0$ then, in this case x^* solves system (3.26) and the Jacobian of $F = (F_i)$ $(i = 1, 2, \ldots, n)$</u>

<u>(where, $F_i(x) = \Phi(|f_i(x) - x_i|) - \Phi(f_i(x)) - \Phi(x_i)$) at x^* is nonsingular.</u>

This result is proved in [A 191].

We note that a differentiable strictly increasing function $\Phi: R \to R$ which can be used in <u>Theorem 3.13</u> such that F is global differentiable is, $\Phi(r) = r|r|$.

In this case system (3.26) becomes,

(3.27):
$$\left\| \begin{array}{l} [f_i(x) - x_i]^2 - f_i(x)|f_i(x)| - x_i|x_i| = 0 \\ i = 1, 2, \ldots, n. \end{array} \right.$$
□

CHAPTER 4

EXITENCE THEOREMS

We consider in this chapter the General Complementarity Problem or some special forms of this problem and our aim is the study of several existence theorems.

The General Complementarity Problem will be considred in R^n, in Hilbert spaces, in Banach spaces and in locally convex spaces.

Also, for this problem, we will be interested to know some remarkable properties of the solution set.

We note that, the space R^n will be considered with its euclidean structure, that is, the hilbertian structure defined by the inner-product $\langle x, y \rangle = \sum_{i=1}^{n} x_i y_i$; $x = (x_i)$, $y = (y_i) \in R^n$, $(i = 1, 2 \ldots n)$.

4.1 Boundedness of the solution set

Let $\langle E, E^* \rangle$ be a dual system of Banach or locally convex spaces and let $K \subset E$ be a closed convex cone.

Given an arbitrary mapping $f: K \to E^*$ the General Complementarity Problem associated to f and K is,

$$\text{G.C.P.}(f,K): \left\| \begin{array}{l} \text{find } x_* \in K \text{ such that,} \\ f(x_*) \in K^* \text{ and } \langle x_*, f(x_*) \rangle = 0 \end{array} \right.$$

For problem G.C.P.(f,K) the feasible set is, $F = \{x \in K \mid f(x) \in K^*\}$.

Obviously, if we denote by S the solution set of problem G.C.P.(f,K) we have, $S \subset F$.

We say that problem G.C.P.(f,K) is solvable if S is nonempty and feasible if F is nonempty.

If F is nonempty then an element $x \in F$ is called a feasible solution.

If problem G.C.P.(f,K) is given we are not sure that S is nonempty.

Generally speaking, the structure of the feasible set F is not so simple.

Suppose for example in problem G.C.P.(f,K) that f has the following structure:

$$f(x) = T(x) + b; \text{ for each } x \in K,$$

where T is a linear operator from E into E^* and $b \in E^*$.

If we denote, $\mathcal{F}_o = \{x \in K \mid T(x) \in K^* \setminus \{0\}\}$ we have the following result.

<u>Proposition 4.1.1</u>

<u>If $\mathcal{F}_o$ and $\mathcal{F}$ are nonempty then $\mathcal{F}$ is unbounded.</u>

<u>Proof</u>

Let $0 \neq x_0 \in \mathcal{F}_o$ and $x_1 \in \mathcal{F}$. Then for every $\lambda \in R_+$ we have, $x_1 + \lambda x_0 \in K$ and

$f(x_1 + \lambda x_0) = T(x_1 + \lambda x_0) + b = T(x_1) + \lambda T(x_0) + b = (T(x_1) + b) + \lambda T(x_0) \in K^* + K^* \subset K^*$, which implies that $\mathcal{F}$ is unbounded. □

Since $\mathcal{F}$ can be unbounded it is important to know when S is bounded, supposing that it is nonempty.

This problem is very important of both practical and theoretical interest.

We study now the boundednes of S when the problem G.C.P.(f,K) is the Linear Complementarity Problem in a Hilbert space.

Let $(H, \langle , \rangle)$ be a real Hilbert space and let $K \subset H$ be a closed convex cone.

The collection of all linear continuous operators from H to H will be denoted by L(H).

Given $T \in L(H)$ and $b \in H$ we consider the problem:

$$\text{L.C.P.}(T,b,K): \quad \left\| \begin{array}{l} \text{find } x_* \in K \text{ such that,} \\ T(x_*) + b \in K^* \text{ and } \langle T(x_*) + b, x_* \rangle = 0 \end{array} \right.$$

and for the next results we denote by S the solution set of this problem.

Denoting by T^* the adjoint of T we recall that T is self-adjoint if and only if $T = T^*$. We observe that $T + T^*$ is self-adjoint.

<u>Lemma 4.1.1</u>

<u>For every $x \in H$ we have that $\frac{1}{2} \langle (T + T^*)(x), x \rangle = \langle T(x), x \rangle$.</u>

<u>Proof</u>

Indeed, we have,

$$\frac{1}{2} \langle (T + T^*)(x), x \rangle = \frac{1}{2} \langle T(x), x \rangle + \frac{1}{2} \langle T^*(x), x \rangle =$$

$$= \frac{1}{2} \langle T(x), x \rangle + \frac{1}{2} \langle x, T(x) \rangle = \frac{1}{2} \langle T(x), x \rangle + \frac{1}{2} \langle T(x), x \rangle = \langle T(x), x \rangle \quad \square$$

Consider the mapping Φ defined by, $\Phi(x) = \frac{1}{2} \langle (T + T^*)(x), x \rangle + \langle b, x \rangle$; for all $x \in H$.

<u>Remark</u>

From <u>Lemma 4.1.1</u> we deduce that, $\Phi(x) = \langle T(x) + b, x \rangle$, for all $x \in H$.

As in the introduction of this chapter we denote by $\mathcal{F}$ the feasible set of problem L.C.P.(T,b,K), that is,

$$\mathcal{F} = \{x \in K \mid T(x) + b \in K^*\}$$

From the precedent remark and the definition of K^* we deduce easily the following result.

Proposition 4.1.2

For every $x \in \mathcal{F}$, $\Phi(x) \geq 0$. Moreover, if x_0 is a solution of problem L.C.P.(T,b,K) then x_0 is a global minimum for Φ on $\mathcal{F}$ and $0 = \phi(x_0) = \text{global} \min_{x \in \mathcal{F}} \Phi(x)$.

We will study the boundedness of S using the numerical range of an operator. Let $(H, \langle , \rangle)$ be a Hilbert space over C.

Definition 4.1.1

The numerical range $\omega(T)$ of $T \in L(H)$ is defined by, $\omega(T) = \{\langle T(x), x \rangle \mid x \in H, \|x\| = 1\}$.

The following results on the numerical range are necessary for this section. The reader finds the proofs and other details on this concept in [C43].

Since we can prove that if T is self-adjoint then $\langle T(x), x \rangle$ is a real number for all $x \in H$ we obtain that for every self-adjoint operator $T \in L(H)$, the numerical range $\omega(T)$ is a subset of R. Moreover $\omega(T)$ is a bounded subset of R.

If we denote by $\sigma(T)$ the spectrum of T, that is, the set of $\lambda \in C$ such that $\lambda I - T$ is either not injective or not surjective then we have the following result.

Theorem 4.1.1

Let $(H, \langle , \rangle)$ be a Hilbert space over C.

If $T \in L(H)$ is self-adjoint then, $\sigma(T) \subset Cl[\omega(T)]$. □

Definition 4.1.2

Let $(H, \langle , \rangle)$ be a Hilbert space over C.

If $T \in L(H)$ is self-adjoint, then we set, $M(T) = \sup \omega(T)$ and $m(T) = \inf \omega(T)$.

In [C43] is proved the following result.

Theorem 4.1.2

Let $(H, \langle , \rangle)$ be a Hilbert space over C.

If $T \in L(H)$ is self-adjoint, then $M(T) \in \sigma(T)$ and $m(T) \in \sigma(T)$. □

Remark

If dim $H <+\infty$ and $T \in L(H)$ is self-adjoint, then a classical result is that in this case T has a finite number of eigenvalues $\{\lambda_k\}_{k=1,\dots m}$ and $\sigma(T) = \{\lambda_k | k=1,2,\dots,m\}$.

In this case if $\lambda_1 \le \lambda_2 \le \dots \le \lambda_m$ we have $m(T) = \lambda_1$ and $M(T) = \lambda_m$. [C43].

If $(H, < , >)$ is a complex Hilbert space and $T \in L(H)$ is a self-adjoint operator, then we say that T is semi-positive definite if and only if $< T(x), x > \ge 0$ for every $x \in H$.

If T is semi-positive definite and moreover, $< T(x), x > > 0$ for every $x \in H \setminus \{0\}$ then we say that T is positive definite.

We can prove that if $T \in L(H)$ is positive definite and invertible then T^{-1} is also positive definite, and if T is self-adjoint and invertible then T^{-1} is self-adjoint too. [C43].

From the definition of m(T) we deduce easily that if $T \in L(H)$ is self-adjoint then it is semi-positive definite if and only if $m(T) \ge 0$.

Proposition 4.1.3

If $T \in L(H)$ is self-adjoint and $m(T)>0$ then T is positive definite and invertible.

Proof

Obviously, T is postive definite since $m(T) > 0$ and $<T(x), x> \ge m(T)\|x\|^2$, for every $x \in H$.

We have, $m(T)\|x\|^2 \le <T(x), x> \ge \|T(x)\|\cdot\|x\|$; for all $x \in H$, which implies,

$m(T)\|x\| \le T(x)$; for all $x \in H\setminus\{0\}$

and hence T is injective.

We prove now that T is surjective.

Indeed, since T from H into T(H) is an isomorphism, T(H) is complete, which implies that T(H) is closed.

Denoting by $[\ \ ]^{o}$ the polar we have, $[T(H)]^{o} = \{0\}$ since for every y $[T(H)]^{o}$ we have $<T(y), y> = 0$ and because $<T(y), y> \geqq m(T)\|y\|^2$, we deduce that $y = 0$.

We have,

$T(H) = \overline{T(H)} = [[T(H)]^{o}]^{o} = \{0\}^{o} = H$

and hence T is bijective which implies (because H is a Hilbert space) that T is invertible and $T^{-1} \in L(H)$. □

Remark

The converse is also true.

We say that $T \in L/H)$ is <u>negative definite</u> if and only if T is self-adjoint and $\langle T(x), x\rangle < 0$, for all $x \in H\backslash\{0\}$.

<u>We note that T is negative definite if and only if -T is positive definite.</u>

<u>Theorem 4.1.3</u>

<u>Let $T \in L(H)$ be a self-adjoint operator.</u>

<u>If $m(T) > 0$ then T is invertible, T^{-1} is self-adjoint, positive definite and we have,</u>

(i): $\underline{M(T^{-1}) = [m(T)]^{-1}}$

(ii): $\underline{m(T^{-1}) = [M(T)]^{-1}}$.

<u>Proof</u>

From <u>Proposition 4.1.3</u> we have that T is positive definite and invertible.

Moreover, T^{-1} is self-adjoint and positive definite.

First we prove formula (i). Indeed, from the definition of m(T) we have,

(1): $m(T)\|x\|^2 \leq \langle T(x), x\rangle$; for all $x \in H\backslash\{0\}$, which implies for every $x \in H\backslash\{0\}$,

(2): $$\frac{1}{\langle T(x),x\rangle} \leq \frac{1}{m(T)\|x\|^2}.$$

Since T is positive definite, we deduce multiplying (2) by $\langle T(x),x\rangle^2$,

(3): $$\langle T(x), x\rangle = \frac{\langle T(x),x\rangle^2}{\langle T(x),x\rangle} \leq \frac{\langle T(x),x\rangle^2}{m(T)\|x\|^2} \leq \frac{\|T(x)\|^2\cdot\|x\|^2}{m(T)\|x\|^2} = \frac{1}{m(T)}\|T(x)\|^2.$$

Denoting $y = T(x)$ in (3) we obtain,

(4): $\langle y, T^{-1}(y)\rangle \leq \frac{1}{m(T)}\|y\|^2$; for every $y \in H\backslash\{0\}$.

But formula (4) implies,

(5): $$M(T^{-1}) \leq \frac{1}{m(T)}.$$

We prove now that $M(T^{-1}) \geq \frac{1}{m(T)}$.

Indeed, since T is self-adjoint and positive definite the mapping, $(x, y) \to \langle T(x), y\rangle$ is an inner product on H and the Schwartz inequality with respect to this inner-product is true and hence we have,

(6): $$|\langle T(x),y\rangle|^2 \leq \langle T(x), x\rangle \langle T(y), y\rangle\ ;\ \forall\ x, y \in H.$$

If we consider $y = T^{-1}(x)$ in (6) we obtain,

(7): $$\|x\|^4 \leq \langle T(x), x\rangle \cdot \langle x, T^{-1}(x)\rangle\ ;\ \forall\ x \in H.$$

Using (7) we obtain for every $x \in H\backslash\{0\}$ such that $\|x\| = 1$,

(8): $$1 \leq \langle T(x), x\rangle\ \langle T^{-1}(x), x\rangle.$$

We remark now that for every $\varepsilon > 0$, sufficiently small, there exists $\delta > 0$ such that,

(9): $$\frac{1}{m(T) + \delta} > \frac{1}{m(T)} - \varepsilon.$$

From the definition of $m(T)$ we have that there exists x_o with $\|x_o\| = 1$ such that,

$$< T(x_o), x_o > \leq m(T) + \delta \text{ or,}$$

(10): $$\frac{1}{<T(x_o),x_o>} \geq \frac{1}{m(T) + \delta}.$$

Now, from (10), (8) and (9) we obtain,

$$<T^{-1}(x_o),x_o> \geq \frac{1}{<T(x_o),x_o>} \geq \frac{1}{m(T) + \delta} > \frac{1}{m(T)} - \varepsilon,$$

that is, for every $\varepsilon > 0$ we have,

$$M(T^{-1}) > \frac{1}{m(T)} - \varepsilon,$$

which implies,

(11): $$M(T^{-1}) \geq \frac{1}{m(T)}.$$

Considering (5) and (11) we have that, $M(T^{-1}) = \frac{1}{m(T)}$, that is (i) is true.

To prove (ii) we remark that,

(12): $m(T^{-1})\|y\|^2 \leq <T^{-1}(y),y>$; for all $y \in H\setminus\{0\}$, which implies,

(13): $$\frac{1}{<T^{-1}(y),y>} \leq \frac{1}{m(T^{-1})\|y\|^2}; \text{ for all } y \in H\setminus\{0\}.$$

Multiplying (13) by $<T^{-1}(y), y>^2$ and using the Schwartz inequality with respect to the inner-product $< , >$ we deduce,

(14): $<x, T(x)> \leq \frac{1}{m(T^{-1})}\|x\|^2$; for all $x \in H\setminus\{0\}$, and we have, $M(T) \leq \frac{1}{m(T^{-1})}$.

Now, by a similar calculus as in the proof of formula (i) but for $m(T^{-1})$ we obtain that $M(T) \geq \frac{1}{m(T^{-1})}$ and finally, $M(t) = \frac{1}{m(T^{-1})}$. □

We apply now the concept of <u>numerical range</u> of an operator to the study of the Linear Complementarity Problem.

Let $(H, < , >)$ be a real Hilbert space and let $K \subset H$ be a closed convex cone. We say that K is <u>self-adjoint</u> if and only if $K = K^*$.

If $T \in L(H)$ we denote $T_o = T + T^*$ and we know that T_o is self-adjoint.

The theory of self-adjoint cones in Hilbert spaces is a very nice and very interesting theory with interesting applications in Physics.

Theorem 4.1.4

Let $(H, <\,,\,>)$ be a real Hilbert space and let $K \subset H$ be a self-adjoint closed convex cone.

Suppose that $T \in L(H)$ is a self-adjoint operator such that $m(T) > 0$.

Then for every solution x of Problem L.C.P. (T,b,K) we have,

(15): $\quad \|x\| \leq M(T)\, \|x_b\|\, \|T^{-1}\|$

where $x_b = -T^{-1}(b)$ and $\|T^{-1}\| = [m(T)]^{-1}$.

Proof

To prove this theoroem we consider the Complementarity Problem,

(16): $\quad \left\| \begin{array}{l} \text{find } v \in K \text{ such that,} \\ T^{-1}(v) + x_b \in K^* \text{ and} \\ <v,\, T^{-1}(v) + x_b >= 0 \end{array} \right.$

We prove now that v is a solution of Problem (16) if and only if $x = T^{-1}(v-b)$ is a solution of Problem L.C.P.(T, b, K).

Indeed, if v is a solution of Problem (16) then $v \in K$, $T^{-1}(v-b) \in K^*$ and $<v, T^{-1}(v-b) > = 0$ which imply,

$$x = T^{-1}(v-b) \in K^* \subseteq K,\ T(x) + b = v - b + b = v \in K \subseteq K^*$$

and $<x, T(x) + b > = 0$.

Conversely, if x is a solution of Problem L.C.P. (T, b, K) then we can prove (since K is self-adjoint) that $v = T(x) + b$ is a solution of Problem (16).

Hence, every solution of Problem L.C.P. (T, b, K) has a representation of the form, $x = T^{-1}(v-b)$, where v is a solution of Problem (16).

Obviously, if v is a solution of Problem (16) then we have,

(17): $\quad <v-b, T^{-1}(v-b)> = -<b, T^{-1}(v-b)> = <-T^{-1}(b), v-b>$

(since T^{-1} is a self-adjoint operator).

We suppose now that $v \neq b$, that is, $\|v-b\| \neq 0$.

From formula (17) we deduce,

(18): $\quad m(T^{-1})\|v-b\| \leq <v-b, T^{-1}(v-b)> = <-T^{-1}(b), v-b>$

and from Theorem 4.1.3 we have,

$$\frac{1}{M(T)} \|v-b\|^2 \leq \|x_b\|\|v-b\|,$$

which implies,

(19): $\|v-b\| \le M(T)\|x_b\|$.

Let x be a solution of Problem L.C.P. (T,b,K).

From (19), since $x = T^{-1}(v-b)$ (where v is a solution of Problem (16)) we have,

(20): $\|x\| \le \|v-b\|\cdot\|T^{-1}\| \le M(T)\|x_b\|\|T^{-1}\|$.

If $v = b$ then $x = T^{-1}(v-b) = 0$ is a solution of Problem L.C.P.(T,b,K) and inequality (20) is also satisfied and the theorem is proved since in this case $\|T^{-1}\| = [m(T)]^{-1}$. □

Remark

In Theorem 4.1.4 the assumption $m(T) > 0$ (which is equivalent with the fact that T is positive definite) is essential.

Example

Consider $H = R^2$, $K = R^2_+$ and $\langle x, y\rangle = \sum_{i=1}^{2} x_i y_i$.

The Complementarity Problem considered in this case is,

(21): $$\left\| \begin{array}{l} \text{find } x \in R^2_+ \text{ such that} \\ Tx + b \in R^2_+ \text{ and } \langle x, Tx + b\rangle = 0 \end{array} \right.$$

where $T = \begin{bmatrix} 1 & 1 \\ 1 & 0 \end{bmatrix}$ and $b = \begin{bmatrix} 1 \\ 0 \end{bmatrix}$.

We can prove that $x_* = \begin{bmatrix} 0 \\ 1 \end{bmatrix}$ is a solution of Problem (21).

Since the eigenvalues of T are, $\lambda_1 = \frac{1-\sqrt{5}}{2}$ and $\lambda_2 = \frac{1+\sqrt{5}}{2}$ we have that $M(T) = \frac{1+\sqrt{5}}{2}$.

T is invertible,

$$T^{-1} = \begin{bmatrix} 0 & 0 \\ 1 & -1 \end{bmatrix} \quad \text{and} \quad x_b = -T^{-1}b = \begin{bmatrix} 0 \\ -1 \end{bmatrix}$$

We observe also that T is self-adjoint but it is not positive definite.

Since for every $\lambda > 0$, λx_* is a solution of Problem (21) we have that formula (15) is not satisfied.

For the following theorems we denote, $B(0, r) = \{x \in H \mid \|x\| \le r\}$.

Theorem 4.1.5

Let $T \in L(H)$ be an arbitrary operator and $b \in H$. If $M(T_o) < 0$ then the solution set S of Problem L.C.P. (T, b, K) is a subset of $B\left(0, \frac{2\|b\|}{|M(T_o)|}\right) \cap K$.

Proof

From the definition of the numerical range of T_o we have for every $x \in H\setminus\{0\}$,

$$\Phi(x) = \frac{1}{2} < T_o(x), x > + < b, x > \le \frac{1}{2} M(T_o)\|x\|^2 + \|b\|\|x\| =$$

$$= \left(\frac{1}{2} M(T_o)\|x\| + \|b\|\right)\|x\|.$$

Since $\Phi(x) = < T(x) + b, x >$, the feasibility requires $\Phi(x) \ge 0$ and we obtain that x is infeasible if $\Phi(x) < 0$, that is, if $\frac{1}{2} M(T_o)\|x\| + \|b\| < 0$.

Hence x is infeasible if $\|x\| > \frac{2\|b\|}{|M(T_o)|}$.

Finally, we deduce that the solution set S must be contained in $B\left(0, \frac{2\|b\|}{|M(T_o)|}\right) \cap K$ since $S \subset F$.

Remark

If $M(T_o) \le 0$, then the feasible set F of Problem L.C.P. (T, B, K) can be bounded or unbounded.

Examples [C61]

1°) We consider the problem L.C.P. (T, b, K) where:

$H = R^2$ (with the euclidean structure),

$K = R^2_+$, $T = \begin{bmatrix} -1 & 1 \\ 1 & -1 \end{bmatrix}$ and $b = \begin{bmatrix} -1 \\ 2 \end{bmatrix}$

In this case $T_o = T + T^* = \begin{bmatrix} -2 & 2 \\ 2 & -2 \end{bmatrix}$ has the eigenvalues $\lambda_1 = -2$ and $\lambda_2 = 0$.

For this problem the feasible set

$$F = \left\{ \begin{bmatrix} x_1 \\ x_2 \end{bmatrix} \in R^2_+ \;\middle|\; \begin{matrix} -x_1 + x_2 - 1 \ge 0 \\ x_1 - x_2 + 2 \ge 0 \end{matrix} \right\} \text{ is unbounded.}$$

2°) In the same space ordered by the same cone as in example 1, we consider the problem L.C.P. (T, b, K), where,

$T = \begin{bmatrix} -1 & 0 \\ 4 & -4 \end{bmatrix}$ and $b = \begin{bmatrix} 2 \\ -2 \end{bmatrix}$.

The eigenvalues of $T_o = \begin{vmatrix} -2 & 4 \\ 4 & -8 \end{vmatrix}$ are $\lambda_1 = -10$ and $\lambda_2 = 0$ but the feasible set,

$$F = \left\{ \begin{bmatrix} x_1 \\ x_2 \end{bmatrix} \in R^2_+ \;\middle|\; \begin{matrix} -x_1 + 2 \ge 0 \\ 4x_1 - 4x_2 + 2 \ge 0 \end{matrix} \right\} \text{ is bounded.}$$

Theorem 4.1.6

Let $T \in L(H)$ be an arbitrary operator and $b \in H$. If $m(T_o) > 0$ then the solution set S of Problem L.C.P. (T,b,K) is a subset of $B\left(0, \frac{2\|b\|}{m(T_o)}\right) \cap K$.

Proof

Since $m(T_o) > 0$ then the definition of the numerical range of T_o implies for every $x \in H \setminus \{0\}$,

$$\Phi(x) = \frac{1}{2} \langle T_o(x), x \rangle + \langle b, x \rangle \geq$$

$$\geq \frac{1}{2} m(T_o)\|x\|^2 - \|b\|\|x\|$$

$$\geq \left(\frac{1}{2} m(T_o)\|x\| - \|b\|\right)\|x\|.$$

If $\frac{1}{2} m(T_o)\|x\| - \|b\| > 0$, that is, if $\|x\| > \frac{2\|b\|}{m(T_o)}$, then $\Phi(x) > 0$ and therefore x cannot be a solution of Problem L.C.P. (T,b,K).

Hence, if $x \in K$ is a solution of Problem L.C.P. (T,b,K) we must have

$$\|x\| \quad \frac{2\|b\|}{m(T_o)} \text{ and } S \subset B\left(0, \frac{2\|b\|}{m(T_o)}\right) \cap K. \quad \square$$

Proposition 4.1.4

Let $T \in L(H)$ be an arbitrary operator such that $T_o = T + T^*$ is invertible and let $b \in H$ be an element.

If x is a solution of Problem L.C.P. (T,b,K) and $x_b = -T_o^{-1}(b)$ then we have,

(22): $\langle x - x_b, T_o(x - x_b)\rangle = \langle b, T_o^{-1}(b)\rangle$.

Proof

Let x be a solution of Problem L.C.P. (T,b,K) and let z be an arbitrary element of H.

We have,

$$\frac{1}{2} \langle z-x, T_o(z-x) \rangle + \langle z, T_o(x) \rangle + \langle b,x \rangle + \langle b, z \rangle =$$

$$= \frac{1}{2} \langle z, T_o(z) \rangle - \frac{1}{2} \langle x, T_o(z) \rangle - \frac{1}{2} \langle z, T_o(x)\rangle +$$

$$+ \frac{1}{2} \langle x, T_o(x)\rangle + \langle z, T_o(x)\rangle + \langle b, x\rangle + \langle b, z\rangle =$$

$$= \frac{1}{2} \langle z, T_o(z)\rangle + \langle b, z\rangle + \langle x, T_o(x) + b\rangle =$$

$$= \frac{1}{2} \langle z, T_o(z)\rangle + \langle b, z\rangle = \Phi(z).$$

that is we have,

(23): $\frac{1}{2} <z-x, T_o(z-x)> + <z, T_o(x)> + <b,x> + <b,z> = \Phi(z).$

Let $z = x_b$ in formula (23). We obtain,

$$\frac{1}{2} < x_b-x, T_o(x_b-x)> + <x_b, T_o(x)> + <b,x> + <b, x_b> =$$

$$= \Phi(x_b) = - <b, T_o^{-1}(b)> ,$$

which implies,

$$\frac{1}{2} <x_b - x, T_o(x_b - x)> + <- T_o^{-1}(b), T_o(x)> +$$

$$+ <b, x >+ <b, - T_o^{-1}(b)> = - \frac{1}{2} <b, T_o^{-1}(b)> ,$$

and finally,

$$\frac{1}{2} <x_b - x, T_o(x_b - x)> = <b, T_o^{-1}(b)> . \quad \square$$

Theorem 4.1.7

Let $T \in L(H)$ be an arbitrary opertor and let $b \in H$ be an element different from zero.

If $m(T_o) > 0$ then any solution of Problem L.C.P. (T,b,K), $x \neq x_b = -T_o^{-1}(b)$ satisfies,

$$\frac{\|b\|}{M(T_o)} \leq \|x - x_b\| \leq \frac{b}{m(T_o)} .$$

Proof

First, since $m(T_o) > 0$ we have that T_o is invertible.

From formula (22), the definition of the numerical range of T_o and Theorem 4.1.3 we have,

$$\frac{1}{M(T_o)} \|b\|^2 \leq \|x - x_b\|^2 M(T_o),$$

which implies,

$$\frac{\|b\|}{M(T_o)} \leq \|x - x_b\|.$$

Using again Proposition 4.1.4 and Theorem 4.1.3 we deduce,

$$m(T_o) \|x - x_b\|^2 \leq M(T_o^{-1})\|b\|^2 = \frac{1}{m(T_o)} \|b\|^2,$$

that is, $\|x - x_b\| \leq \frac{\|b\|}{m(T_o)}$.

Finally we have,

$$\frac{\|b\|}{M(T_o)} \le \|x - x_b\| \le \frac{\|b\|}{m(T_o)}$$

and the theorem is proved. □

Theorem 4.1.8

Let $T \in L(H)$ be an arbitrary operator and let $b \in H$ be an element different from zero.

If $M(T_o) < 0$ then any solution of Problem L.C.P. (T,b,K), $x \neq x_b = -T_o^{-1}(b)$ satisfies,

$$\frac{\|b\|}{|m(T_o)|} \le \|x - x_b\| \le \frac{\|b\|}{|M(T_o)|}.$$

Proof

First, we can prove that T_o is invertible since $M(T_o) < 0$.

Because $M(T_o) < 0$ we have that, $m(-T_o) = |M(T_o)|$, $M(-T_o) = |m(T_o)|$ and $-T_o$ is positive definite and self-adjoint.

In this case formula (22) becomes for the operator $-T_o$,

(24): $\langle x - x_b, -T_o(x - x_b)\rangle = \langle b, (-T_o)^{-1}(b)\rangle$

with the same x_b as in Proposition 4.1.4.

Now, using formula (24) and a similar calculus as in the proof of Theorem 4.1.7 we obtain,

$$\frac{\|b\|}{|m(T_o)|} \le \|x - x_b\| \le \frac{\|b\|}{|M(T_o)|}. \qquad \square$$

Remark

If $H = R^n$ with the euclidean structure and the eigenvalues of T (or respective T_o) are $\lambda_1 \le \lambda_2 \le \dots \le \lambda_n$, then $\lambda_1 = m(T)$ (resp. $m(T_o)$) and $\lambda_n = M(T)$(resp. $M(T_o)$) and the bounds defined in Theorems 4.1.4-4.1.8 can be computed using λ_1 and λ_n.

The Complementarity Problem over the complex field was considered in 1977 by McCallun and in 1974 by Berman [see references [A25] in connection with the Mathematical Programming over the complex field as considered by Levison, Hanson and Mond, Abrams and Ben-Israel [A25].

Let $(H, \langle , \rangle)$ be a complex Hilbert space.

In this case a convex cone in H is a subset $K \subset H$ such that,

1°) $K + K \subset K$,

2°) $(\forall \lambda \in R_+)(\lambda K \subset K)$.

If $K \subset H$ is a convex cone, the dual of K is,

$K^* = \{y \in H | \text{Re} <y, x> \ \geq 0;\ \forall x \in K\}$.

Given $T \in L(H)$ and $b \in H$ the Linear Complementarity Problem over the complex field associated to T, b and K is,

$$(25): \quad \left\| \begin{array}{l} \text{find } z \in K \text{ such that,} \\ T(z) + b \in K^* \text{ and} \\ \text{Re} <z, T(z) + b> = 0. \end{array} \right.$$

If we denote,

$$\Phi(z) = \frac{1}{2} <z, (T + T^*)(z)> + \text{Re} <z, b>$$

we have that,

$$(26): \quad \Phi(z) = \text{Re} <z, T(z) + b>.$$

Indeed, formula (26) is true since we have,

$$\frac{1}{2} <z, (T + T^*)(z)> = \frac{1}{2} <z, T(z)> + \frac{1}{2}<z, T^*(z)> =$$

$$= \frac{1}{2} <z, T(z)> + \frac{1}{2} <T(z), z> = \frac{1}{2}<z, T(z)> + \frac{1}{2} \overline{<z, T(z)>} =$$

$$= \text{Re} <z, T(z)>.$$

Formula (26) implies that if $\Phi(z) < 0$ then z is infeasible. [We use that

$\text{Re} <T(z) + b, z> = \text{Re} \overline{<T(z) + b, z>} = \text{Re} <z, T(z) + b>$].

From the theory of numerical range we have that if $T \in L(H)$ is self-adjoint then $\omega(T) \subset R$ and $m(T)$ [resp. $M(T)$] is well defined.

We recall that if $T \in L(H)$ then $T_o = T + T^*$ is self-adjoint.

Now, using the definition of $\Phi(z)$ and the fact that $|\text{Re} <z,b>| \leq |<z,b>|$, we obtain by the same proofs as for Theorems 4.1.5 and 4.1.6 the following results.

Theorem 4.1.9

Let $T \in L(H)$ be an arbitrary operator and $b \in H$. If $M(T_o) < 0$ then if $x_* \in K$ is a solution of Problem (25) we have, $\|x_*\| \leq \dfrac{2\|b\|}{|M(T_o)|}$. □

Theorem 4.1.10

Let $T \in L(H)$ be an arbitrary operator and $b \in H$. If $m(T_o) > 0$ then if $x_* \in K$ is a solution of Problem (25) we have, $\|x_*\| \leq \dfrac{2\|b\|}{m(T_o)}$. □

We consider now the Nonlinear Complementarity Problem.

Let $(H, < , >)$ be a real Hilbert space and let $K \subset H$ be a closed convex cone.

Let $T:K \to H$ be an operator not necessary linear and $\hat{S}:K \to H$ a nonlinear operator.

We say that T is homogeneous of degree $\rho > 0$ if $T(\lambda x) = \lambda^{\rho} T(x)$, for every $x \in K$ and every $\lambda \in R_+$.

Let $\varphi : R_+ \to R_+$ be a mapping such that $\varphi(t) > 0$ for every $t \geq \gamma$, where $\gamma \in R_+$.

Definition 4.1.3

We say that $\hat{S}$ is φ-asymptotically bounded if there exist $r, c \in R_+ \setminus \{0\}$ such that $r \leq \|x\|$, $(x \in K)$ implies that, $\|\hat{S}(x)\| \leq c\varphi(\|x\|)$.

We denote, $\omega(T) = \{<T(x), x> \mid x \in K, \|x\| = 1\}$ and we say that T is K-range bounded if $\omega(T)$ is a bounded subset of R.

If T is K-range bounded then $M_K(T) = \sup \omega(T)$ and $m_K(T) = \inf \omega(T)$ are finite real numbers.

If T is homogeneous of degree $\rho \quad 0$ and K-range bounded, then for every $x \in K \setminus \{0\}$ we have,

$$m_K(T)\|x\|^{\rho+1} \leq \ <T(x), x> \ \leq M_K(T)\|x\|^{\rho+1}.$$

Assuming T homogeneous of degree $\rho > 0$ and K-range bounded and $\hat{S}$ φ-asymptotically bounded we consider the following Nonlinear Complementarity Problem:

$$(27): \left\| \begin{array}{l} \text{find } x \in K \text{ such that,} \\ \\ T(x) + \hat{S}(x) \in K^* \text{ and } <T(x) + \hat{S}(x), x> = 0. \end{array} \right.$$

Theorem 4.1.11

Let $(H, <,>)$ be a real Hilbert space and let $K \subset H$ be a closed convex cone. Suppose that $T:K \to H$ is a not necessary linear operator homogeneous of degree $\rho > 0$ and $\hat{S}:K \to H$ a nonlinear operator φ-asymptotically bounded.

If:

(i): T is K-range bounded and $M_K(T) < 0$

(ii): $\lim\limits_{\|x\| \to \infty} \dfrac{\varphi(\|x\|)}{\|x\|^{\rho}} = 0$,

then the solution set of Problem (27) is bounded.

Proof

Indeed, for every $x \in K \setminus \{0\}$ such that $r \leq \|x\|$ we have,

(28): $\quad <T(x) + \hat{S}(x), x> = <T(x), x> + <\hat{S}(x), x> \leq$

$\leq M_K(T)\|x\|^{\rho+1} + \|\hat{S}(x)\|\|x\| \leq$

$\leq M_K(T)\|x\|^{\rho+1} + c\varphi(\|x\|)\ \|x\| = [M_K(T)\|x\|^{\rho} + c\varphi(\|x\|)]\ \|x\|.$

Since from assumption (i) we have, $-M_K(T) > 0$, we obtain using (ii) that there exists $a \geq r$ such that, $\dfrac{c\varphi(\|x\|)}{\|x\|^{\rho}} < -M_K(T)$, for every $x \in K\setminus\{0\}$ such that $a < \|x\|$.

Hence from (28) we deduce that every $x \in K\setminus\{0\}$ such that $a < \|x\|$ is infeasible for Problem (27).

So, we obtain that if $x \in K\setminus\{0\}$ is a solution of Problem (27), then every solution is feasible, it is necessary to have $\|x\| \leq a$ and the theorem is proved.

__Theorem 4.1.12__

__Let $(H <, >)$ be a real Hilbert space and let $K \subset H$ be a closed convex cone.__

__Let $T:K \to H$ be a not necessary linear operator homogeneous of degree $\rho > 0$ and $\hat{S}:K \to H$ a nonlinear operator φ-asymptotically bounded.__

If:

(i): __T is K-range bounded and $m_K(T) > 0$__

(ii): $\displaystyle\lim_{\|x\|\to\infty} \frac{\varphi(\|x\|)}{\|x\|^{\rho}} = 0,$

__then the solution set of Problem (27) is bounded.__

__Proof__

Indeed, using assumption (i), for every $x \in K\setminus\{0\}$ such that $r \leq \|x\|$ we have,

(29): $\quad <T(x) + \hat{S}(x), x> \geq m_K(T)\|x\|^{\rho+1} - c\varphi(\|x\|)\|x\| =$

$= [m_K(T)\ \|x\|^{\rho} - c\varphi(\|x\|)]\ \|x\|.$

Since $m_K(T) > 0$ we obtain from assumption (ii) that there exists $a \geq r$ such that,

(30): $\quad \dfrac{c\varphi(\|x\|)}{\|x\|^{\rho}} < m_K(T)$; fore every $x \in K\setminus\{0\}$ such that, $a < \|x\|$.

Obviously, using now (30) and (29) we deduce that every $x \in K\setminus\{0\}$ satisfying $a < \|x\|$ cannot be a solution for Problem (27) and hence every solution $x \in K\setminus\{0\}$ of Problem (27) satisfies, $\|x\| \leq a$. □

For the linear Complementarity Problem in a Hilbert space we have another situation when the solution set is bounded.

Let $(H, <, >)$ be again a real Hilbert space and let $K \quad H$ be a closed pointed convex cone.

We say that $T \in L(H)$ is __copositive plus__ if:

1°) <u>$x \in K$ implies $<T(x), x> \geq 0$</u>.

2°) <u>$x \in K$ and $<T(x), x> = 0$ imply $(T + T^*)(x) = 0$</u>.

This concept was defined in R^n by Lemke and studied in Hilbert space by Gowda.

We say that K is <u>well-based</u> if and only if there exists a bounded convex set B such that $0 \notin \bar{B}$ and $K = \bigcup_{\lambda \geq 0} \lambda B$.

We can prove that K is well-based if and only if there exists a linear continuous functional h on H such that $\|x\| \leq h(x)$, for every $x \in K$.

<u>Theorem 4.1.13</u>

<u>If the following assumptions are satisfied</u>:

1°) <u>T is copositive plus on K</u>,

2°) <u>The map $x \to <T(x), x>$ is weak lower semicontinuous on K</u>,

3°) <u>K is well-based</u>,

4°) <u>$\{x \in K | T(x) \in K^*, <T(x), x> = 0 \text{ and } <q, x> = 0\} = \{0\}$</u>,

<u>then the solution set S of problem</u>

$$\text{L.C.P.}(T,q,K): \left\| \begin{array}{l} \text{find } x_o \in K \text{ such that} \\ T(x_o) + q \in K^* \text{ and } <x_o, T(x_o) + q> = 0 \end{array} \right.$$

<u>is bounded (if it is nonempty)</u>.

<u>Proof</u>

We suppose S nonempty and unbounded.

Then there exists a sequence $\{x_n\}_{n\in N} \subset S$ such that $\lim_{n\to\infty} \|x_n\| = +\infty$.

Since for every $n \in N$, $x_n \in S$ we have,

$$(31): \left\| \begin{array}{ll} x_n \in K \text{ for every } n \in N & \\ < T(x_n) + q, x > \geq 0 & \text{for every } x \in K \\ < T(x_n) + q, x_n > = 0 & \text{for every } n \in N. \end{array} \right.$$

Because K is well-based there exists a linear continuous functional h on H such that,

$$(32): \quad \|x\| \leq h(x) \text{ for every } x \in K.$$

From (32) we have that $\lim_{n\to\infty} h(x_n) = + \infty$.

We can suppose that $h(x_n) \neq 0$ for every $n \in N$ and from (31) we deduce,

$$(33): \quad <T\left(\frac{x_n}{h(x_n)}\right) + \frac{q}{h(x_n)}, \frac{x_n}{h(x_n)} > = 0 \text{ for every } n \in N.$$

Since $\frac{x_n}{h(x_n)} \in B = K \cap \{x | h(x) = 1\}$ for every $n \in N$ and since B is weakly compact we have that the sequence $\{\frac{x_n}{h(x_n)}\}_{n\in N}$ has a subsequence $\{\frac{x_{n_i}}{h(x_{n_i})}\}_{n\in N}$

weakly convergent to an element $d \in K$.

From (33) and assumption 2°) we have,

$< T(d), d > \leq 0.$

Since T is copositive plus on K we have $< T(d), d > \geq 0$ and hence,

(34): $< T(d), d > = 0,$

which implies (using assumption 1°)) that

(35): $T(d) = -T^*(d).$

From (31) we get,

$< q, x_n > = - < T(x_n), x_n > \leq 0$, for every $n \in N$,

or

$< q, \frac{x_{n_i}}{h(x_{n_i})} > \leq 0$, for every $i \in N$ and finally,

(36): $< q,d > \leq 0.$

Because, $< T(x_{n_i}) + q, x > \geq 0$, for every $x \in K$ we deduce,

$< T(\frac{x_{n_i}}{h(x_{n_i})}) + \frac{q}{h(x_{n_i})}, x > \geq 0$, for every $x \in K$ which implies,

$< T(d), x > \geq 0$, for every $x \in K$,

that is, $T(d) \in K^*$.

Let x_o be an element of S.

We have, $< T(x_o) + q, d > \geq 0$ which implies,

$$< q, d > \geq - < T(x_o), d > = - < x_o, T^*(d) > = < x_o, T(d) > \geq 0$$

and using (36) we obtain

(37): $< q,d > = 0.$

Hence, $d \in \{x \in K | T(x) \in K^*, < T(x), x > = 0 \text{ and } < q,x > = 0\}$ and using assumption 4° we get, $d = 0$, which is impossible since $d \in B$, B is closed and $0 \notin B$.

This contradiction shows that S must be bounded. □

We finisth this section considering a special nonlinear Complementarity Problem with the property that the solution set is bounded.

We consider $H = R^n$ with the euclidean structure, $K_+ = R^n$, $D \subset R^n$ a subset such that $R^n_+ \subset D$ and $f:D \to R^n$.

We recall [Chapter 1] that f is <u>monotone</u> if $<f(x^1) - f(x^2), x^1 - x^2 > \geq 0$; for all $x^1, x^2 \in D$.

Supposing f monotone we consider the Complementarity Problem,

$$(38): \quad \left\| \begin{array}{l} \text{find } x_* \in R^n_+ \text{ such that,} \\ \\ f(x_*) \in R^n_+ \text{ and } \langle x_*, f(x_*) \rangle = 0. \end{array} \right.$$

We suppose now that Problem (38) <u>is feasible</u> and there exist p (with $p \geq 1$) feasible points $x^1, x^2, \ldots, x^p$ such that there exist $\lambda_j > 0$ $(j = 1, 2, \ldots, p)$ with $\sum_{j=1}^{p} \lambda_j = 1$ and such that the components of $\hat{V} = \sum_{j=1}^{p} \lambda_j V^j$ are strictly positive, where $V^j = f(x^j)$; $j = 1, 2, \ldots, p$.

If x_* is an arbitrary solution of Problem (38) then we have,

$$\langle V^j - V_*, x^j - x_* \rangle \geq 0; \text{ where } V_* = f(x_*), \text{ which implies,}$$

$$\langle V^j, x^j \rangle \geq \langle V^j, x_* \rangle + \langle V_*, x^j \rangle \; ; \; j = 1, 2, \ldots, p$$

or,

$$\sum_{j=1}^{p} \lambda_j \langle V^j, x^j \rangle \geq \sum_{j=1}^{p} \lambda_j \langle V^j, x_* \rangle + \sum_{j=1}^{p} \lambda_j \langle V_*, x^j \rangle$$

and finally (since $\sum_{j=1}^{p} \lambda_j \langle V_*, x^j \rangle \geq 0$),

$$(39): \quad \sum_{j=1}^{p} \lambda_j \langle V^j, x^j \rangle \geq \left\langle \sum_{j=1}^{p} \lambda_j V^j, x_* \right\rangle \geq \|x_*\|_1 \left(\min_{i=1,\ldots,n} \hat{V}_i\right),$$

where $\hat{V}_i$ $(i = 1, 1, \ldots, n)$ are the components of $\hat{V}$.

Hence, if $x_* \in R^n_+$ is an arbitrary solution of Problem (38) we have,

$$\|x_*\|_1 \leq \frac{\sum_{j=1}^{p} \lambda_j \langle V^j, x^j \rangle}{\min_{i=1,\ldots,n} \hat{V}_i}.$$

4.2 Feasibility and solvability

Let $\langle E, E^* \rangle$ be a dual system of locally convex spaces and let $K \subset E$ be a closed convex cone.

Given $f: K \to E^*$ we consider again the General Complementarity Problem,

$$G.C.P.(f,K): \quad \left\| \begin{array}{l} \text{find } x_* \in K \text{ such that} \\ \\ \quad f(x_*) \in K^* \text{ and } \langle x_*, f(x_*) \rangle = 0. \end{array} \right.$$

We say that G.C.P.(f,K) is feasible if its feasible set $F = \{x \in K \mid f(x) \in K^*\}$ is nonempty.

Obviously, every solvable Complementarity Problem is feasible, while a feasible Complementarity Problem is not necessary solvable.

Example

Consider $E = E^* = R^2$ with the euclidean structure, $K = R^2_+$ and

$$f\left(\begin{bmatrix} x_1 \\ x_2 \end{bmatrix}\right) = \begin{bmatrix} ax_1x_2 - ax_2 + 1 \\ -ax_1^2 + 2ax_1 - a \end{bmatrix} \quad \text{where } a \in R_+ \setminus \{0\}.$$

The Complementarity Problem considered in this case is,

$$(1): \quad \left\| \begin{array}{l} \text{find } \begin{bmatrix} x_1 \\ x_2 \end{bmatrix} \in R^2_+ \text{ such that,} \\ \\ f\left[\begin{pmatrix} x_1 \\ x_2 \end{pmatrix}\right] \in R^2_+ \text{ and } \left\langle \begin{bmatrix} x_1 \\ x_2 \end{bmatrix}, f\left(\begin{bmatrix} x_1 \\ x_2 \end{bmatrix}\right)\right\rangle = 0. \end{array} \right.$$

In this case,

$$F = \left\{\begin{bmatrix} x_1 \\ x_2 \end{bmatrix} \in R^2_+ \mid x_1 = 1 \text{ and } x_2 \geq 0\right\}$$

which is nonempty and since

$\left\langle \begin{bmatrix} x_1 \\ x_2 \end{bmatrix}, f\left(\begin{bmatrix} x_1 \\ x_2 \end{bmatrix}\right)\right\rangle = 1$ for every $\begin{bmatrix} x_1 \\ x_2 \end{bmatrix} \in F$ we deduce that problem (1) is feasible but not solvable.

A conclusion of this example is that, it is important to know when a feasible Complementarity Problem is solvable.

A. The finite dimensional case.

We consider $E = E^* = R^n$ endowed with the euclidean structure or with a Banach structure and $\langle R^n, R^n \rangle$ is the natural duality.

If $f: R^n \to R^n$ is a mapping we denote by f_i $(i = 1, 2 \ldots, n)$ the components of f.

If $K \subset R^n$ is a convex cone and $f: R^n \to R^n$ a mapping we consider the Complementarity Problem,

$$(2): \quad \left\| \begin{array}{l} \text{find } x_* \in K \text{ such that} \\ \\ f(x_*) \in K^* \text{ and } \langle x_*, f(x_*) \rangle = 0 \end{array} \right.$$

and we recall that problem (2) is equivalent to the variational inequality,

$$(3): \quad \left\| \begin{array}{l} \text{find } x_* \in K \text{ such that} \\ \langle x - x_*, f(x_*) \rangle \geq 0, \text{ for all } x \in K. \end{array} \right.$$

Lemma 4.2.1

Let $D \subset R^n$ be a convex set. If $u \in R^n$ is given then $v \in D$ satisfies
(4): $< x - v, u > \geq 0$; for all $x \in D$ if and only if there exists a constant $a > \|v\|$ for which (4) holds for all $x \in D_a = \{x \in D \mid \|x\| \leq a\}$.

Proof

We prove only the sufficiency since obviously the condition is necessary.

Indeed, if $x \in D$ we choose $0 < \lambda < 1$ sufficiently small such that $w = \lambda x + (1-\lambda)v \in D_a$. We have, $0 \leq < w - v, u > = \lambda < x - v, u >$ and v satisfies (4).

Theorem 4.2.1

Let $f: K^n \to R$ be a continuous mapping.

If there is a $u_o \in K$ and a constant $r > \|u_o\|$ such that,

$<x - u_o, f(x)> \geq 0$, for all $x \in K$

with $\|x\| = r$ then problem (2) is solvable and it has a solution x_* such that $\|x_*\| \leq r$.

Proof

By Theoreme 3.1 [Ch. 3] [Hartman-Stampacchia] there exists $x_* \in K_r = \{x \in K \mid \|x\| \leq r\}$ solution for inequality (3) but for all $x \in K_r$.

If $\|x_*\| < r$ then Lemma 4.2.1 with $u = f(x_*)$ implies that x_* is a solution of problem (3).

If $\|x_*\| = r$ then $< x_* - u_o, f(x_*) > \geq 0$ and since $< x - x_*, f(x_*) > \geq 0$ for all $x \in K_r$ it follows that $< x - u_o, f(x_*) > \geq 0$ for all $x \in K_r$.

But $\|u_o\| < r$ and thus Lemma 4.2.1 implies that $x - u_o, f(x_*) > \geq 0$ for all $x \in K$ and adding to $<u_o - x_*, f(x_*) > \geq 0$ the theorem is proved. Obviously x_* is a solution of problem (2) and $\|x_*\| \leq r$.

We say that the convex cone $K \subset R^n$ is rectangular if $K = \prod_{i=1}^{n} I_i$, where I_i is either $\{0\}$, $[0, +\infty)$, $(-\infty, 0]$ or $(-\infty, +\infty)$.

We denote, $\|x\|_\infty = \max \{|x_i| \mid 1 \leq i \leq n\}$ the ℓ^∞ norm on R^n.

Theorem 4.2.2

Assume that $f: R^n \to R^n$ is a continuous function on the rectangular cone $K_o \subset R^n$.

If there is a $u \in K_o$ and a constant $r > \|u\|_\infty$ such that,

(5): $\max_{1 \le i \le n} \{(x_i - u_i) f_i(x)\} > 0$, for all $x \in K_o$ with $\|x\|_\infty = r$,

then the problem G.C.P. (f, K_o) is solvable.

Proof

Using Hartman-Stampacchia Theorem [Theorem 3.1 Ch. 3] we obtain an element $x_* \in K_{or} = \{x \in K_o \mid \|x\|_\infty \le r\}$ such that,

(6): $< x - x_*, f(x_*) > \ge 0$; for all $x \in K_{or}$.

If $\|x_*\|_\infty = r$ we define, $\hat{x} \in R^n$ by, $\hat{x}_i = u_i$ and $\hat{x}_j = x_{*j}$ for $j \ne i$, we put $\hat{x} = (x_i)$ and $x_* = (x_{*i})$.

Since K_{or} is the cartesian product of n intervals we have, $\hat{x} \in K_{or}$ and from (6) we deduce,

$[(u_i - x_*)f_i(x_*)] \ge 0$, for each $i = 1, 2, \ldots, n$.

This contradicts (5) and hence $\|x_*\|_\infty < r$.

The theorem now follows from Lemma 4.2.1 and the equivalence between the Complementarity Problem and variational inequality (4) associated to f and K_o.

Corollary

Assume $f: R^n \to R^n$ is continuous on R^n_+.

If there is a $u \in R^n_+$ and a constant $r > \|u\|_\infty$ such that, $\max_{1 \le i \le n} \{(x_i - u_i) f_i(x)\} > 0$, for all $x \in R^n_+$ with $\|x\|_\infty = r$, then the problem,

(7): $$\left\| \begin{array}{l} \text{find } x_* \in R^n_+ \text{ such that,} \\ f(x_*) \in R^n_+ \text{ and } <x_*, f(x_*)> = 0 \end{array} \right.$$

has a solution x_* with $\|x_*\|_\infty \le r$.

Let $K \subset R^n$ be a convex cone. We say that K is solid if it has nonempty interior relative to R^n.

If K is a pointed solid closed convex cone then K^* also has the same properties and $(K^*)^* = K$. We recall that such cones generate partial orderings on R^n which are denoted by "$\le_K$" and "$<_K$" respectively as follows: $x \le_K y \iff y - x \in K$ and $x <_K y \iff y - x \in \text{Int } K$.

In particular, $0 \leq_K x$ $(0 <_K x)$ if and only if $x \in K$ $(x \in \text{Int } K)$. The partial orderings "$\leq_{K^*}$" and "$<_{K^*}$" generated by K^* are defined similarly.

Given $K \subset R^n$ a pointed solid closed convex cone and $f: K \to R^n$ we say that the problem G.C.P. (f, K) is <u>strictly feasible</u> if there exists and element $x_o \in K$ such that $0 <_{K^*} f(x_o)$.

We note that when $K = R^n_+$ we denote "$\leq_{R^n_+}$" by "$\leq$" and "$<_{R^n_+}$" by "$<$" (in case $(R^n_+)^* = R^n_+$).

<u>Theorem 4.2.3</u>

<u>If $f: R^n \to R^n$ is a continuous monotone mapping on R^n_+, then the problem G.C.P. (f, R^n_+) is solvable if it is strictly feasible.</u>

<u>Proof</u>

By assumption there exists an element $u \in R^n_+$ such that $0 < f(u)$ (since "$<$" is the order defined by Int. R^n_+, every component of $f(u)$ is strictly positive).

Since f is monotone we have,

$$< x - u, f(x)> \geq <x - u, f(u)> = <x, f(u)> - <u, f(u)>$$

and because $0 < f(u)$ there exists an $r > \|u\|_\infty$ such that $<x, f(u)> > <u, f(u)>$, for all $x \geq 0$ with $\|x\|_\infty = r$.

Now, the theorem is a consequence of the Corollary of Theorem 4.2.2.

<u>Theorem 4.2.4</u>

<u>Let $f: R^n \to R^n$ be a continuous function on R^n_+ such that for each $x \neq y$ in R^n_+ and any index k with $|x_k - y_k| = \|x - y\|_\infty$ it follows that,</u>

(8) $\qquad (x_k - y_k)[f_k(x) - f_k(y)] \geq 0.$

<u>In this case the problem G.C.P. (f, R^n_+) is solvable if it is feasible.</u>

<u>Proof</u>

We suppose the problem G.C.P. (f, R^n_+) feasible. In this case there exists $u \in R^n_+$ such that $0 \leq f(u)$.

If $0 < f(u)$ then we choose $r > 2\|u\|_\infty$ and we take any $x \geq 0$ with $\|x\|_\infty = r$. Then $x \neq u$ and $|x_k - u_k| = \|x - u\|_\infty$ implies that $x_k > u_k$. Thus,

$$\max_{1 \leq i \leq n} \{(x_i - u_i)f_i(x)\} \geq (x_k - u_k)f_k(x) \geq (x_k - u_k)f_k(u) > 0,$$

and we obtain the solvability of problem G.C.P. (f, R^n_+) by the Corollary of Theorem 4.2.2.

If $0 \leq f(u)$ then we consider the mapping h defined by, $h_i(x) = f_i(x) + \lambda$ for any $\lambda > 0$ and $h = (h_i)$, $i = 1, 2, \ldots, n$.

Since $h(u) > 0$ there is an $x_*(\lambda) \geq 0$ with $\|x_*(\lambda)\| \leq r$ for which,

(9): $\quad h(x_*(\lambda)) \geq 0$ and $\langle x_*(\lambda), h(x_*(\lambda))\rangle = 0$.

But since $\|x_*(\lambda)\| \leq r$ there is a sequence $\{\lambda_m\} \to 0$ such that $\{x_*(\lambda_m)\}_{m \in N}$ is convergent to an element $x_* \geq 0$.

By continuity from (9) it follows that x_* is a solution of the problem G.C.P. (f, R^n_+).

Definition 4.2.1

We say that a function $f: R^n \to R^n$ is off-diagonally antitone on $D \subset R^n$ if for each $u \geq v$ in D it follows from $u_k = v_k$ that $f_k(u) \leq f_k(v)$.

Theorem 4.2.5

Let $f: R^n \to R^n$ be a continuous off-diagonally antitone function on R^n_+.

Then the problem G.C.P. (f, R^n_+) is solvable if it is feasible.

Proof

If we assume that the problem G.C.P. (f, R^n_+) is feasible then there exists an element $u \in R^n_+$ such that $f(u) \geq 0$.

We construct a sequence $\{x^k\}$ such that $f(x^k) \geq 0$, where $0 \leq x^{k+1} \leq x^k$ and

(10): $\quad x_i^{k+1} \cdot f_i(x_1^k, \ldots, x_{i-1}^k, x_i^{k+1}, x_{i+1}^k, \ldots, x_n^k) = 0,$

for $i = 1, 2, \ldots, n$ and $k = 0, 1, 2, \ldots$.

To do this we put $x^o = u$ and supposing that x^k is defined, we define x^{k+1} by the following construction.

For any index i set $x_i^{k+1} = 0$ if

$$f_i(x_1^k, \ldots, x_{i-1}^k, 0, x_{i+1}^k, \ldots, x_n^k) \geq 0,$$

otherwise, since $f_i(x^k) \geq 0$, there is an $x_i^{k+1} \in [0, x_i^k]$ which satisfies (10).

In either case,

$$f_i(x_1^k, \ldots, x_{i-1}^k, x_i^{k+1}, x_{i+1}^k, \ldots, x_n^k) \geq 0.$$

Thus, $0 \leq x^{k+1} \leq x^k$ and since f is off-diagonally antitone, $f(x^{k+1}) \geq 0$.

If the sequence is constructed we observe that $\{x^k\}$ converges to some $x_* \in R_+^n$ (since $0 \leq x^{k+1} \leq x^k$, for every $k = 0, 1, 2, \ldots$).

Since f is continuous we obtain from (10) that x_* is a solution of the problem G.C.P. (f, R_+^n).

The concept of off-diagonally antitone mapping is a generalization of the concept of Z-matrix.

A real n by n matrix is called Z-matrix if its off-diagonal elements are non-positive.

A particular case of Theorem 4.2.5 is the following result.

Theorem 4.2.6

If $f:R^n \to R^n$ has the form, $f(x) = Ax + b$, where A is a Z-matrix and $b \in R^n$, then the problem G.C.P. $(Ax + b, R_+^n)$ is solvable if it is feasible.

Definition 4.2.2

Let $K \subset R^n$ be a convex cone. A mapping $f:K \to R^n$ is pseudomonotone if, for every pair of points $x, y \in K$ such that $x \neq y$, we have

$$\langle x - y, f(y)\rangle \geq 0 \Longrightarrow \langle x - y, f(x)\rangle \geq 0.$$

Remarks 4.2.1

1°) Every monotone map is pseudomonotone but the converse is not true.

Example

The mapping $f:R_+ \to R$ given by, $f(x) = \frac{a}{a + x}$, where $a \in R_+\setminus\{0\}$ is pseudomonotone but not monotone.

2°) From Definition 4.2.2 we deduce that if $f:K \to R^n$ is pseudomonotone then for every distinct points $x, y \in K$ we have

$$<x - y, f(y)> > 0 \Longrightarrow <x - y, f(x)> > 0.$$

Indeed, if $<x - y, f(y)> > 0$ and $<x - y, f(x)> = 0$ then we have,

$$0 = <x - y, f(x)> \Longrightarrow <y - x, f(x)> = 0 \Longrightarrow$$

$$\Longrightarrow <y - x, f(y)> \geq 0, \text{ that is } x - y, f(y) > \leq 0 \text{ which is impossible.}$$

Let D be a nonempty open subset of R^n (eventually $K \subset D$) and let $f:D \to R$ be a differentiable function.

We denote by $\nabla f(x_o)$ the gradient of f at the point x_o.

Definition 4.2.3

The function f is pseudoconvex on D if for every pair of distinct points x, y D we have,

$$<x - y, \nabla f(y)> \geq 0 \Longrightarrow f(x) \geq f(y).$$

Remark 4.2.2

Every convex function is pseudoconvex but the converse is not true.

The relation between the concept of pseudomonotone mapping and pseudoconvex function is given by the following result (well known in Convex Analysis).

Theorem 4.2.7

Let D be a nonempty, open convex subset of R^n. Let $f:D \to R$ be a differentiable function on D. Then f is pseudoconvex on D if and only if ∇f is pseudomonotone on D.

Lemma 4.2.2

Let $K \subset R^n$ be a pointed closed convex cone with nonempty interior ($\text{Int } K \neq \Phi$).

If $v \in R^n$ then we have:

i) $0 <_{K^*} v$ if and only if for all $x \in K \setminus \{0\}$ we have $0 < <x, v>$;

ii) $0 <_{K^*} v$ if and only if for any real $\alpha > 0$, the set $K_{v,\alpha} = \{x \in K \mid 0 \leq <x, v> \leq \alpha\}$ is compact.

Proof

i) Let $v \in R^n$ such that $0 <_{K^*} v$. Assume that there exists x_o satisfying,

(11) $\quad x_o \in K \setminus \{0\}$ and $<x_o, v> = 0$.

Since $0 <_{K^*} v$, there exists $\varepsilon > 0$ such that $v - \varepsilon x_o \in K^*$.

From (11) we have, $0 \le \langle x_o, v - \varepsilon x_o \rangle = \langle x_o, v\rangle - \varepsilon \langle x_o, x_o\rangle < 0$ which is a contradiction.

Hence, $\langle x, v\rangle > 0$ for all $x \in K\setminus\{0\}$.

Conversely, let $0 < \langle x, v\rangle$ for all $x \in K\setminus\{0\}$ and assume that $v \notin \text{Int } K^*$.

Then for every $\varepsilon > 0$ the closed ball $B(v; \varepsilon)$ contains a point $y(\varepsilon) \notin K^*$.

For every such $y(\varepsilon)$ there exists an element $x(\varepsilon) \in K$ such that,

(12) $\langle x(\varepsilon), y(\varepsilon)\rangle < 0.$

Since K is a cone, we may choose $x(\varepsilon) \in K$ in (12) with $\|x(\varepsilon)\| = 1$.

Now let $\{\varepsilon_n\}_{n\in N}$ be a sequence of real numbers such that $\varepsilon_n > 0$ for every $n \in N$ and $\lim_{n\to\infty} \varepsilon_n = 0$.

From (12) we have

(13) $\langle x(\varepsilon_n), y(\varepsilon_n)\rangle < 0$ for all $n \in N$,

(14) $x(\varepsilon_n) \in K$, $\|x(\varepsilon_n)\| = 1$ for all $n \in N$,

(15) $\|y(\varepsilon_n) - v\| \le \varepsilon_n$ for all $n \in N$.

Since the set $\{x \in K \mid \|x\| = 1\}$ is compact there exists a subsequence of $\{x(\varepsilon_n)\}_{n\in N}$ convergent to an element $x_o \in K$ with $\|x_o\| = 1$.

Also from (15) we have that the corresponding subsequence of $\{y(\varepsilon_n)\}_{n\in N}$ converges to v.

But from (13) we have that $\langle x_o, v\rangle \le 0$ which is a contradiction and we deduce that it is necessary to have $v \in \text{Int } K$, that is $0 <_{K^*} v$.

ii) Let $0 <_{K^*} v$. Since $K_{v,\alpha}$ is closed it is sufficient to show that it is bounded.

Indeed, if $K_{v,\alpha}$ is unbounded then there exists a sequence $\{x_n\}_{n\in N} \subset K_{v,\alpha}$ such that $\lim_{n\to\infty} \|x_n\| = +\infty$ and we can suppose that $x_n \ne 0$ for every $n \in N$.

Let $y_n = \frac{x_n}{\|x_n\|}$ (which is an element of K for every $n \in N$).

The set $\{y_n\}_{n\in N}$ being in a compact set, has a subsequence $\{y_{n_i}\}_{i\in N}$ convergent to an element $y_o \in K$ with $\|y_o\| = 1$.

Since $v \in \text{Int } K$ we have from part i)

$$0 < \langle y_{n_i}, v\rangle = \frac{1}{\|x_{n_i}\|} \langle x_{n_i}, v\rangle \le \frac{\alpha}{\|x_{n_i}\|}.$$

Passing to the limit as $i \to \infty$ we have $0 \le \langle y_o, v\rangle \le 0$, which is a contradiction.

Therefore, $K_{v,\alpha}$ is a compact set.

Conversely, we suppose $K_{v,\alpha}$ compact and we assume that $v \notin \text{Int } K^*$.

It follows from part i) that there exists $x_o \in K\setminus\{0\}$ such that $<x_o, v> = 0$.

Hence $\lambda x_o \in K_{v,\alpha}$ for all scalars $\lambda > 0$.

Taking a sequence $\{\lambda_n\}_{n\in N}$ with $\lambda_n > 0$ and $\lim_{n\to\infty} \lambda_n = +\infty$ we obtain that $K_{v,\alpha}$ is unbounded and we have a contradiction. Thus $v \in \text{Int } K^*$ and the proof is finished.

Theorem 4.2.8

Let $K \subset R^n$ be a pointed closed convex cone with nonempty interior and let $f:K \to R^n$ be a continuous psudomonotone mapping.

If the problem G.C.P. (f,K) is strictly feasible then it has a solution.

Proof

From chapter 3 we know that the problem G.C.P. (f, K) is equivalent to the following variational inequality.

$$(16)\quad \left\|\begin{array}{l}\text{find } x_* \in K \text{ such that} \\ <x - x_*, f(x_*)> \ \geq 0, \text{ for all } x \in K.\end{array}\right.$$

Let x_o be a strictly feasible point for the problem G.C.P. (f, K).

If $<x_o, f(x_o)> = 0$ then x_o is itself a solution. Therefore, we assume that $<x_o, f(x_o)> > 0$.

From Lemma 4.2.2, we know that

$$(17)\qquad D = \{x \in K \mid <x, f(x_o)> \ \leq \ < x_o, f(x_o)>\}$$

is a compact subset of K.

For every $u \in K$, define

$$D_u = \{x \in D \mid <u - x, f(x)> \ \geq 0\}.$$

It is clear that any $x_* \in D$ would solve the problem G.C.P. (f, K) if and only if $x_* \in \bigcap_{u\in K} D_u$.

To show that such an x_* exists, consider an arbitrary set of points $\{u_i \mid i = 1, 2, \ldots, m\}$ in K and let

$$C = \text{conv } [D \cup \{u_1, u_2, \ldots, u_m\}]$$

It is well known that C is a compact convex subset of K.

Since $f:C \to R^n$ is continuous on C it follows from Theorem 3.1 [Hartman-Stampacchia] that there exists $\hat{x} \in C$ such that

$$(18)\qquad <x - \hat{x}, f(\hat{x})> \ \geq 0 \text{ for all } x \in C.$$

In particular we have,

$$<u_i - \hat{x}, f(\hat{x})> \ \geq 0 \text{ for } i = 1, 2, \ldots, m.$$

Since $x_o \in C$ it follows from (18), the pseudomonotonicity of f and (17) that $\hat{x} \in D$.

Thus, we have shown that the intersection of any finite number of the collection $\{D_u\}_{u \in K}$ of closed subsets of D is nonempty.

Since D is compact it follows that $\bigcap_{u \in K} D_u$ is nonempty and the theorem is proved.

Definition 4.2.4

A monotone mapping $f:K \to R^n$ is said to be proper at a point $x_o \in K$ if the set $D(x_o) = \{x | x_o \leq_K x, \ f(x_o) \leq_{K^*} f(x), \ <x - x_o, f(x) - f(x_o)> = 0\}$ is bounded.

Theorem 4.2.9

Let $K \subset R^n$ be a pointed closed convex cone with nonempty interior and let $f:K \to R^n$ be a continuous monotone mapping.

If the problem G.C.P. (f, K) has a feasible point x_o where f is proper, then it has a solution.

Proof

Choose a $u \in \text{Int } K^*$ and for every real $r > 0$ let $C_r = \{x_o | x_o \leq_K x, \ <x - x_o, u> = r\}$.

Since C_r is nonempty, compact convex and since $f(x) - f(x_o)$ is continuous on C_r it follows from Theorem 3.1 [Hartman-Stampacchia] that for every $r > 0$ there exists $x_r \in C_r$ satisfying,

$$<x, f(x_r) - f(x_o)> \ \geq \ <x_r, f(x_r) - f(x_o)> \text{ for all } x \in C_r$$

and from the monotonicity of f we have,

$$<x - x_o, f(x_r) - f(x_o)> \ \geq \ <x_r - x_o, f(x_r) - f(x_o)> \ \geq 0$$

for all $x \in C_r$, which implies $f(x_o) \leq_{K^*} f(x_r)$.

Since f is proper at x_o, there exists a $\hat{r} > 0$ for which

(19) $$<x - x_o, f(x_{\hat{r}}) - f(x_o)> \ \geq \ <x_{\hat{r}} - x_o, f(x_{\hat{r}}) - f(x_o)> 0 \text{ for all } x \in C_r$$

From (19) and Lemma 4.2.2 we deduce $0 \leq_{K^*} f(x_o) <_{K^*} f(x_{\hat{r}})$ and since every monotone mapping is psudomonotone the existence of a solution to the problem G.C.P. (f, K) follows from Theorem 4.2.8.

Corollary

Let $K \subset R^n$ be a pointed closed convex cone with nonempty interior and let $f:K \to R^n$ be a continuous strictly monotone mapping.

If the problem G.C.P. (f, K) is feasible then it has a unique solution.

Proof

Obviously, f is proper at each point $x_o \in K$ since $D(x_o) = \{x_o\}$.

Hence from Theorem 4.2.9 the problem G.C.P. (f, K) has a solution.

In this case the problem G.C.P. (f, K) has a unique solution since if we suppose that $x_1 \neq x_2$ are solutions we have,

$$0 < \langle x_1 - x_2, f(x_1) - f(x_2)\rangle = - \langle x_1, f(x_2)\rangle - \langle x_2, f(x_1)\rangle \leq 0,$$

which is a contradiction.

The concept of derivative with respect to a convex cone has been intensively used in Nonlinear Analysis.

A mapping $f:K \to R^n$ is said to be differentiable at $x_o \in K$ with respect to K if there exists a linear mapping $f^1(x_o);K)$ (called the derivative of f at x with respect to K) and a mapping $\omega(x_o,\cdot):K \to R^n$ such that for all $h \in K$ we have,

$$f(x_o + h) = f(x_o) + f'(x_o;K)h + \omega(x_o,h)$$

with $$\lim_{\|h\| \to 0} \frac{\|\omega(x_o,h)\|}{\|h\|} = 0.$$

Definition 4.2.5

A linear mapping $A:K \to R^n$ is said to be copositive with respect to the cone K if $\langle x, Ax\rangle \geq 0$ for all $x \in K$.

It is said to be strictly copositive if $\langle x, Ax\rangle > 0$ for all $x \in K \setminus \{0\}$.

Lemma 4.2.3

Let $f:K \to R^n$ be differentiable at $x_o \in K$ with respect to K. Then:

i) $f'(x_o;K)$ is copositive with respect to K if and only if,

$$\langle x - x_o, f(x) - f(x_o)\rangle \geq 0 \text{ for all } x_o \leq_K x,$$

ii) if $f'(x_o; K)$ is strictly copositive with respect to K then,

$$\langle x - x_o, f(x) - f(x_o)\rangle > 0 \text{ for all } x_o \leq_K x,\ x \neq x_o.$$

Proof

Consequences of Definition 4.2.5.

Theorem 4.2.10

Let $K \subset R^n$ be a pointed closed convex cone with nonempty interior and let $f:K \to R^n$ be a continuous pseudomonotone mapping. Let x_0 be a feasible point to the problem G.C.P. (f, K).

If f is differentiable with respect to K and $f'(x_0;K)$ is strictly copositive with respect to K then the problem G.C.P. (f, K) has a solution.

Proof

We consider the set C_r and the element x_r defined in the proof of Theorem 4.2.9 for a real number $r > 0$.

Since we have,

$$\langle x - x_o, f(x_r) - f(x_o)\rangle \geq \langle x_r - x_o, f(x_r) - f(x_o)\rangle \geq 0$$

for all $x \in C_r$, $x_r \neq x_0$ and $f'(x_0;K)$ is strictly copositive we deduce that $0 \leq_{K*} f(x_0) <_{K*} f(x_r)$, that is the problem G.C.P. (f, K) is strictly feasible and we apply Theorem 4.2.8.

Let $K \quad R^n$ be a closed convex cone and let $f:K \to R^n$ be a point-to-set mapping.

We denote by G.M.C.P. (f, K) the Generalized Multivalued Complementarity Problem defined in Chapter 1, that is,

G.M.C.P. (f, K): find $x_0 \in K$ and $y_0 \in R^n$ such that,

$$y_0 \in f(x_0) \cap K^* \text{ and } \langle x_o, y_o \rangle = 0$$

We say that f is upper semicontinuous on K if for every $x_0 \in K$ and every open set $V \subset R^n$ such that $f(x_0) \subset V$ there exists a neighborhood $U(x_0)$ of x_0 such that for every $x \in U(x_0) \cap K$ we have $f(x) \subset V$.

Under this condition and with f(x) compact for every x we have that if D is compact then $f(D) = \bigcup_{x \in D} f(x)$ is also compact [C. Berge: Topological spaces, MacMillan, New York (1963)].

A subset D of R^n is called contractible if there is a continuous mapping $h:D \times [0, 1] \to D$ such that $h(x, 0) = x$ and $h(x, 1) = x_0$ for some $x_0 \in D$.

We note that if D is convex, it is contractible since for any $x_0 \in D$ the mapping $h(x, t) = tx_o + (1 - t)x$.

With $x \in D$ and $t \in [0, 1]$ would satisfy the above property.

Also, a set starshaped at x_0 is also contractible to x_0.

The problem G.M.C.P. (f, K) is equivalent to a fixed point problem.

Indeed, let $D \subset K$ be a convex set.

Define, for each $y \in R^n$,

$$P_D(y) = \{x \mid x \text{ solution of the problem } \min_{u \in D} \langle u, y \rangle\}$$

D^* = the class of all nonempty subsets of D.

We remark that $P_D : R^n \to D^* \cup \{\Phi\}$ and if D is compact then $P_D(y)$ is nonempty and compact for each $y \in R^n$.

Also P_D is an upper semicontinuous mapping and $P_D(y)$ is convex for each y and thus contractible.

Let $P = P_K$ and define,

$T : K \times R^n \to (K^* \cup \{\Phi\}) \times R^{n*}$ by $T(x, y) = (P(y), f(x))$.

We denote by F the set of all fixed point of T, that is,

$F = \{(x, y) \in R^{2n} \mid (x, y) \in T(x, y)\}$.

<u>Proposition 4.2.1</u>

<u>Every fixed point of T are a solution of G.M.C.P. (f, K) and conversely.</u>

<u>Proof</u>

Let (x, y) be a fixed point of T.

Then $y \in f(x)$ and $\langle x, y \rangle \le \langle u, y \rangle$ for all $u \in K$.

Since $0 \in K$, $\langle x, y \rangle \le 0$. Also $x \in K$ implies $\lambda x \in K$ for $\lambda \ge 0$, thus $(\lambda - 1)\langle x, y \rangle \ge 0$ and $\lambda - 1$ changes sign, we must have $\langle x, y \rangle = 0$ and thus $\langle u, y \rangle \ge 0$ for all $u \in K$ which implies that $y \in K^*$ and we have that x solves the problem G.M.C.P. (f, K).

Now let x solve the problem G.M.C.P. (f, K).

Hence $\langle u, y \rangle \ge \langle x, y \rangle = 0$ for all $u \in K$ and thus $x \in P(y)$. Also $y \in f(x)$ implies $(x, y) \in T(x, y)$.

We say that f satisfies the Karamardians's condition with respect to K if:

(K.C.): <u>there exists a compact set $C \subset K$ such that for every $x \in K \backslash C$, there is a $u \in C$ for which $\langle u - x, y \rangle < 0$ for all $y \in f(x)$.</u>

<u>Proposition 4.2.2</u>

<u>Let f be upper semicontinuous, f(x) be a compact and contractible for each $x \in K$.</u>

If f satisfies the condition (K.C.) for some compact subset C of K then there are compact, convex subsets $D \subset K$ and $E \subset R^n$ and an upper semicontinuous mapping $T^1: D \times E \to D^* \times E^*$ with $T^1(x, y)$ contractible and compact, such that the fixed point set of T^1 is equal to the fixed point set of T.

Proof

Define D to be any compact convex set $C \subset D \subset K$ such that for all $x \in C$, $y \in K \backslash D$ there is a $\lambda > 0$ such that $(1 - \lambda) x + \lambda y \in D$.

Define E to be any compact convex set containing $f(D) = \bigcup_{x \in D} f(x)$.

Also, let $T^1: D \times E \to D^* \times E^*$ be the mapping $(x, y) \to (P_D(y), f(x))$ and let F^1 be the fixed point set of T^1 and F of T.

We show that $F^1 = F$. First, observe that the condition (K.C.) implies both T and T^1 have fixed points, if any, in $C \times R^n$. To see this, let (x, y) $T(x, y)$ with $x \notin C$. Since $x \in P(y)$, $\langle x, y\rangle \leq \langle z, y$ for all $z \in K$. But from (K.C.) we have that there is $u \in C$ such that $\langle x, y\rangle > \langle u, y\rangle$ for all $y \in f(x)$, which is a contradiction, since $C \subset K$.

A similar argument can be used to show the result for T^1.

Since $x \in P(y) \cap C$ implies $x \in P_D(y)$, we have $F \subset F^1$. We now show the converse.

Let $(x, y) \in T^1(x, y)$ or $x \in P_D(y) \cap C$ and $y \in f(x)$. Thus $\langle x, y\rangle \leq \langle z, y\rangle$ for all $z \in D$.

Let $u \in K \backslash D$ and by the choice of D there is a $\lambda > 0$ such that $(1 - \lambda)x + \lambda u \in D$ and thus $\langle x, y\rangle \leq (1 - \lambda) \langle x, y\rangle + \lambda \langle u, y\rangle$ or $\lambda \langle x, y\rangle \leq \lambda \langle u, y\rangle$. Thus $x \in P(y)$ and $(x, y) \in T(x, y)$ or $F^1 \subset F$ and the proof is finished.

We now give the classical Eilenberg-Montgomery fixed point theorem.

Let X be a topological space. We denote by $H_n(X)$ the n-dimensional singular homology group of the space X.

If we denote by $S_n(X)$ the free abelian group with generators all singular n-simpleces of X, it is known tha the homomorphism $\varepsilon: S_o(X) \to Z$ (the ring of integers) which carries each singular 0-simplex into $1 \in Z$ induces an epimorphism $\varepsilon_*: H_o(X) \to Z$.

A space X is called acyclic if $H_n(X) = 0$ for $n > 0$ and ε_* is an isomorphism $H_o(X) \approx Z$.

The following is a classical result.

Proposition 4.2.3

Any convex set C in a Euclidean space is acyclic.

Proof

See: S. MacLane: Homology. Academic Press (1963), Prop. 8.1, p. 58. □

Proposition 4.2.4

Each space which is contractible is acyclic.

Proof

See: K. Borsuk: Theory of retracts. Warszama (1967), p. 43. □

Suppose X is a topological space and $r:X \to M$ is a continuous mapping with $M \subseteq X$. The mapping r is called a retraction if and only if $r(x) = x$ for all $x \in M$. In that case, the set M is called a retract of X.

A set D in a topological space X is called a neighborhood retract if and only if D is a retract of some one of its neighborhoods.

An absolute neighborhood retract space is a compact metric space M with the universal property that every homeomorphic image of M in a separable metric space is a neighborhood retract.

Every compact convex set in a Euclidean space is an absolute neighborhood retract and acyclic.

Theorem 4.2.11 [Eilenberg-Montgomery].

Let M be an acyclic absolute neighborhood retract and $T:M \to M$ an upper semicontinuous point-to-set mapping such that for every $x \in M$ the set $T(x)$ is acyclic. Then T has a fixed point.

Proof

See: S. Eilenberg and D. Montgomery: Fixed point theorems for multi-valued transformations. Amer. J. Math. 6;8(1946, 214-222. □

Theorem 4.2.12

Let f be an upper semicontinuous point-to-set mapping, with $f(x)$ nonempty compact and contractible for each $x \in K$.

If f satisfies the condition (K.C.) for some compact set $C \subset K$ then the problem G.M.C.P. (f, K) has a solution.

Proof

Let T be the point-to-set mapping used in Proposition 4.2.12 and let D, E, T^1 be as in Proposition 4.2.2.

The theorem is proved if we show that the fixed point set of T is nonempty.

But using Proposition 4.2.2 it is sufficient to prove that the fixed point set of T^1 is nonempty.

Indeed this is true since $T^1(x, y)$ is contractible and T^1 is upper semicontinuous from $D \times E$ into $D^* \times E^*$, where $D \times E$ is compact and convex. Hence we can apply the Eilenberg-Montgomery theorem and we obtain that T^1 has a fixed point. □

In Chapter 1 (1.1) we defined the concept of monotone mapping for multivalued mappings.

We now define the concept of pseudomonotone mapping for point-to-set (multivalued) mappings.

The point-to-set mapping $f:K \to R^n$ is called pseudomonotone if, for every pair of vectors x and y in K and every $x^* \in f(x)$ and $y^* \in f(y)$, $\langle x - y, y^* \rangle \geq 0$ implies $\langle x - y, x^* \rangle \geq 0$.

We can show that every monotone multivalued mapping is pseudomonotone.

The converse is not true.

The feasible set for the problem G.M.C.P. (f, K) is

$F = \{x_o \in K | f(x_o) \cap K^* \neq \Phi\}$.

If Int (K^*) is nonempty the problem G.M.C.P. (f, K) is called strictly feasible if there exists an element $x_o \in K$ such that $f(x_o) \cap \text{Int}\ (K^*) \neq \Phi$.

In this case we say that x_o is a strict feasible solution.

Theorem 4.2.13

Let $K \subset R^n$ be a solid closed convex cone and let $f:K \to R^n$ be a pseudomonotone upper semicontinuous multivalued mapping, such that $f(x)$ is compact and contractible for each $x \in K$.

Then the problem G.M.C.P. (f, K) has a solution if it is strictly feasible.

Proof

By assumption we have that there exists a strict feasible solution $x_o \in K$.

Let $u^* \in f(x_o) \cap \text{Int}\ (K^*)$. Denoting $C = \{x \in K | \langle x - x_o, u^* \rangle \leq 0\}$ we have (because $u^* \in \text{Int}\ (K^*)$) that C is compact.

We show that f satisfies condition (K.C.) for C. Indeed, let $x \in K \setminus C$.

Then, $\langle x - x_o, u^* \rangle > 0$ which implies (as in remark 2 of Definition 4.2.2) that $\langle x - x_o, v^* \rangle > 0$ for all $v^* \in f(x)$.

Thus $x_o \in C$ works uniformly for all $x \in K\backslash C$ to satisfy condition (K.C.).

Hence the theorem follows from Theorem 4.2.12. □

We finish this section with a classical and important result about the Linear Complementarity Problem with respect to R^n_+ .

Definition 4.2.6

We say that a matrix $M \in \mathcal{M}_{nxn}(R)$ is copositive plus on R^n_+ if:

i) $x \in R^n_+$ implies $\langle Mx, x \rangle \geq 0$,

ii) $x \in R^n_+$ and $\langle Mx, x\rangle = 0$ imply $(M + M^*)\, x = 0$

(where M^* is the adjoint of M).

Remarks 4.2.3

1°) The class of copositive plus matrices includes positive-semidefinite matrices.

2°) Positive linear combinations of copositive plus matrices are copositive plus.

3°) As observed in [A174] if M_1 and M_2 are copositive plus then $\begin{bmatrix} M_1 & -A^* \\ A & M_2 \end{bmatrix}$ are copositive plus for every A.

We denote by L.C.P. (M, R^n_+, q) the Linear Complementarity Problem L.C.P. (f, K) where $K = R^n_+$ and $f(x) = Mx + q$ with $q \in R^n$.

Theorem 4.2.14 [Lemke].

Let M be a n x n-copositive plus matrix on R^n_+.

If L.C.P. (M, R^n_+, q) is feasible then it is solvable.

Proof

We obtain this result by a detailed analysis of Lemke's algorithm as presented in [A174]. □

B. The infinite dimensional case.

We begin this section with a generalization of Theorem 4.2.8 in infinite dimensional spaces.

Let $E(\tau)$ be a locally convex space and let E^* the topological dual of E.

We denote by $\langle , \rangle$ the duality between E and E^*.

Let $K \subset E$ be a pointed closed convex cone.

A convex set B is said to be a <u>base</u> for K if for each $x \in K\setminus\{0\}$, there exist λ $\lambda \in R_+\setminus\{0\}$ and $b \in B$ such that $x = \lambda b$ and the decomposition is unique.

For instance, each closed pointed cone in a separable Banach space has a base.

A classical result is that a pointed closed convex cone is locally compact if and only if it has a compact base.

We can prove that if K has a compact base then every base of K is also compact. We note that this result is not true for bounded bases.

We say that a linear functional ψ on E is <u>strictly positive</u> with respect to K if $\psi(x) > 0$ for every $x \in K\setminus\{0\}$,

A convex set $B \subset K$ is a base for K if and only if there exists a linear functional ψ strictly positive with respect to K such that $B = K \cap \psi^{-1}(1)$.

We say that K is <u>well based</u> if there exists a bounded convex set A such that

$0 \notin \overline{A}$ and $K = \bigcup_{\lambda \geq 0} \lambda A$.

If the topology τ of E is defined by a <u>sufficient family of seminorms</u> $\{p_\alpha\}_{\alpha \in A}$) then we have the following result:

<u>"the closed convex cone $K \subset E$ is well based if and only if there exists a linear continuous functional $\psi: E \to R$ such that,</u>

(20): $(\forall \alpha \in A)(\exists c_\alpha > 0)(\forall x \in K)(c_\alpha p_\alpha(x) \leq \psi(x))$"

In this case since the family $\{p_\alpha\}_{\alpha \in A}$ is sufficient we obtain from (20) that ψ is strictly positive with respect to K and the set

$B = \{x \in K \mid \psi(x) = 1\}$ is a closed base for K.

If the mapping $f: K \to E^*$ is defined we consider the Complementarity Problem:

$$\text{G.C.P.}(f,K): \left\| \begin{array}{l} \text{find } x_* \in K \text{ such that} \\ f(x_*) \in K^* \text{ and } \langle x_*, f(x_*)\rangle = 0 \end{array} \right.$$

We denote (if nonempty)

$K^{\circledast} = \{y \in E^* \mid y \text{ strictly positive with respect to } K\}$.

If K is locally compact or well based then $K^{\circledast}$ is nonempty. Supposing E a separable Banach space and K a closed pointed convex cone in E we can prove that $K^{\circledast}$ is nonempty.

<u>Definition 4.2.7</u>

<u>The problem G.C.P. (f, K) is called strongly feasible if and only if:</u>

1°) <u>$K^{\circledast}$ is nonempty,</u>

2°) $F_s = \{x_o \in K \mid f(x_o) \in K^{\circledast}\}$ is nonempty.

As in definition 4.2.2, we say that <u>f is pseudomonotone if for every pair of points $x, y \in K$ such that $x \neq y$ we have $\langle x - y, f(y)\rangle \geq 0 \Longrightarrow \langle x - y, f(x)\rangle \geq 0$.</u>

<u>Theorem 4.2.15</u>

<u>Let $K \subset E$ be a pointed locally compact convex cone and let $f:K \to E^*$ be a continuous pseudomonotone mapping.</u>

<u>If the problem G.C.P. (f, K) is strongly feasible then it has a solution.</u>

<u>Proof</u>

We consider the variational inequality:

$$(21): \quad \left\| \begin{array}{l} \text{find } x_* \in K \text{ such that} \\ \langle x - x_*, f(x_*)\rangle \geq 0 \text{ for all } x \in K. \end{array} \right.$$

This variational inequality is equivalent to the problem G.C.P. (f, K).

From assumptions we have that there exists an element $x_o \in K$ such that $f(x_o)$ is a continuous linear functional strictly positive on K.

If $x_o = 0$ then it is itself a solution.

We assume $x_o \in K\setminus\{0\}$. In this case we have $\alpha = \langle x_o, f(x_o)\rangle > 0$ and the set $B = \{x \in K| \langle x, \frac{f(x_o)}{\alpha}\rangle = 1\}$ is a compact base for K.

We can prove that the convex set

$$D = \{x \in K| \langle x, \frac{f(x_o)}{\alpha}\rangle \leq 1\}$$

is compact.

Indeed, it is sufficient to show that D is bounded.

Let p be a continuous seminorm on E.

Since B is bounded there exists a real number $r_p > 0$ such that for every $x \in B$ we have $p(x) \leq r_p$.

If $x \in D$ and $p(x) = 0$ obviously we have $p(x) \leq r_p$.

If $x \in D$ is such that $p(x) > 0$ then there exist $\lambda_x > 0$ and $b_x \in B$ such that $x = \lambda_x \cdot b_x$, which implies $p(x) = \lambda_x p(b_x) \leq r_p \lambda_x$.

Finally we have,

$1 \geq \langle x, \frac{f(x_o)}{\alpha}\rangle = \lambda_x \langle b_x, \frac{f(x_o)}{\alpha}\rangle = \lambda_x \geq \frac{p(x)}{r_p}$ and hence $p(x) \leq r_p$ for every $x \in D$.

So, D is bounded and closed and K being locally compact we deduce that D is compact.

For every $u \in K$, now we define

$$D_u = \{x \in D \mid \langle u - x, f(x)\rangle \geq 0\}.$$

It is clear that any $x_* \in D$ would solve the problem G.C.P. (f,K) if and only if $x_* \in \bigcap_{u\in K} D_u$.

The theorem is proved if we show that $\bigcap_{u\in K} D_u$ is indeed nonempty.

Let $\{u_i \mid i = 1, 2, \ldots, m\}$ be an arbitrary set of points in K and let

$$C = \text{conv } [D \cup \{u_1, u_2, \ldots, u_m\}].$$

Since $f:C \to E^*$ is continuous and C is compact it follows from Theorem 3.1. [Hartman-Stampacchia] that there exists an element $\hat{x} \in C$ such that

(22): $\langle x - \hat{x}, f(\hat{x})\rangle \geq 0$ for all $x \in C$.

In particular we have

$$\langle u_i - \hat{x}, f(\hat{x})\rangle \geq 0 \text{ for } i = 1, 2, \ldots, m.$$

Since $x_o \in C$ it follows using (22), the pseudomonotonicity of f and the definition of D that $\hat{x} \in D$.

Thus, we have shown that the intersection of any finite number of the collection $\{D_u\}_{u\in K}$ of closed subsets of D is nonempty and since D is compact it follows that $\bigcap_{u\in K} D_u$ is nonempty. □

Let $(H, \langle ., . \rangle)$ be a real Hilbert space, A is a continuous linear operator on H, K is a closed convex cone in H and b is an element of H.

We consider the Generalized Linear Complementarity Problem associated to the triplet (A, K, b), that is the problem:

$$\text{G.L.C.P.}(A,K,b): \left\| \begin{array}{l} \text{find } x \in K \text{ such that} \\ A(x) + b \in K^* \text{ and} \\ \langle A(x) + b, x \rangle = 0. \end{array} \right.$$

The adjoint A^* of A is the (unique) bounded linear operator on H defined by $\langle A^*(x), y \rangle = \langle x, A(y)\rangle$; $(\forall\, x, y \in H)$.

When $H = R^n$, A^* is exactly the transpose of A. If $A^* = A$, we say that A is a self-adjoint operator.

Definiton 4.2.8

We say that A is copositive plus on K if:

a) $x \in K$ implies $\langle A(x), x\rangle \geq 0$

b) $x \in K$ and $\langle A(x), x\rangle = 0$ imply $(A + A^*)(x) = 0$.

A convex cone $K \subset H$ is said to be polyhedral if there exists $u_1, u_2, \ldots u_n \in K$ such that

$$K = \{x \in H | x = \sum_{i=1}^{n} \lambda_i u_i, \ \lambda_i \geq 0, \ i = 1, 2, \ldots, n\}$$

We say in this case that $u_1, u_2, \ldots, u_n$ are the generators of K.

A real valued function $\phi(x)$ is lower semi-continuous (l.s.c. for short) on K with respect to the weak topology, if $\liminf_{y \to x (weak)(y \in K)} \phi(y) \geq \phi(x)$.

We prove now the Lemke's Theorem for an arbitrary polyhedral cone in an arbitrary Hilbert space.

Theorem 4.2.16

Let K be polyhedral and let A be copositive plus on K. If G.L.C.P. (A,K,b) is feasible then it is solvable.

Proof

If we suppose that $\dim H < +\infty$ then, by using an inner product preserving transformation, we can assume that $H = R^n$ for some $n \in N$. (R^n is endowed with the usual euclidean structure).

Since K is polyhedral, there is a positive integer m and a linear positive transformation $B: R^m \to R^n$ such that $B(R^m_+) = K$.

We can show that $A_* = B^* A B$ is copositive plus on R^m_+.

Since G.L.C.P. (A, K, b) is feasible, there is some $x_o \in K$ such that $< A(x_o) + b, x > \geq 0$; $(\forall x \in K)$.

Let u_o be an element in R^m_+ such that $B(u_o) = x_o$. Then

$$< A_*(u_o) + B^*(b), x > = < B^*(A(x_o) + b), x > =$$

$$= < A(x_o) + b, B(x) > \geq 0; \ (\forall x \in R^m_+).$$

Thus G.L.C.P. $(A_*, R^m_+, B^*(b))$ is feasible.

By Theorem 4.2.14 (Lemke's Theorem) there exists an element $u \in R^m_+$ such that

$$(23): \quad \begin{aligned} &< A_*(u) + B^*(b), x > \geq 0; \ (\forall x \in R^m_+) \text{ and} \\ &< A_*(u) + B^*(b), u > = 0, \end{aligned}$$

which implies:

$< A(B(u)) + b, B(x) > \geq 0$; $(\forall x \in R_+^m)$ and

$< A(B(u)) + b, B(u) > = 0$.

So, we have that $B(u)$ solves G.L.C.P. (A, K, b).

For the general case let E be a finite dimensional subspace of H containing K.

Let P be the (orthogonal) projection from H into E and set $A_o = PA$.

Then $A_o : E \to E$ is copositive plus on K and $< A_o(x_o) + P(b), x > = < A(x_o) + b, x > \geq 0$ for each $x \in K$.

Thus G.L.C.P. $(A_o, K, P(b))$ is feasible in E. By the previous case, there exists $x_* \in K$ such that

$$(24): \quad \begin{cases} < A_o(x_*) + P(b), x > \geq 0; \ (\forall x \in K) \text{ and} \\ < A_o(x_*) + P(b), x_* > = 0. \end{cases}$$

Since $<A_o(x_*) + P(b), x> = <A(x_*) + b, x>$ for each $x \in K$, (24) shows that x_* solves G.L.C.P. (A, K, b) and the theorem is proved. □

This result was proved by Gowda and Seidman [C20].

Definition 4.2.9

We say that a closed convex cone $K \subseteq H$ is a Galerkin cone if there exists a family of convex subcones $\{K_n\}_{n \in N}$ of K such that,

1°) K_n is locally compact for every $n \in N$,

2°) if $n \leq m$ then $K_n \leq K_m$,

3°) $K = \overline{\bigcup_{n \in N} K_n}$

We denote a Galerkin cone by $K(K_n)_{n \in N}$ and we say that $\{K_n\}_{n \in N}$ is a Galerkin approximation of K.

Examples

1°) If H is a separable Hilbert space and if H (or K) admits a Schauder base then K is a Galerkin cone.

2°) In an arbitrary Hilbert space H a closed convex cone is a Galerkin cone if K admits a Galerkin approximation by the finite element method.

If $K(K_n)_{n \in N}$ is a Galerkin cone such that for every $n \in N$ K_n is a polyhedral cone we say that K is a Galerkin cone with a polyhedral approximation.

Let H be a Hilbert space ordered by a pointed closed convex cone $K \subset H$.

We say that H is a normed vector lattice if and only if:

1°) with respect to the order defined by K, H is a vector lattice (i.e. for every $x, y \in H$ there exist $x \vee y := \sup(x, y)$ and $x \wedge y := \inf(x, y)$,

2°) for every $x, y \in H$ such that $|x| \leq |y|$ we have $\|x\| \leq \|y\|$ (where $|x| = x \vee (-x)$ and $|y| = y \vee (-y)$).

From the proof of Proposition 4.3 of paper [A133] we deduce the following result.

Proposition 4.2.5

If H is an infinite-dimensional Hilbert space and $K \subset H$ a closed separable well based cone such that with respect to the order defined by K, H is a normed vector lattice then K is a Galerkin cone with a polyhedral approximation.

In a general Banach space E if $C \subset E$ is a convex set, then we say that a continuous operator P defined on all of E is a projection on C if $P(E) = C$ and $P(x) = x$, for $x \in C$.

The following result was proved by Krasnoselskii and Zabreiko. [M.A. Krasnoselskii and P.P. Zabreiko: Geometrical Methods of Nonlinear Analysis, Springer-Verlag, (1984)].

Theorem 4.2.17

Let K be a closed convex cone in a Banach space E. For $\alpha > 0$ there exists a projection P_α onto K such that,

$$\|x - P_\alpha(x)\| \leq (1 + \alpha)\, d(x, K); \text{ for every } x \in E, \text{ with } d(x, K) = \inf_{y \in K} \|x - y\|.$$

Let $K(K_n)_{n \in N}$ be a Galerkin cone in a Hilbert space H.

In this case we can suppose that, for every $n \in N$ there exists a projection P_n onto K_n such that, for every $x \in K$, $\{P_n(x)\}_{n \in N}$ is convergent to x (with respect to the norm $\|\ \|$ of H).

Indeed, since H is a Hilbert space, using Theorem 4.2.17 for an arbitrary real number $\alpha \geq 0$, we obtain for every $n \in N$ a continuous projection P_n onto K_n such that

$$\|x - P_n(x)\| \leq (1 + \alpha)\, d(x, K_n), \text{ for every } x \in K.$$

In particular, for each $x \in K$ and each $\varepsilon > 0$ there exists $n \in N$ with $\|x - x_n\| < \varepsilon$ where $x_n \in K_n$.

Hence, for $m \geq n$ we get,

$$\|x - P_m(x)\| \leq (1 + \alpha)\, d(x, K_n) \leq (1 + \alpha)\, d(x, K_n) < (1 + \alpha)\varepsilon$$

and hence, $\lim_{n \to \infty} P_n(x) = x$, for every $x \in K$.

Theorem 4.2.18

Suppose that H is a Hilbert space, $T: H \to H$ a continuous linear operator, $b \in H$ and $K \subset H$ a closed convex cone such that:

1) T is copositive plus on K,

ii) <u>the map $x \to \langle T(x), x\rangle$ is weak l.s.c,</u>

iii) <u>K is a well based Galerkin cone with a polyhedral approximation $\{K_n\}_{n\in N}$,</u>

iv) <u>$\{x \in K | T(x) \in K^*, \langle T(x), x\rangle = 0$ and $\langle b, x\rangle = 0\} = \{0\}$.</u>

<u>If G.L.C.P. (T, K, b) is feasible then it is solvable.</u>

<u>Proof</u>

Let x_o be a feasible point for G.L.C.P. (T, K, b).

We can assume that $x_o \neq 0$, since if $x_o = 0$ then x_o is a solution for G.L.C.P. (T, K, b).

If we add x_o as generator at every K_n we can suppose that $x_o \in K_n$ for every $n \in N$ and we obtain another polyhedral Galerkin approximation for K and in addition T is copositive plus on K_n for each $n \in N$.

We note that for every $n \in N$ there exists a projection P_n onto K_n such that for every $x \in K$, $\lim_{n\to\infty} P_n(x) = x$.

Now, fix an $n \in N$. Since x_o is feasible for G.L.C.P. (T, K_n, b), <u>Theorem 4.2.16</u> gives an $x_n \in K_n$ such that

(25): $\langle T(x_n) + b, x\rangle \geq 0; \ \forall x \in K_n$ and $\langle T(x_n) + b, x_n\rangle = 0$.

We claim that the sequence $\{x_n\}_{n\in N}$ is bounded.

Suppose to the contrary one were to have $\|x_n\| \to \infty$ (as $n \to \infty$).

Since K is well based there exists a continuous linear functional Φ such that $B = \{x \in K | \phi(x) = 1\}$ is a bounded base for K and $\|x\| \leq \phi(x)$ for every $x \in K$.

Hence $\phi(x_n) \to \infty$ (as $n \to \infty$) and we can suppose $\phi(x_n) \neq 0$ for every $n \in N$.

Then from (25) we would have

(26): $\langle T\left(\frac{x_n}{\phi(x_n)}\right) + \frac{b}{\phi(x_n)}, \frac{x_n}{\phi(x_n)}\rangle = 0$; for every $n \in N$.

Since $\frac{x_n}{\phi(x_n)} \in B$ for every $n \in N$ we have that $\{\frac{x_n}{\phi(x_n)}\}$ (or a subsequence) converges weakly to some element $x_* \in K$.

From assumption (ii), formula (26) and since $\{\frac{x_n}{\phi(x_n)}\}_{n\in N}$ is bounded we deduce, $\langle T(x_*), x_*\rangle \leq 0$.

Because T is copositive plus on K, one has $\langle T(x_*), x_*\rangle \geq 0$ so $\langle T(x_*), x_*\rangle = 0$.

This implies by (i)

(27): $T(x_*) = - T^*(x_*)$.

Now (26) implies that

$< b, x_n > = - <T(x_n), x_n > \leq 0$ (by (i)).

This gives

(28): $< b, x_* > \leq 0$.

For any $k \in K$ we denote $k_n = P_n(k)$ and we have $k = \lim_{n\to\infty} k_n$ and $k_n \in K_n$.

If we denote $k' = \phi(x_n)k_n$ we have

$< T(x_n) + b, k' > \geq 0$ which implies

$< T\left(\frac{x_n}{\phi(x_n)}\right) + \frac{b}{\phi(x_n)}, k_n > \geq 0$ and since

$< T\left(\frac{x_n}{\phi(x_n)}\right), k_n > + < \frac{b}{\phi(x_n)}, k_n > \geq 0$ implies,

$< \frac{x_n}{\phi(x_n)}, T^*(k_n) > + < \frac{b}{\phi(x_n)}, k_n > \geq 0$

we deduce, $< x_*, T^*(k) > \geq 0$; for every $k \in K$, that is

(29): $T(x_*) \in K^*$.

The feasibility of G.L.C.P. (T, K, b) implies that $< T(x_o) + b, x_* > \geq 0$ whence

(30): $< b, x_* > \geq - <T(x_o), x_* > = - <x_o, T^*(x_*) > =$

$= <x_o, T(x_*) > \geq 0$ (by (29)).

From (28) and (30) we get $< b,x_*> = 0$. Thus $x_* \in \{x \in K | T(x) \in K^*, <T(x),x > = 0$ and $<b, x> = 0\}$ and by assumption (iv) we deduce $x_* = 0$, that is $0 \in$ weak closure (B) = closure (B) = B which is impossible.

It follows that $\{x_n\}_{n\in N}$ is bounded.

We may then assume that $\{x_n\}_{n\in N}$ (or a subsequence) converges weakly to some element $u_o \in K$.

Now (25) gives by assumption (ii) that

(31): $< T(u_o) + b, u_o > \leq \liminf_{(x_n \to u_o)(\text{weak})} < T(x_n) + b, x_n > = 0$.

Since for every $k \in K$ we have $k = \lim_{n\to\infty} k_n$ where $k_n = P_n(k) \in K_n$ we obtain using again (25),

(32): $< T(u_o) + b, k > \geq 0$ for every $k \in K$.

Finally (31) and (32) imply that u_o solves G.L.C.P. (T, b, K). □

Remark 4.2.4

Theorem 4.2.18 is similar to a result proved by Gowda and Seidman [C20] since we can prove that a "thin cone" in Gowda and Seidman's sense is exactly a well based cone. Our proof is based on the idea used by Gowda and Seidman.

In Theorem 4.2.18 the assumption that K is well based is essential.

Example [Gowda and Seidman]

Consider the space

$$H = \ell_2 = \{x = (x_n)_{n\in N} | x_n \in R \text{ and } \sum_{n=1}^{\infty} x_n^2 < +\infty\}$$

which is a Hilbert space with respect to the inner product $< x, y > = \sum_{n=1}^{\infty} x_n y_n$.

Let $K = \ell_2^+ = \{x = (x_n)_{n\in N} \in \ell_2 | x_n \geq 0; \forall n \in N\}$, K is separable and it has a polyhedral Galerkin approximation.

Consider the elements

$$v = \alpha\left(1, \frac{1}{2^2}, \frac{1}{3^2}, \ldots\right),$$

$$u = \beta\left(\gamma, \frac{1}{2}, \frac{1}{3}, \ldots\right)$$

where α, β, γ are chosen so that $\alpha, \beta > 0$, $\|u\| = \|v\| = 1$ and $< u, v > = 0$.

Consider the operator $P: \ell_2 \to \ell_2$ defined by $P(x) = < x, u > u + < x, v > v$.

Since P is a projection on ℓ_2 we have that P is monotone on ℓ_2 and hence copositive plus on $K(= \ell_2^+)$.

The monotonicity of P implies the weak lower semi-continuity of $x \to <P(x), x>$.

In the theory of ordered vector spaces is known that ℓ_2^+ is not well based.

Put $b = -u$. For every $n < 1$, e_n denotes the element in ℓ_2^+ with 1 as the nth entry and zero elsewhere.

$$P\left(\frac{n}{\beta} e_n\right) = < \frac{n}{\beta} e_n, u > u + < \frac{n}{\beta} e_n, v > v = u + \frac{\alpha}{\beta} \frac{1}{n} v.$$

Since $v \in K^*$ (= K), we have

$$P\left(\frac{n}{\beta} e_n\right) + b = \frac{\alpha}{\beta}\frac{1}{n} v \in K^*.$$

Thus G.L.C.P. (P, K, b) is feasible.

Hence we have verified all the conditions in Theorem 4.2.18 except the fact that K is well based.

Now, we show that G.L.C.P. (P, K, b) is not solvable.

Suppose, there was an $x_* \in \ell_2^+$ with

$< P(x_*) + b, x > \geq 0$ for $x \in \ell_2^+$ and

$< P(x_*) + b, x_* > = 0$.

The inequality $< P(x_*) + b, \frac{n}{\beta} e_n > \geq 0$,

(true for every $n \in N$) gives $< x_*, u > + \frac{\alpha}{\beta}\frac{1}{n} < x_*, v > - 1 \geq 0$

for $n = 2, 3, \ldots$ and hence $< x_*, u > \geq 1$.

The equality,

$<x_*, u>^2 + < x_*, v >^2 - < x_*, u > = < P(x_*) + b, x_* > = 0$

and the inequality $< x_*, u > \geq 1$ imply $< x_*, v > = 0$.

Since $x_* \in \ell_2^+$ and all entries of v are positive, one must then have $x_* = 0$.

This implies that

$< -u, x > = < b, x > = < P(x_*) + b, x > \geq 0$ for all $x \in \ell_2^+$.

But $< -u, e_2 > = -\frac{\beta}{2} < 0$.

Thus G.L.C.P. (P, ℓ_2^+, β) is not solvable. □

We present now a result in Banach spaces.

Let $(E, \| \|)$ be a Banach space and let $K \subset E$ be a closed convex cone.

If $f:K \to E^*$ is an arbitrary mapping we consider the General Complementarity Problem:

$$\text{G.C.P.}(f,K): \left\| \begin{array}{l} \text{find } x_* \in K \text{ such that} \\ \\ f(x_*) \in K^* \text{ and } <x_*, f(x_*)> = 0 \end{array} \right.$$

where $< .,. >$ is the natural duality between E and its topological dual E^*.

<u>Theorem 4.2.19</u>

<u>Let $(E, \| \|)$ be a reflexive Banach space and let $K \subset E$ be a closed convex cone.</u>

<u>Suppose $f:K \to E^*$ to be a bounded (not necessary linear) hemicontinuous strictly monotone operator.</u>

<u>If the problem G.C.P. (f, K) is feasible then it is uniquely solvable.</u>

<u>Proof</u>

Since the problem G.C.P. (f, K) is feasible there exists $x_o \in K$ such that

$f(x_o) \in K^*$. We can suppose $x_o \neq 0$ since if $x_o = 0$ we observe that $x_o = 0$ is a solution of G.C.P. (f, K).

For every $u \in K^*$ and $r > 0$ we denote

$$K_r(u) = \{x \in K \mid 0 \leq < x, u > \leq r\}.$$

Obviously $K_r(u)$ is convex. Since the mapping $\Phi(x) = <x, u>$ is continuous and $K_r(u) = \Phi^{-1}([0, r])$ we have that $K_r(u)$ is closed.

By Mosco's Theorem [Chapter 1] we obtain for every $r > 0$ and $u \in K^*$ a unique element $x_r \in K_r(u)$ such that

(33): $\quad < z - x_r, f(x_r) > \geq 0$ for every $z \in K_r(u)$.

Since $0 \in K_r(u)$ we obtain from (33) that $<x_r, f(x_r)> \leq 0$ for every $r > 0$.

We have the following two cases.

i) <u>There exist $u \in K^*$ and $r > 0$ such that $x_r \in K_r^<(u) = \{x \in K \mid 0 \leq <x, u> < r\}$</u>.

In this case there exists $\lambda > 1$ such that $\lambda x_r \in K_r(u)$.

Indeed, if $x_r, u = 0$ we can take an arbitrary $\lambda > 1$, but if $0 \leq <x_r, u>$ there exists $\lambda > 1$ such that $\lambda x_r \in K_r^=(u) = \{x \in K \mid <x, u> = r\} \subset K_r(u)$.

From (33) we have

(34): $\quad <x_r, f(x_r) > \leq <\lambda x_r, f(x_r)> = \lambda <x_r, f(x_r)>$.

But since $<x_r, f(x_r) > \leq 0$ and $\lambda > 1$ it is necessary to have $<x_r, f(x_r) > = 0$.

Hence, in this case if we show that $f(x_r) \in K^*$ we obtain that x_r is a solution of G.C.P. (f, K).

Let x be an arbitrary element of K.

There exists $\rho > 0$ such that $\rho x \in K_r(u)$ and from (33) we have,

$<\rho x, f(x_r)> = <\rho x - x_r, f(x_r)> \geq 0$ that is $<x, f(x_r) > \geq 0$ for every $x \in K$.

ii) <u>For every $u \in K^*$ and every $r > 0$ we have $x_r \in K_r^=(u)$</u>.

We show now that this case is impossible.

If we put $u = f(x_o)$ we consider $r > 0$ such that $0 < < x_o, f(x_o) > < r$.

We have $x_o \in K_r^<(f(x_o))$ and since f is monotone we obtain

(35): $\quad < z - x_o, f(z) > \geq < z - x_o, f(x_o) > > 0$

for every $z \in K_r^=(f(x_o))$.

Since $x_r \in K_r^=(f(x_o))$ we deduce,

(36): $\quad < x_r - x_o , f(x_r) > > 0$

But because $x_o \in K^{<}_r(f(x_o)) \subset K_r(f(x_o))$ formula (33) implies,

$$< x_o - x_r , f(x_r) > \geq 0, \text{ that is } < x_r - x_o , f(x_r) > \leq 0$$

which contradicts (36). So case ii) is impossible and the existence is proved.

The uniqueness is a consequence of the assumption that f is strictly monotone.□

. 4.3 General existence theorems

In this section we suppose that $< E,E^* >$ is a dual system of locally convex spaces or of Banach spaces.

Given a closed convex cone $K \subset E$ and $f:K \to E^*$ a mapping we will consider some general existence theorems for the complementarity problem

G.C.P. (f,K): find $x_* \in K$ such that $f(x_*) \in K^*$ and $< x_*, f(x_*) > = 0$.

If A is a subset of E we denote by conv(A) the convex hull of A.

Let $D \subset E$ be an arbitrary nonempty subset.

Definition 4.3.1.

We say that a point-to-set map $f:D \to E$ is a KKM-map if $\text{conv}(\{x_1, x_2, \ldots, x_n\}) \leq \leq \bigcup_{i=1}^{n} f(x_i)$ for each finite subset $\{x_1, x_2, \ldots, x_n\} \subset D$.

The next theorem is fundamental for some of our results.

Theorem 4.3.1 [Ky Fan]

Let D be an arbitrary nonempty set in a Hausdorff topological vector space E.
Let $f:D \to E$ be a KKM-map.
If all the sets f(x) are closed in E and if one is compact, the $\bigcup_{x \in D} f(x) \neq \Phi$.

Proof

See: Ky Fan: A generalization of Tychonoff's fixed point theorem. Math Ann 142 (1961), 305-310. □

Theorem 4.3.2.

Let $E(\tau)$ be a topological Hausdorff vector space, $D_o \subset E$ a nonempty convex subset and $f:D_o \to E^*$ a mapping such that $y \to <f(y), y-x>$ is lower semicontinuous on D_o for every $x \in D_o$.

If there exists a nonempty convex compact subset $K_o \subset D_o$ such that

$$(\forall y \in D_o \setminus K_o)(\exists x \in K_o)(< f(y), y-x > > 0$$

then there exists a vector $\hat{y} \in K_o$ such that

$$< f(\hat{y}), \hat{y}-x > \leq 0, \text{ for all } x \in D_o.$$

Proof

First, we remark that the function $\Psi(x,y) = < f(y), y-x >$ has the following properties:

1°). $\Psi(x,x) = 0$, for all $x \in D_o$,

2°). for every $x \in D_o$ the function $y \to \Psi(x,y)$ is lower semicontinuous,

3°). for every $y \in D_o$ the set $\{x \in D_o | \Psi(x,y) > 0\}$ is convex,

4°). there exists a nonempty convex compact subset $K_o \subset D_o$ such that

$$(\forall y \in D_o \setminus K_o)(\exists x \in K_o)(\Psi(x,y) > 0).$$

For every $x \in D_o$ we consider the set $K(x) = \{y \in K_o | \Psi(x,y) \leq 0\}$ and from property 2°) we obtain that for every $x \in D_o$ the set $K(x)$ is closed.

The family $\{K(x)\}_{x\in D_o}$ has the finite intersection property.

Indeed, let $x_1, x_2, \ldots, x_m \in D_o$ be arbitrary element and denote

$$C = \text{conv}(K_o \cup \{x_1, x_2, \ldots, x_m\}).$$

Obviously, C is a compact convex subset of D_o. For every $x \in D_o$ we set

$$F(x) = \{y \in C | \Psi(x,y) \leq 0\}$$

From property 1°) we have that for every $x \in C$ the set $F(x)$ is nonempty (since $x \in F(x)$) and from property 2°) we have that $F(x)$ is closed and consequently $F(x)$ is compact.

We prove now that F is a KKM-map.

Indeed, if we suppose that F is not a KKM-map then there exist $\{u_1, u_2, \ldots, u_n\} \subset C$ and $\alpha_j \geq 0$; $1 \leq j \leq n$ with $\sum_1^n \alpha_j = 1$ and such that $\sum_{j=1}^n \alpha_j u_j \notin \bigcup_{i=1}^n F(u_i)$, that is, $\Psi(u_i, \sum_{j=1}^n \alpha_j u_j) > 0$, for every $i = 1, 2, \ldots, n$.

But property 3°) implies

$$\Psi(\sum_{j=1}^n \alpha_j u_j, \sum_{j=1}^n \alpha_j u_j) > 0$$

which is a contradiction of property 1°).

Hence applying Theorem 4.3.1. we obtain that $\bigcap_{x\in C} F(x) \neq \Phi$, which implies that there exists $y \in C$ such that $\Phi(x,y) \leq 0$; for all $x \in C$.

So, property 4°) implies now that $\bar{y} \in K_o$ and particularly we have $\bar{y} \in \bigcap_{i=1}^{m} K(x_i)$, that is the family $\{K(x)\}_{x \in D_o}$ has the finite intersection property.

Using the fact athat K_o is compact we have that $\bigcap_{x \in D_o} K(x) \neq \Phi$.

If $\hat{y} \in \bigcap_{x \in D_o} K(x)$ then we have that $\Phi(x,\hat{y}) \leq 0$; for all $x \in D_o$ and the theorem is proved. □

As direct consequences of Theorem 4.3.2 we obtain two important existence theorem for the problem G.C.P.(f,K).

Theorem 4.3.3.

Let $< E,E^* >$ be a dual system of locally convex spaces, $K \subset E$ a closed convex cone and $F: K \to E^*$ such that

1°). for every $y \in K$ the mapping $x \to < x-y, f(x) >$ is lower semicontinuous on K,

2°). there exists a nonempty compact convex set $D \subset K$ such that for every $x \in K \setminus D$ there exists $y \in D$ with $< x-y, f(x) > > 0$,

then the problem G.C.P.(f,K) has a solution x_* and $x_* \in D$. □

Remark 4.3.1.

Theorem 4.3.3. was initially proved by Allen [A12], but with assumption 1°) supposing to be:

"the mapping $x \to < f(x), x >$ is lower semicontinuous on K".

In our paper [C27] we showed that this Allen's theorem is incorrect and the correct form is Theorem 4.3.3.

Theorem 4.3.4. [Karamardian][A157]

Let $< E,E^* >$ be a dual system of locally convex spaces, $K \subset E$ a closed convex cone and $f: K \to E^*$ a mapping such that the mapping $(x,y) \to < x,f(y) >$ is continuous on K x K.

If there exists a nonempty convex compact set $D \subset K$ such that $(\forall x \in K \setminus D)(\exists y \in D)(< x-y, f(x) > > 0)$ then the problem G.C.P.(f,K) has a solution x_* and $x_* \in D$. □

Remark 4.3.2

The condition, "there exists a nonempty convex compact set $D \subset K$ such that $(\forall x \in K \setminus D)(\exists y \in D)(< x-y, f(x) > > 0)$" will be named "Karamardian's condition"

Karamardian's theorem implies some interesting results in finite dimensional spaces.

Indeed, let $E = R^n$, $K \subset R^n$ a closed convex cone and $f: K \to R^n$.

The space R^n is considered with euclidean structure.

Definition 4.3.2.

We say that f is strongly K-monotone if there exists a constant $M > 0$ such that for every $x, y \in K$ satisfying $x - y \in K$ we have $< f(x) - f(y), x - y > \geq M\|x-y\|^2$.

Definition 4.3.3.

We say that f is strongly K-copositive if there exists a constant $M > 0$ such that for every $x \in K$ we have $< f(x) - f(0), x > \geq M \|x\|^2$.

Remarks 4.3.3.

1°). Every strongly K-monotone mapping is strongly K-copositive.
2°). Every differentiable function satisfying the following property:

$$(\theta : \left\| \begin{array}{l} \text{there exists } M > 0 \text{ such that} \\ \text{for all } x, y \in K \text{ we have} \\ < x, \dot{J}_f(y)(x) > \geq M \|x\|^2 \end{array} \right.$$

(where $J_f(y)$ is the Jocobian of f at the point y) is strongly K-=copositive.

Indeed, for every $x \in K$ we consider the function $\Phi : [0, 1] \to R$ defined by $\Phi(t) = < x, f(tx) >$.

By differentiation we have $\Phi'(t) = < x, J(tx)(x) >$ and from (Θ) we obtain $\Phi'(t) \geq M \|x\|^2$, which implies $< x, f(x) - f(0) > = \Phi(1) - \Phi(0) =$

$$= \int_0^1 \Phi'(t)\, dt \geqslant \int_0^1 M \|x\|^2 dt = M \|x\|^2$$

Theorem 4.3.5.

If $f: K \to R^n$ is a strongly K-coposition continuous function then the problem G.C.P.(f, K) has a solution.

Proof

If $f(0) = 0$ then $x_* = 0$ is a solution of the problem G.C.P.(f,K).

We suppose now $f(0) \neq 0$. Since f is strongly K-copositive there exists $M > 0$ such that

(1): $\langle x, f(x) \rangle \geq \langle x, f(0) \rangle + M \|x\|^2$; $\forall x \in K$

We put $\rho = \|f(0)\| / M$ and we define $D = \{x \in K \mid \|x\| \leq \rho\}$.

We have that D is nonempty convex compact and for every $x \in K \setminus D$,

(2): $M \|x\|^2 > \|x\| \|f(0)\|$.

From (1), (2) and Schwartz's inequality we obtain,

$\langle x, f(x) \rangle > \langle x, f(0) \rangle + \|x\|.\|f(0)\| \geq 0$; $\forall x \in K \setminus D$

which implies Karamardian's condition.

Hence, from Theorem 4.3.4. we obtain that the problem G.C.P.(f,K) has a solution.

By a similar proof as for Theorem 4.3.5. but using Theoremd 4.3.3. we have the following result.

Theorem 4.3.6.

Let $(E, \| \|)$ be a Banach space, $K \subset E$ a locally compact convex cone and $f: K \to E^*$ a strongly K-copositive mapping.

If for every $y \in K$ the mapping $x \to \langle x - y, f(x) \rangle$ is lower semicontinuous then the problem G.C.P.(f,K) has a solution.

In the next results we will use the concepts of α-monotone operator of maximal monotone operator and other concepts defined in chapter 1.

Bazaraa, Goode and Nashed proved in 1972 [A17] the following result.

Theorem 4.3.7.

Let $\langle E, E^* \rangle$ be a dual system of reflexive Banach spaces and let $K \subset E$ be a closed convex cone.

If $f : K \to E^*$ is a bounded α-monotone and hemicontinuous operator then the problem G.C.P.(f,K) has a solution and this solution is unique.

The initial proof of this theorem is very long but we will prove now a more general result.

We remark also that the next result shows that in Theorem 4.3.7 the assumption that f is bounded is not necessary.

If $D \subset E$ is a convex set we denote by Ψ_D the indicatrix of D [see chapt. 1].

Theorem 4.3.8 [Luna] [A181]

Les $\langle E, E^* \rangle$ be a dual system of reflexive Banach spaces and let $K \subset E$ be a closed convex cone.

If the point-to-set mapping $f : E \to E^*$ satisfies the following assumptions:

1°) $f + \partial\Psi_K$ is maximal monotone,

2°) there exists $\beta \subset 0$ such that $x \in D(f) \cap K$, $x^* \in f(x)$ and $\|x\| > \beta$ imply $\langle x, x^* \rangle \geq 0$, then there exist $x_o \in K$ and $x^* \in f(x_o)$ such that $x^* \in K^*$ and $< x_o, x^*_o > = 0$.

Proof

To prove this theorem we will apply Rockafellar's Theorem [chap. 1].

Let $x^* \in (f + \partial\Psi_K)(x)$ be an element such that $x \in D(f) \cap K$ and $\|x\| > \beta$.

Then $x^* = y^* + z^*$, where $y^* \in f(x)$ and $z^* \in \partial\Psi_K(x)$ which implies

$$< x, x^* > = < x, y^* + z^* > = < x, y^* > + < x, z^*) >$$

and since $z^* \in \partial\Psi_K(x)$ and K is a convex cone we have $< x, z^* > =)$ and hence we have $< x, x^* > = < x, y^* >$.

From assumption 2°) we deduce $< x, x^* > \geq 0$ and applying Rockafellar's Theorem we have that there exists $x_o \in K$ such that $0 \in (f + \partial\Psi_K)(x_o)$, which implies that there exist $x^*_o \in f(x_o)$ and $y^*_o \in \partial\Psi_K(x_o)$ such that $0 = x^*_o + y^*_o$.

Hence, $0 = < x_o, 0 > = < x_o, x^*_o + y^*_o > = < x_o, x^*_o >$ and the proof is finished.

The next corrollay is a generalization ofTheorem 4.3.7.

Corollary

Let $\langle E, E^* \rangle$ be a dual system of reflexive Banach space and let $K \subset E$ be a closed convex cone.

If $f : E \to E^*$ is a hemicontinuous α-monotone operator then the problem G.C.P.(f,K) has a solution and this solution is unique.

Proof

From Browder-Stampacchia-Rockafellar's Theorem [chap. 1] we have that $f + \partial\Psi_K$ is a maximal monotone operator and because f is α-monotone we get,

$< x, f(x) > \geq \|x\| \alpha(\|x\| + < x, f(o) > \geq \|x\| [\alpha(\|x\|) - \|f(0)\|]$.

If $\beta > 0$ is sufficiently big such that $\|x\| > \beta$ implies $\alpha(\|x\|) \geq \|f(0)\|$ then we can apply Theorem 4.3.8 and we obtain an element $x_o \in K$ such that

$f(x_o) \in K^*$ and $< x_o, f(x_o) > = 0$.

If we suppose that there exists another element $x_1 \in K$ such that $f(x_1) \in K^*$ and $<x_1, f(x_1) > = 0$ then we have, $0 \geq < x_o - x_1, f(x_o) - f(x_1) > \geq$

$\geq \|x_o - x_1\| \alpha(\|x_o - x_1\| \geq 0$, which implies $x_o = x_1$. □

To apply Theorem 4.3.8. it is important to know some maximality tests.

In this sense we can use the results sproved in: R.T. Rockafellar: On the maximality of sums of nonlinear monotone operators. Trans. Amer. Math. Soc. Vol. 149(1970, 75-88.

For example we have the following tests.

The operator $f + \partial\Psi_K$ is maximal monotone if one of the following conditions is satisfied:

i°) $\dim E < +\infty$ and $(\mathrm{ri}\, D(f)) \cap K \neq \Phi$ (where ri C is the interior of C with respect to the affine hull of C)

ii°) $D(f) \cap \mathrm{Int}\, K \neq \Phi$,

iii°) $\mathrm{Int}\, D(f) \cap K \neq \Phi$.

Remark 4.3.4

The Corollary of Theorem 4.3.8 is not true if the condition "α-monotone" is replaced by "strictly monotone".

Example

$$E = R,\ K = R_+,\ f : K \to R,\ f(x) = -\frac{1}{1+x}.$$

In this case f is strictly monotone but $\langle x, f(x)\rangle = 0$ implies $x = 0$ while $f(0) = -1 \notin K^* = R$.

We consider now the following problem: given $r > 0$ when does the problem G.C.P. (f, K) have a solution $x_* \in K$ such that $\|x_*\| \leq r$?

This problem is important in practice or when we approximate solutions of the general complementarity problem by some numerical methods based on the global optimization.

Let $\langle E, E^* \rangle$ be a dual system of Banach spaces and let $K \subset E$ be a closed convex cone.

Given two operators $T_1, T_2 : K \to E^*$ we consider the problem G.C.P. (f,K) with $f(x) = T_1(x) - T_2(x)$.

This case seems to be frequently used in practical problems.

First, we need to introduce the operators of class $(S)^1_+$.

In this sense we denote by "(w)-lim" the limit with respect to the weak topology.

Definition 4.3.4.

We say that a mapping $T : E \to E^*$ satisfies condition $(S)^1_+$ if, for any sequence $\{x_n\}_{n\in N} \subset E$ with $(w)\text{-}\lim_{n\to\infty} x_n = x_*$, $(w)\text{-}\lim_{n\to\infty} T(x_n) = u \in E^*$ and $\limsup_{n\to\infty} \langle x_n, T(x_n)\rangle \leq \langle x_*, u\rangle$ we have that $\{x_n\}_{n\in N}$ is norm convergent to x_*.

We consider on E* the dual norm of the norm $\|\ \|$ given on E and we denote it by

We say that a continuous and strictly increasing function $\Phi : R_+ \to R_+$ is a weight if $\Phi(0) = 0$ and $\lim_{r\to+\infty} \Phi(r) = +\infty$.

Given an arbitrary weight Φ a duality mapping on E associated to Φ is a mapping $\dot{J} : E \to 2^{E^*}$ such that $\dot{J}(x) = \{x^* \in E^* \mid \langle x,x^* \rangle = \|x\|.\|x^*\|_* \text{ and } \|x^*\|_* = \Phi(\|x\|)\}$

A consequence of Hahn-Banach theorem is the fact that for every $x \in E$ that $\dot{J}(x)$ is nonempty.

Examples

1°). If E is a Hilbert space and $\Phi(r) = r$, for every $r \in R_+$ then the duality mapping associated to Φ is $J(x) = x$, for every $x \in E$.

2°). If $E = L^p(\Omega)$, $\|u\| = (\int_\Omega |u|^p dx)^{1/p} = \|u\|_{L^p(\Omega)}$ and $\Phi(r) = r^{p-1}$ then $\dot{J}(u) = |u|^{p-2}. u$

3°). If $E = W_o^{1,p}(\Omega)$, $\|u\| = (\sum_{i=1}^n \|D_i u\|^p_{L^p(\Omega)})^{1/p}$ and $\Phi(r) = r^{p-1}$ then

$$\dot{J}(u) = -\sum_{i=1}^n \frac{\partial}{\partial x_i} \left(\left| \frac{\partial u}{\partial x_i} \right|^{p-2} \frac{\partial u}{\partial x_i} \right).$$

A duality mapping is a monotone operator and it is strictly monotone if E is strictly convex.

About duality mappings we have the following classical results.

A). If $(E, \| \|)$ is a reflexive Banach space with $(E^*, \| \|_*)$ strictly convex then a duality mapping associated to a weight function Φ is a hemicontinuous point-to-point mapping. □

B). If $(E, \| \|)$ is a Banach space then a duality mapping on E is a point-to-point mapping and norm continuous if and only if the norm of E is Frêchet differentiable. □

We recall that a Banach space $(E, \| \|)$ is locally uniformly convex if for every $\varepsilon > 0$ and x with $\|x\| = 1$ there exists $\delta(\varepsilon,x) > 0$ such that the inequality $\|x-y\| \geq \varepsilon$ implies $\|x+y\| \leq 2 [1 - \delta(\varepsilon, x)]$ for every $y \in E$ with $\|y\| = 1$.

Certainly, every uniformly convex Banach space is locally uniformly convex and reflexive.

Every locally uniformly convex Banach space is strictly convex.

We note that Sobolev spaces $W_o^{m,p}(\Omega)$ are locally uniformly convex.

As consequence of some classical results proved by Lindenstrauss, Asplund and Troyanski we have the following result.

If $(E, \| \|)$ is a reflexive Banach space, then there exists on E an equivalent norm $\| \|_1$ such that $(E, \| \|_1)$ and $(E^*, \| \|_{1*})$ are locally uniformly convex. Moreover the norm $\| \|_1$ and $\| \|_{1*}$ are Frêchet diffentiable.

We say that a Banach space $(E, \| \|)$ is Kadeç if for each sequence $\{x_n\}_{n\in N}$ in E which converges weakly to x_* with $\lim_{n\to\infty} \|x_n\| = \|x_*\|$ we have $\lim_{n\to\infty} \|x_n - x_*\| = 0$.

Each L^p space $(1 < p < \infty)$ has this property as does $l_1(S)$ and any locally uniformly convex Banach space in particular every Hilbert space.

Proposition 4.3.1

Let $(E, \| \|)$ be a Banach space which is Kadeç and such that E^* is strictly convex.

If $\dot{J}$ is a duality mapping on E associated to a weight Φ, then $\dot{J}$ satisfies condition $(S)^1_+$.

Proof

Since E^* is strictly convex we have that $\dot{J}$ is a point-to-point mapping.

Consider a sequence $\{x_n\}_{n\in N} \subset E$ such that $(w) - \lim_{n\to\infty} x_n = x_*$, $(w) - \lim_{n\to\infty} \dot{J}(x_n) = u$ and $\lim_{n\to\infty} \sup < x_n, \dot{J}(x_n > \leq > x_*, u >$.

From the definition of $\dot{J}$ we have

$< x_n - x_*, \dot{J}(x_n) - \dot{J}(x_*) > = < x_n, \dot{J}(x_n) > - < x_*, \dot{J}(x_*) > - < x_n, \dot{J}(x_*) > +$
$< x_*, \dot{J}(x_*) > = [\Phi(\|x_n\|) - \Phi(\|x_*\|)] \cdot [\|x_n\| - \|x_*\|] + [\|\dot{J}(x_*)\|_* \cdot \|x_n\| -$
$- < x_n, \dot{J}(x_*) >] + [\|\dot{J}(x_n)\|_* \| x_*\| - < x_*, \dot{J}(x_n) >]$, that is,

(3): $< x_n - x_*, \dot{J}(x_n) - \dot{J}(x_*) > \geq [\Phi(\|x_n\|) - \Phi(\|x_*\|)] \cdot [\|x_n\| - \|x_*\|] \geq 0$,

which implies

$0 \leq \lim_{n\to\infty} \inf [\Phi(\|x_n\|) - \Phi(\|x_*\|] \cdot [\|x_n\| - \|x_*\|] \leq$

$\leq \lim_{n\to\infty} \sup [\Phi(\|x_n\|) - \Phi(\|x_*\|)] [\|x_n\| - \|x_*\|] \leq$

$\leq \lim_{n\to\infty} \sup < x_n, \dot{J}(x_n) > - \lim_{n\to\infty} < x_*, \dot{J}(x_n) > -$

$- \lim_{n\to\infty} < x_n - x_*, \dot{J}(x_*) > \leq < x_*, u > - < x_*, u > = 0$,

that is, we have

(4): $\lim_{n\to\infty} [\Phi(\|x_n\|) - \Phi(\|x_*\|] \cdot [\|x_n\| - \|x_*\|] = 0$

We show now that (4) implies that $\{\|x_n\|\}_{n\in N}$ is convergent to $\|x_*\|$.

To show that $\{\|x_n\|\}_{n\in N}$ is convergent to $\|x_*\|$ we show that every subsequence of $\{\|x_n\|\}_{n\in N}$ has a subsequence convergent to $\|x_*\|$.

Indeed, let $\{\|x_{n_k}\|\}_{k\in N}$ be a subsequence of $\{\|x_n\|\}_{n\in N}$. The sequence $\{x_{n_k}\}_{k\in N}$ is bounded since $\{x_{n_k}\}$ is weakly convergent to x_*.

Hence $\{\|x_{n_k}\|\}_{k\in N}$ has a convergent subsequence. We denote this last subsequence by $\{\|x_i\|\}_{i\in N}$

The sequence $\{\|x_i\|\}_{i\in N}$ must be convergent to $\|x_*\|$.

Indeed, if we suppose the contrary we have $\lim_{i\to\infty} \|x_i\| = \|x_*\| + c$. with $c \neq o$.

The mapping Φ being continuous we have, $\lim_{i\to\infty} \Phi(\|x_i\|) = \Phi(\|x_*\| + c)$ and $\Phi(\|x_*\| + c) \neq \Phi(\|x_*\|)$ since Φ is strictly increasing.

So, $\lim_{i\to\infty} [\Phi(\|x_i\|) - \Phi(\|x_*\|) = \alpha \neq 0$ and hence

$\lim_{i\to\infty} [\Phi(\|x_i\|) - \Phi(\|x_*\|)]\,[\|x_i\| - \|x_*\|] = \alpha\, c \neq 0$, which is a contradiction of (4).

Hence $\{x_n\}_{n\in N}$ is weakly convergent to x_*, $\{\|x_n\|\}_{n\in N}$ is convergent to $\|x_*\|$ and since E is Kadeç we obtain that $\{x_n\}_{n\in N}$ is convergent to x_* and finally we have that $\dot{J}$ satisfies $(S)_+^1$. □

In nonlinear analysis was much used another condition similar to $(S)_+^1$ denoted by $(S)_+$.

<u>Definition 4.3.5</u>

<u>A mapping $T:E \to E^*$ is said to satisfy condition $(S)_+$ if for any sequence $\{x_n\}_{n\in N} \subset E$ which converges weakly to x_* in E and for which</u> $\limsup_{n\to\infty} < x_n - x_*, T(x_n) > \leq 0$ <u>we have the norm convergence of</u> $\{x_n\}_{n\in N}$ <u>to</u> x_*

It is important to remark that one can verify this property under suitable concrete hypotheses for the maps of a Sobolev space $W_o^{m,P}(\Omega)$ into conjugate space $W^{-m,p'}(\Omega)$ (where $p^1 = \frac{p}{p-1}$) obtained from an elliptic operator in generalized divergence form $T(u) = \sum_{\alpha\in m} (-1)^{|\alpha|} D^\alpha T_\alpha(x, u, \ldots, D^m u)$.

But these mappings are not necessarily duality mappings.

The next result shows also that condition $(S)_+$ can be satisfied for operators which are not necessarily duality mappings.

We say that $f:E \to E^*$ is strongly ρ-monotone if there exists a continuous strictly increasing function $\rho: R_+ \longrightarrow R_+$ such that $< x-y, f(x)-f(y) > \geq \rho(\|x-y\|)$.

Proposition 4.3.2.

Each strongly ρ-monotone mapping $T:E \to E^*$ satisfies condition $(S)_+$.

Proof

Let $\{x_n\}_{n\in N}$ be a sequence weakly convergent to x_* in E and such that $\limsup_{n\to\infty} < x_n - x_*, T(x_n) > \leq 0$.

Since $\rho(\|x_n-x_*\|) \leq < x_n-x_*, T(x_n) - T(x_*) > = < x_n-x_*, T(x_n) > - < x_n-x_*, T(x_*)>$ we obtain

$0 \leq \liminf_{n\to\infty} \rho(\|x_n-x_*\|) \leq \limsup_{n\to\infty} \rho(\|x_n-x_*\| \leq$

$\leq \limsup_{n\to\infty} < x_n-x_*, T(x_n) > - \lim_{n\to\infty} < x_n-x_*, T(x_*) > \leq 0$, which implies $\lim_{n\to\infty} \rho(\|x_n-x_*\|) = 0$ and since ρ is strictly increasing and continuous we can show that $\|x_n-x_*\|$ is convergent to zero. □

We remark that the class of operators satisfying condition $(S)_+$ is invariant under conpact perturbations that is, if $T_1:E \to E^*$ satisfies $(S)_+$ and $T_2:E \to E^*$ is compact, then $T_1 + T_2$ satisfies $(S)_+$.

In particular, when $E = E^*$ is a Hilbert space then the class of operators satisfying condition $(S)_+$ contains Leray-Schauder operators.

Proposition 4.3.3.

If a mapping $T:E \to E$ satisfies condition $(S)_+$ then it satisfies $(S)_+^1$.

Proof

Let $\{x_n\}_{n\in N}$ be weakly convergent to x_* in E, such that $\{T(x_n)\}_{n\in N}$ is weakly convergent to u E* and $\limsup_{n\to\infty} < x_n, T(x_n) > \leq < x_*, u >$. We have $< x_n - x_*, T(x_n) > =$ $= < x_n, T(x_n) > - < x_*, T(x_n) >$, which implies,

$\limsup_{n\to\infty} < x_n-x_*, T(x_n) > = \limsup_{n\to\infty} < x_n, T(x_n) > - \lim_{n\to\infty} < x_*, T(x_n) > \leq < x_*, u > -$
$- < x_*, u > = 0.$

Since $(S)_+$ is satisfied for T then we have $\{x_n\}_{n\in N}$ norm convergent to x_*. □

Let $< E, E^* >$ be a dual system of Banach spaces, $K \subset E$ a closed convex cone and $T_1, T_2 : K \to E^*$ two mappings.

Definition 4.3.6.

We say that T_2 satisfies Altman's condition with respect to T_1 for $r > 0$ if for every $x \in K$ with $\|x\| = r$ we have $< x, T_2(x) > \leq < x, T_1(x) >$

If E is a Hilbert space, $K = E$ and $T_1(x) = x$, for every $x \in E$ then we obtain from Definition 4.3.6 the classical Alatman's condition for T_2 used in the fixed point theory.

Theorem 4.3.9

If $K \subset E$ is a locally compact convex cone in a Banach space E and $T_1, T_2 : K \to E^*$ two mappings and the following assumptions are satisfied:

1°) T_1 and T_2 are continuous,

2°) T_2 satisfies Altman's condition with respect to T_1 for $r > 0$,

then the problem G.C.P.$(T_1 - T_2, K)$ has a solution x_* with $\|x_*\| \leq r$.

Proof

Since K is locally compact the set $D = \{x \in K \mid \|x\| \leq r\}$ is convex and compact

By Theorem 3.1 [Hartman-Stampacchia] there exists and element $x_* \in D$ such that

(5) : $< x - x_*, T_1(x_*) - T_2(x_*) > \geq 0$, for all $x \in D$.

We show not that x_* is a solution of the problem G.C.P. $(T_1 - T_2, K)$.

Indeed, about x_* we have two possible cases:

i) $\|x_*\| < r$. Then for every $x \in K$ there exists $\lambda \in (0,1)$ such that $u = \lambda x + (1-\lambda)x_* \in D$.

Using the element u in inequality (5) we get,

$< x - x_*, T_1(x_*) - T_2(x_*) > \geq 0$, for all $x \in K$, which is equivalent to the fact that x_* solves the problem G.C.P.$(T_1 - T_2, K)$.

ii) $\|x_*\| = r$. In this case from (5) we have

$< 0 - x_*, T_1(x_*) - T_2(x_*) > \geq 0$, that is,

$< x_*, T_1(x_*) - T_2(x_*) > \leq 0$ or

$< x_*, T_1(x_*) > \leq < x_*, T_2(x_*) >$ and using assumption 2°) we obtain

(6) $< x_*, T_1(x_*) - T_2(x_*) > = 0$.

The proof is finished if we show that $T_1(x_*) - T_2(x_*) \in K^*$.

Indeed, from (5) and (6) we have $< x, T_1(x_*) - T_2(x_*) > \geq 0$, for all $x \in D$.

Scaling leads to $< x, T_1(x_*) - T_2(x_*) > \geq 0$, for all $x \in K$, that is, we have $T_1(x_*) - T_2(x_*) \in K^*$. □

Corollary

Let K be a locally compact convex cone in a Banach space.

If $T:K \to E^*$ is a continuous mapping and for some r 0 we have $x, T(x)$ 0, for all $x \in K$ with $\|x\| = r$, then the problem G.C.P.(T, K) has a solution x with $\|x_*\| \leq r$.

Proof

We consider $T_1 = t$, $T_2 = 0$ and we apply Theorem 4.3.9. □

The next result is an extension of Theorem 4.3.9 to a general Galerkin cone [See Definition 4.2.9) in an arbitrary reflexive Banach space.

Theorem 4.3.10

Let $< E, E^* >$ be a dual system of reflexive Banach spaces and Let $K(K_n)_{n\in N}$ be a Galerkin cone in E.

Suppose given two continuous mappings $T_1, T_2:K \to E^*$.

If the following assumptions are satisfied:

1°) T_1 is bounded and satisfies condition $(S)_+^1$ with respect to K,

2°) T_1 is a compact operator,

3°) T_2 satisfies Altman's condition with respect to T_1 for $r > 0$,

then the Problem G.C.P.(T_1-T_2, K) has a solution x_* such that $\|x_*\| \leq r$.

Proof

We consider the sequence of problems $\{G.C.P.(T_1 - T_2, K_n)\}_{n\in N}$.

Since the all conditions of Theorem 4.3.9. are satisfied we obtain a sequence $\{x_n\}_{n\in N}$ in K such that for every $n \in N$, x_n is a solution of the problem G.C.P. ($T_1 - T_2$, K_n), that is, we have

(7) $< x_n, T_2(x_n) > = < x_n, T_1(x_n) >$; $\forall n \in N$.

From Theorem 4.3.9. and hypothesis 3°) we have that $\|x_n\| \leq r$; $\forall n \in N$, that is $\{x_n\}_{n\in N}$ is bounded.

Since E is reflexive, $\{x_n\}_{n\in N}$ has a subsequence denoted also by $\{x_n\}_{n\in N}$ which is weakly convergent to $x_* \in K$.

The sequence $\{x_n\}_{n\in N}$ being bounded and T_1 a bounded operator, we have that $\{T_1(x_n)\}_{n\in N}$ is norm bounded in E*.

Because E* is also reflexive we have that $\{T_1(x_n)\}_{n\in N}$ has a subsequence weakly convergent to an element $u \in E^*$.

By assumption 2°) (eventually considering a subsequence we may also suppose that $\{T_2(x_n)\}_{n\in N}$ is norm convergent to an element $v \in E^*$.

From (7) we have

(8): $\lim_{n\to\infty} < x_n, T_1(x_n) > = \lim_{n\to\infty} < x_n, T_2(x_n) > = < x_*, v >$.

Let $\{P_n\}_{n\in N}$ be a sequence of projections such that for every $n \in N$, P_n is projection on K_n and for every $x \in K$, $\lim_{n\to\infty} P_n(x) = x$. (This sequence exists since $K(K_n)_{n\in N}$ is a Galerkin-cone).

We set $\hat{x}_n := P_n(x_*)$.

Since for every $n \in N$, x_n solves the problem G.C.P. (T_1-T_2, K_n), which is equivalent to a variational inequality [see ch.3] and since denoting $z_n := \hat{x}_n + (1 + \frac{1}{n}) x_n$ we have that $z_n \in K_n$ (for every $n \in N$) we obtain

$$0 \leq < z_n - x_n, T_1(x_n) - T_2(x_n) > = < \hat{x}_n + \frac{x_n}{n}, T_1(x_n) - T_2(x_n) > =$$

$$= < \hat{x}_n, T_1(x_n) - T_2(x_n) > + \frac{1}{n} < x_n, T_1(x_n) - T_2(x_n) > = < \hat{x}_n, T_1(x_n) - T_2(x_n) > ,$$

which implies $< \hat{x}_n, T_2(x_n) > \leq < \hat{x}_n, T_1(x_n) >$ and computing the limit in the last inequality we deduce,

(9): $< x_*, v > \leq < x_*, u >$.

From (8) and (9) we obtain $\lim_{n\to\infty} < x_n, T_1(x_n) > \leq < x_*, u >$.

Since T_1 satisfies condition $(S)^1_+$ we have that $\{x_n\}_{n\in N}$ is norm convergent to x_*.

The proof is finished if we show that x_* is a solution of the problem G.C.P. $(T_1 - T_2, K)$.

Indeed, let $z \in K$ be an arbitrary element.

If we denote $z_n := P_n(z)$ we have $\lim_{n\to\infty} (z_n - x_n) = z - x_*$.

Since $z_n \in K$, for every $n \in N$ and x_n solves the problem G.C.P. $(T_1 - T_2, K_n)$, we obtain (using again that G.C.P. $(T_1 - T_2, K_n)$ is equivalent to a variational inequality)

(10): $< z_n - x_n, T_1(x_n) - T_2(x_n) > \geq 0$.

Taking the limit in (10) as n tends to $+\infty$ we obtain,

$< z - x_*, T_1(x_*) - T_2(x_*) > \geq 0$; $\forall z \in K$,

that is, x_* solves the problem G.C.P. $(T_1 - T_2, K)$ and by construction we have also $\|x_*\| \leq r$. □

The last theorem has several interesting consequences.

Corollary 1

Let $(E, \| \|)$ be a reflexive Banach space and let $K(K_n)_{n\in N}$ be a Galerkin cone in E.

Suppose given two continuous mappings $T_1, T_2 : K \to E^*$.

If the following assumptions are satisfied:

1°) T_1 is bounded and strongly ρ-monotone,

2°) T_2 is compact and satisfies Altman's condition with respect to T_1 for $r > 0$,

then the problem G.C.P. $(T_1 - T_2, K)$ has a solution x_* such that $\|x_*\| \leq r$.

Proof

Consequence of Proposition 4.3.2 and Theorem 4.3.10. □

Corollary 2.

Let $E = H$ be a Hilbert space and let $K(K_n)_{n\in N}$ be a Galerkin cone in H.

If $T:K \to H$ is continuous compact and there exists $r > 0$ such that for every $x \in K$ with $\|x\| = r$ we have $< x, T(x) > \leq \|x\|^2$, then the problem G.C.P. $(I-T, K)$ has a solution x_* such that $\|x_*\| \leq r$.

Proof

Consequence of Theorem 4.3.10. □

Let $(H, <,>)$ be a Hilbert space and $K \subset H$ a closed convex cone.

We say that $T:K \to H$ is monotone decreasing on rays with respect to K if $\Psi(s) = < x, T(sx) >$ is a monotonically decreasing function of the positive real variable s for all $x \in K$ and large enough s.

Suppose $T:K \to H$ to be bounded and monotonically decreasing on rays on K.

Consider the function $h:R_+\setminus\{0\} \to R$ defined by:

$h(s) = \sup \{ <u, T(su)> \mid u \in K, \|u\| = 1\}$.

Since T is bounded the function h is well defined.

Suppose $t \le s$ large enough and consider a sequence $\{u_n\}_{n\in N}$ of elements of K of norm one such that $h(s) = \lim_{n\to\infty} < u_n, T(su_n)>$.

Because T is monotonically decreasing on rays on K we have,

$< u_n, T(su_n) > \le < u_n, T(tu_n) > \le h(t)$, that is we have $h(s) \le h(t)$.

So, there exists a positive value of s for example $s = r$ such that $h(r) \le r$.

If $x \in K$ is such that $\|x\| = r$, we put $x = ru$, where $\|u\| = 1$ and we have,

$< x, T(x) > = r < u, T(u) > \le rh(r) \le \|x\|^2$, that is T satisfies Altman's condition with respect to I.

So, we have the following result.

Corollary 3.

Let H be a Hilbert space and let $K(k_n)_{n\in N}$ be a Galerkin cone in H.

If $T:K \to G$ is continuous compact and monotonically decreasing on rays with respect to K, then the problem G.C.P.(I-T, K) has a solution.

Proof

Consequence of Corollary 2. □

The next result is a fixed point theorem similar to a fixed point theorem proved by Shinbrot [M. Shinbrot: A fixed point theorem and some applications. Arch. Rat. Mech. Anal. 17(1965), 255-277], but our theorem is with respect to a cone.

Shinbrot's theorem is proved by a long proof and supposing the weak continuity, or by another method using the topological degree.

Our result is a fixed point theorem on a Galerkin cone and it is a consequence of the Complementarity Theory.

Theorem 4.3.11

Let H be a Hilbert space and let $K(K_n)_{n\in N}$ be a Galerkin cone in H.

f $T:K \to K$ is completely continuous and monotonically decreasing on rays with respect to K then T has a fixed point.

Proof

Consequence of Corollary 3 and of the fact that in this case the complementarity problem G.C.P. (I-T, K) is equivalent to the existence of a fixed point for T on K [Chap. 3]. □

We consider now the case $E = R^n$ endowed with euclidean structure.

Corollary 4

Let $K \in R^n$ be a closed pointed convex cone, $G:K \to R^n$ a continuous mapping and $b \in R^n$ an arbitrary element.

If there exists $r > 0$ such that

(11): $< x, G(x) - b > \geq 0$, for every $x \in K$ with $\|x\| = r$ then the problem G.C.P. $(G - b, K)$ has a solution x_* such that $\|x_*\| \leq r$.

Proof

We consider $T_1(x) = x$, $T_2(x) = x - [G(x) - b]$ for every $x \in K$ and we apply Theorem 4.3.9.

Remark 4.3.5.

Condition (11) is satisfied if there exists a constant $a > 0$ such that

(12): $< x, G(x) > \geq a \|x\|^2$, for every $x \in K$.

Indeed, in this case if we choose $r > 0$ such that $\|b\| \leq ar$, we obtain for every $x \in K$ with $\|x\| = r$, $< x, G(x) > \geq a \|x\|^2 = ar^2 \geq r \|b\| \geq < x, b >$.

Finally, we remark that condition (12) is also satisfied if $\lim\limits_{\|x\| \to \infty} \frac{<x, G(x)>}{\|x\|} = +\infty$

Corollary 4 is true if K is a locally compact cone in a Hilbert space.

In the general case, operator T_2 satisfies Altman's condition with respect to T_1 if there exists $\gamma > 0$ such that $< x, T_1(x) > \geq \gamma \|x\|^\rho$, with $\rho > 2$ and if T_2 is linear and continuous.

Indeed, in this case we choose $r > 0$ such that $\|T_2\| \leq \gamma r^{\rho-2}$.

The next result is similar to Theorem 4.3.10 but condition $(S)^1_+$ is replaced by another condition.

Theorem 4.3.12

Let $(E, \| \ \|)$ be a reflxive Banach space and let $K(K_n)_{n\in N}$ be a Galerkin cone in E Suppose given two continuous mappings $T_1, T_2 : K \to E^*$.

If the following assumptions are satisfied:

1°) T_2 satisfies Altman's condition with respect to T_1 for $r > 0$,

2°) $T_1 - T_2$ is sequentially weak-to-weak continuous,

3°) if $\{x_n\}_{n\in N} \subset K$, (w) - $\lim x_n = x_o$ and $< x_n, T_1(x_n) - T_2(x_n) > = 0$, for every $n \in N$ then $< x\ , T\ (x\ ) - T\ (x\ ) > \leq 0$,

then the problem G.C.P.$(T_1 - T_2$, K) has a solution x_* with $\|x_*\| \leq r$.

Proof

From Theorem 4.3.9. we have that for every $n \in N$ the problem G.C.P.$(T_1 - T_2, K)_n$ has a solution x_n with $\|x_n\| \leq r$.

Since E is reflexive the sequence $\{x_n\}_{n\in N}$ has a subsequence $\{x_{n_k}\}_{k\in N}$ weakly convergent to an element $x_* \in K$.

We have $\|x_*\| \leq r$. We denote the sequence $\{x_{n_k}\}_{k\in N}$ again by $\{x_n\}_{n\in N}$.

Since for every $n \in N$, we have that $T_1(x_n) - T_2(x_n) \in K_n{}^*$, we deduce $<x, T_1(x_m) - T_2(x_m)> \geq 0$ for every $x \in K_n$ and every $m \geq n$ which imply $<x, T_1(x_*) - T_2(x_*)> \geq 0$, for every $x \in K_n$ and finally using assumption 2°) and the fact that K is a Galerkin cone, we can show that $< x, T_1(x_*) - T_2(x_*)> \geq 0$, for every $x \in K$, that is $T_1(x\ ) - T_2(x_*) \in K^*$.

The proof is finished if we show that $< x_*, T_1(x_*) - T_2(x_*) > = 0$.

Indeed, because $< x_n, T_1(x_n) - T_2(x_n) > = 0$, for every $n \in N$ and $\{x_n\}$ is weakly convergent to x_* we obtain from assumption 3°), $< x_*, T_1(x_*) - T_2(x_*)> \leq 0$ and since $T_1(x_*) - T_2(x_*) \in K^*$ we have $< x_*, T_1(x_*) - T_2(x_*)> = 0$. □

Remark 4.3.6

We give now a condition which implies that $T_1 - T_2$ is sequentially weak-to-weak continuous.

If $T \in L(E,F^*)$ we denote by T^* the adjoint of T.

We say that $f:K \to E^*$ is Gâteaux differentiable along the convex cone $K \subset E$ if the function f has a linear Gâteaux differential $f'(x) \in L(E,E^*)$ at every $x \in K$.

We suppose E to be a reflexive Banach space and $T_1, T_2:K \to E^*$ two mappings.

We can show that if $T_1 - T_2$ is Gâteaux differentiable along to K and for every bounded sequence $\{x_n\}_{n\in N} \subset K$ there exists a subsequence $\{x_{n_k}\}_{k\in N}$ such that $\bigcup_{k\in K} [(T_1 - T_2)'(x_{n_k})]^*(x)$ is strongly precompact then $T_1 - T_2$ is sequentially weak-to-weak continuous on K.

Let $< E,E^* >$ be a dual system of reflexive Banach spaces. We consider on E^* the strong topology.

Proposition 4.3.4

Let $K \subset E$ be a locally compact convex cone and let $f:K \to E^*$ be a continuous mapping.

If there is an element $u_o \in K$ and a constant $r > \|u_o\|$ such that

(13): $< x - u_o, f(x) > \geq 0$, for all $x \in K$ with $\|x\| = r$,

then the problem G.C.P. (f, K) has a solution x^* such that $\|x_*\| \leq r$.

Proof

By Theorem 3.1. there exists an element $x_* \in K_r = \{x \in K \mid \|x\| \leq r\}$ such that,

(14): $< x - x_*, f(x_*) > \geq 0$; for all $x \in K_r$(since K_r is convex compact).

We have two possibilities.

Case 1: $\|x_*\| < r$. If $x \in K$ then there exists $\lambda \in]0, 1[$ sufficiently small such that $w = \lambda x + (1 - \lambda) x_* \in K_r$ and from (14) we have,

$< w - x_*, f(x_*) > = \lambda \leq x - x_*, f(x_*) > \geq 0$, that is $< x - x_*, f(x_*) > \geq 0$, for all $x \in K$, which implies that x_* is a solution of the problem G.C.P. (f, K).

Case 2: $\|x_*\| = r$. Then we have $< x_* - u_o, f(x_*) > \geq 0$ (from assumption (13)) and since $< x - x_*, f(x_*) > \geq 0$, for all $x \in K_r$, we obtain that $< x - u_o, f(x_*) > =$ $= < x - x_* + x_* - u_o, f(x_*) > = < x - x_*, f(x_*) > + < x_* - u_o, f(x_*) > \geq 0$, that is we have,

(15): $< x - u_o, f(x_*) > \geq 0$, for all $x \in K_r$.

But $\|u_o\| < r$ implies that if $x \in K$ there exists $\lambda \in]\, 0, 1\, [$ such that $v = \lambda x + (1 - \lambda) u_o \in K_r$.

If we put $x = v$ in (15) we have $\lambda < x - u_o, f(x_*) > \geq 0$, that is,

(16): $< x - u_o\ f(x_*) > \geq 0$; for all $x \in K$.

Since $\|u_o\| < r$, then from (14) we obtain,

(17): $< u_o - x_*, f(x_*) > \geq 0$.

Now, from (16) and (17) we deduce $< x - x_*, f(x_*) > \geq 0$; for all $x \in K$, that is x_* is a solution of the problem G.C.P. (f, K) with $\|x_*\| \leq r$. □

We introduce now two conditions.

Let $K(K_n)_{n\in N}$ be a Galerkin cone in E and let $f:K \to E^*$ be a mapping.

Definition 4.3.6

We say that f satisfies the generalized Karamardian's condition (denoted by (GK)) with respect to $K(K_n)_{n\in N}$ if there exists a countable family $\{D_n\}_{n\in N}$ of subsets of K such that:

i) for every $n\in N$, D_n is a convex compact subset of K_n,

ii) for every $x \in K_n \setminus D_n$ there exists $y \in D_n$ such that $\langle x-y, f(x) \rangle > 0$.

Definition 4.3.7

We say that f satisfies the generalized Moré's condition (denoted by (G.M)) with respect to $K(K_n)_{n\in N}$ if for every $n \in N$ there exist $r_n > 0$ and $u_n \in K_n$ such that $\|u_n\| < r_n$ and $\langle x - u_n, f(x) \rangle \geq 0$ for all $x \in K_n$ with $\|x\| = r_n$.

Theorem 4.3.13

Let $(E, \|\ \|)$ be a reflexive Banach space and let $K(K_n)_{n\in N}$ be a Galerkin cone in E.

Suppose given two continuous mappings $T_1, T_2 : K \to E^*$.

If the following assumptions are satisfied

1°) $T_1 - T_2$ satisfies condition (G.K.) or (G.M.),

2°) T_1 is bounded and satisfies condition $(S)^1_+$,

3°) there exists a function $\rho : R_+ \to R_+$ such that $\limsup_{r\to+\infty} \rho(r) = +\infty$ and $\|x\|\,\rho(\|x\|) \leq \langle x, T_1(x) \rangle$, for all $x \in K$,

4° T_2 is compact and φ- asymptotically bounded and $\limsup_{r\to+\infty} \varphi(r) < +\infty$,

then the problem G.C.P. (T_1-T_2, K) has a solution.

Proof

If $T_1 - T_2$ satisfies condition (G.K.) or respective (G.M.) then from Theorem 4.3.4. or respective from Proposition 4.3.4 we have that for every $n \in N$, the problem G.C.P. $(T_1 - T_2, K_n)$ has a solution $x_n \in K_n$ (such that $x_n \in D_n$ or $\|x_n\| \leq r_n$).

The sequence $\{x_n\}_{n\in N}$ is bounded.

Indeed, if we suppose the contrary then we have, $\langle x_n, T_1(x_n) - T_2(x_n)\rangle = 0$ which implies

(18): $\langle x_n, T_1(x_n)\rangle = \langle x_n, T_2(x_n)\rangle$; $\forall n \in N$.

Considering eventually a subsequence we can suppose that $\lim_{n\to\infty} \|x_n\| = +\infty$.

From (18) and using assumptions 3°) and 4°) we obtain,
$\|x_n\| \rho(\|x_n\|) \le c\varphi(\|x_n\|) \|x_n\|$; $\forall n \in N$ such that $\|x_n\| > r$)where r is defined by the assumption tht T_2 is φ- asymptotically bounded).

Hence we have the inequality $\rho(\|x_n\|) \le c\varphi(\|x_n\|$; $\forall n \in N$ such that $\|x_n\| > r$ which is impossible since the properties of ρ and φ.

So, $\{x_n\}_{n\in N}$ is bounded and the proof follows the proof of Theorem 4.3.10. □

We remark that in Theorem 4.3.13 assumption 3°) and assumption that T_2 is - asymptotically bounded with $\limsup_{r\to\infty} (r) < +\infty$,are necessary to show that the sequence $\{x_n\}_{n\in N}$ of solutions of partial problems G.C.P.(T_1-T_2, K) is bounded.

So, if we use in Theorem 4.3.13 a strong condition of equilimitation we obtain the following interesting result.

Corollary 1

Les (E, ‖ ‖) be a reflexive Banach space and let $K(K_n)_{n\in N}$ be a Galerkin cone in E.

Suppose given two continuous mappings $T_1, T_2 : K \to E^*$.

If the following assumptions are satisfied:

1°) $T_1 - T_2$ satisfies condition (G.K.) with a equibounded family $\{D_n\}_{n\in N}$ or condition (G.M.) with a bounded family of elements $\{u_n\}_{n\in N}$,

2°) T_1 is bounded and satisfies condition $(S)^1_+$

3°) T_2 is compact,

then the problem G.C.P. (T_1-T_2, K) has a solution x_* and we can compute $r > 0$ such that $\|x_*\| \le r$. □

Assumption 1°) of the last corollary is satisfied if $T_1 - T_2$ is strongly K-copositive in the sense of Definition 4.3.3 but with respect to the cone K, that is there exists a number $M > 0$ such that for all $x \in K$ we have

(19): $\langle x, T_1(x) - T_2(x) \rangle \ge \langle x, T_1(0) - T_2(0) \rangle + M\|x\|^2$

Certainly, if $T_1 - T_2$ is strongly K-monotone, that is, there exists $m > 0$ such that, $(\forall x, y \in K)(x - y \in K)(\langle x - y, T_1(x) - T_2(x) - T_1(y) + T_2(y) \rangle \ge m\|x - y\|^2$, then $T_1 - T_2$ is strongly K-copositive.

Corollary 2

Let $(E, \| \|)$ be a reflexive Banach space and let $K(K_n)_{n\in N}$ be a Galerkin cone in E.

Suppose given two continuous mappings $T_1, T_2 : K \to E^*$.

If the following assumptions are satisfied

1°) $T_1 - T_2$ *is strongly K-copositive and* $T_1(0) \neq T_2(0)$,

2°) T_1 *is bounded and satisfies conditon* $(S)^1_+$,

3°) T_2 *is compact,*

then the problem G.C.P. $(T_1 - T_2, K)$ *has a solution* x_* *with* $\|x\| \leq \|T_1(0)-T_2(0)\|/M$, *where* $M > 0$ *is the constant defined by assumption 1°).*

Proof

We sobserve that the corollary is a consequence of Corollary 1 of Theorem 4.3.13 if we show that $T_1 - T_2$ satisfies condition (G.K.) with an equibounded family $\{D_n\}_{n\in N}$.

Indeed, since $T_1 - T_2$ is strongly K-copositive then there exists $M > 0$ such that the relation (19) is satisfied.

If we put $r_0 = \dfrac{\|T_1(0) - T_2(0)\|}{M}$ and $K_{r_0} = \{x \in K \mid \|x\| \leq r_0\}$ then for all $x \in K \setminus K_{r_0}$ we have

$$\|x\|^2 > r_0 \|x\| = \frac{\|T_1(0) - T_2(0)\|}{M} \|x\| \text{, that is,}$$

$$M \|x\|^2 > [\, \|T_1(0) - T_2(0)\|] \|x\|, \text{ which implies}$$

$$< x, T_1(x) - T_2(x) > \geq < x, T_1(0) - T_2(0) > + \|x\| \|T_1(0) - T_2(0)\| \geq 0;$$

for all $x \in K \setminus K_{r_0}$.

Thus, if we put $D_n = K_{r_0} \cap K_n$, for all $n \in N$ we observe that condition (G.K.) is satisfied with $y = 0$ for all $x \in K_n \setminus D_n$ and further, the family $\{D_n\}_{n\in N}$ is equibounded. Since, every partial problem G.C.P. $(T_1 - T_2, K_n)$ has a solution x_n satisfying $\|x_n\| \leq r_0$, the corollary is proved. □

Remark 4.3.7

Theorem 4.3.13 and its corollaries can be applied to study the complementarity problem associated to the problem of post-equilibrium state of a thin elastic plate resting without friction on a flat rigid support, that is to the problem defined by the model 2.4.5. (Chapter 2).

Comments

The problem to compute the radius of the ball containing the solution set, when this set is bounded, was first considered by Pardalos and Rosen in finite dimensional spaces in [C61].

The results presented in 4.1. in a general Hilbert space were obtained by Isac [C28].

The relation between feasibility and solvability is an important fact.

Theorems 4.2.2, 4.2.3, 4.2.4, 4.2.5 and 4.2.6 were obtained by Moré [A219], [A220].

The results concerning pseudomonotone mappings were obtained by Karamardian [A156]. The results for multivalued pseudomonotone mappings were proved by Saigal [A262]. The concept of Galerian cone as is used in this chapter was defined by Isac.

Gowda and Seidman proved recently Theorem 4.2.16 [C20] and Theorem 4.2.18 is an improvement of a Gowda and Seidman's result.

Theorem 4.2.19 was initially proved by Dash and Nanda [A73].

Condition $(S)_+^1$ was introduced by Isac.

Theorems 4.3.10, 4.3.11 and 4.3.13 were obtained by Isac and Theorem 4.3.12 by Gowda.

CHAPTER 5
THE ORDER COMPLEMENTARITY PROBLEM

In this chapter we will study the Order Complementarity Problem, defined in section B of Chapter 1.

We begin this study by recalling certain elementary facts about vector lattices. The reader is referred to A.L. Peressini: Ordered Topological Vector Spaces. Harper and Row, New York, (1967).

Let E be a vector space and let $K \subset E$ be a pointed convex cone.

We denote by "$\leq$" the ordering defined by, $x \leq y \iff y - x \in K$.

As defined in Section 1.1 of Chapter 1, if the supremum $\sup\{x, y\} = x \vee y$ and the infimum $\inf\{x\ y\} = x \wedge y$ of every pair $\{x, y\}$ of elements of E exist, then we say that E is a vector lattice.

Suppose now (E, K) to be a vector lattice.

The absolute value of x is $|x| = x \vee (-x)$.

The positive part of x is $x^+ = x \vee 0$ and the negative part is $x^- = (-x)^+ = -(x \wedge 0)$,

Then $|x| = x^+ + x^-$, $x = x^+ - x^-$ and $x^+ \wedge x^- = 0$.

We say that two elements $x, y \in E$ are (lattice) orthogonal if $|x| \wedge |y| = 0$.

Since (E, K) is a vector lattice the following identities are true for every x, y, $z \in E$ and $\lambda \in R_+$.

1°) $x \vee y = -\{(-x) \wedge (-y)\}$,

2°) $z - (x \vee y) = (z - x) \wedge (z - y)$,

3°) $z + (x \vee y) = (z + x) \vee (z + y)$,

4°) $z + (x \wedge y) = (z + x) \wedge (z + y)$,

5°) $\lambda(x \vee y) = (\lambda x) \vee (\lambda y)$,

6°) $\lambda(x \wedge y) = (\lambda x) \wedge (\lambda y)$,

7°) $x + y = x \quad y + x \wedge y$

8°) $x \vee (y \wedge z) = (x \vee y) \wedge (x \vee z)$

9°) $x \wedge (y \vee z) = (x \wedge y) \vee (x \wedge z)$

A least element of a set D in E is an element α of D such that $\alpha \leq x$ for all x in D.

If (E, K) is a vector lattice and at the same time a topological vector space we say that E is a topological vector lattice.

Given a topological vector lattice (E, K) and $T:K \to E$ an arbitrary mapping (not necessary linear), the Order Complementarity Problem associated to T and K is

$$\text{O.C.P. }(T,K): \left\| \begin{array}{l} \text{find } x_o \in K \text{ such that} \\ T(x_o) \wedge x_o = 0 \end{array} \right.$$

It is clear that if x_o is a solution of this problem then $T(x_o) \in K$.

So the feasible set for the problem O.C.P. (T, K) is $F(T) = \{x \in K | T(x) \in K\}$.

If $T(x) = L(x) + q$, where L is a continuous linear operator from E into E (that is $L \in L(E, E)$) and q is an element of E, then we obtain the Linear Order Complementarity Problem denoted by L.O.C.P. (L, q) and in this case the feasible set is

$$F(L, q) = \{x \in K | L(x) + q \in K\}$$

In this chapter we are interested to know:

i) when the problem O.C.P. (T, K) has a solution,

ii) when this solution is the least element of the feasible set.

5.1 The Linear Order Complementarity Problem

Theorem 5.1.1

Let (E, K) be a vector lattice and $L:E \to E$ be linear and satisfying the following property: for each $x \in K\setminus\{0\}$ there exists $u \in K\setminus\{0\}$ with $u < x$ (that is $x - u \in K\setminus\{0\}$) and a strictly positive number $\lambda(u)$ with $L(u) \leq \lambda(u)u$.

Then, whenever L.O.C.P. (L, q) is feasible and admits a least element, that least element solves L.O.C.P. (L, q).

Proof

Suppose (L, q) nonempty and x_* is a least element.

Set $x = (L(x_*) + q) \wedge x_*$.

Then $x \in K$. If $x \neq 0$ we may select an element $u \in K\setminus\{0\}$ such that $L(u) \leq \lambda(u)u$. Set $\lambda = \lambda(u) \vee 1$ and $y = x_* - \lambda^{-1} u$. We have, $y \in K$ and $L(y) + q = L(x_*) + q - \lambda^{-1} L(u) \geq u - \lambda^{-1}(u)L(u) \geq 0$.

Thus y lies in $F(L, q)$ and strictly minorizes x_*. This contradiction shows that $x = 0$ and x_* solves L.O.C.P. (L, q). □

Remark 5.1.1

The property used in Theorem 5.1.1 is satisfied if there exists $\lambda > 0$ such that $L \leq \lambda I$ in the induced operator ordering.

Definition 5.1.1

We say that a linear operator $L:E \to E$ is type (λI) if there exists a linear positive operator $P:E \to E$ that is $P(K) \subseteq K$) such that $L = \lambda I - P$.

Theorem 5.1.2

Suppose that (E, K) is an order complete vector lattice.

If $L:E \to E$ is type (λI) then L.O.C.P. (L, q) has a least element solution, for every $q \in E$, whenever it is feasible.

Proof

We have $L = \lambda I - P$ with P linear positive and $\lambda > 0$.

By assumption we have that $F(L, q)$ is nonempty.

We define,

$$T(x) = \lambda^{-1} (P(x) - q)^+; \ \forall \ x \in E.$$

Let $D = \{x \in K | T(x) \le x\}$.

Since $T(x) \le x$ for each $x \in F(L, q)$ we have that inf D is well defined.

We set $y = \inf D$.

Because P is increasing we deduce that T is increasing.

We have $T(x) \le x$ and $y \le x$ which imply $T(y) \le T(x) \le x$ and finally $T(y)$ is a lower bound for D.

Thus $T(y) \le y$. But now $y \in D$ and $T(y) = y$.

However $T(y) = y$ is equivalent to $(P(y) - q - \lambda y) \vee (-\lambda y) = 0$, which implies that y is a solution of L.O.C.P. (L, q).

The proof is finished since we remark that by construction $F(L, q) \subset D$ and hence y is the least element of $F(L, q)$. □

Definitions 5.1.2

Let $L:E \to E$ be linear.

i) L is type (A) if for $x \in E$

(1): $[L(x)]^+ \wedge x^+ = 0$ implies $x \le 0$.

ii) L is type (P) if for $x \in E$.

(2): $[L(x)] \wedge x \le 0 \le [L(x)] \vee x$ implies $x = 0$.

Remark 5.1.2

I) Condition (2) of Definitions 5.1.2 is equivalent to

(3): $[L(x)]^+ \wedge x^+ = 0 = [L(x)]^- \wedge x^-$ implies $x = 0$.

(3) ==> (2). Indeed, if $[L(x)] \wedge x \le 0 \le [L(x)] \vee x$ then $[L(x)]^+ \wedge x^+ = \{[L(x)] \wedge x\} \vee 0 \le 0 \le [L(x)]^+ \wedge x^+$ and we have $[L(x)]^+ \wedge x^+ = 0$.

Since $0 \le [L(x)] \vee x$ we have $[-L(x)] \wedge (-x) = -[(L(x)) \vee x] \le 0$ which implies $[L(x)]^- \wedge x^- \le 0 \le [L(x)]^- \wedge x^-$.

So, we have $[L(x)]^+ \wedge x^+ = 0 = [L(x)]^- \wedge x^-$ and since (3) is supposed true we obtain $x = 0$, that is (2) is true.

(2) ⟹ (3). We suppose $[L(x)]^+ \wedge x^+ = 0 = [L(x)]^- \wedge x^-$.

In this case we have $\{[L(x)] \wedge x\} \vee 0 = [L(x)]^+ \wedge x^+ = 0$ which implies

(4): $[L(x)] \wedge x \le 0$.

On the other hand we have

$$0 = [L(x)]^- \wedge x^- = [(-L(x)) \vee 0] \wedge [(-x) \vee 0] =$$
$$= [(-L(x)) \wedge (-x)] \vee 0, \text{ which implies } (-L(x)) \wedge (-x) \le 0 \text{ or } -[L(x)) \vee x] \le 0,$$

that is

(5): $[L(x)] \vee x \ge 0$.

But (4) and (5) imply by (2) that $x = 0$. □

II) Every (A) operator is type (P).

Indeed, replacing x by $-x$ in (1) we obtain that $[L(x)]^- \wedge x^- = 0$ implies $x \ge 0$. Using now Remark I we obtain that (A) implies (P).

III) In R^n the class of (P) operators is exactly the class of (P)-matrices.

Definitions 5.1.3

Let $L:E \to E$ be a linear operator.

iii) L is type (Z) if $x \wedge y = 0$ implies $[L(x)] \wedge y \le 0$.

iv) L is type (E) if for x in E $[L(x)] \wedge x \le 0$ and $x \le 0$ imply $x = 0$.

Remark 5.1.3

If L is type (Z) and type (E) then L is type (P).

Indeed, if L is type (Z) and (E) and if $[L(x)]^+ \wedge x^+ = 0$ then since $x^+ \wedge x^- = 0$ we have $[L(x^+)] \wedge x^- \le 0$ (because L is type (Z)) so that $L(x^+) \le 0$, $[L(x^+)]^+ = 0$ and hence $[L(x^+)]^+ \wedge x^+ = 0$ implies $0 = [(L(x^+)) \vee 0] \wedge x^+ = [(L(x^+)) \wedge x^+] \vee (0 \wedge x^+) = [(L(x^+)) \wedge x^+] \vee 0$ whence $[L(x^+)] \wedge x^+ \le 0$.

Since L is type (E) we obtain that $x^+ = 0$, whence L is type (A) and hence type (P).

Theorem 5.1.3

Let $L \in L(E, E)$ be an arbitrary operator.

The problem L.O.C.P. (L, q) has at most one solution for each $q \in E$ if and only if L is type (P).

Proof

Suppose L is type $(\mathcal{P})$ and L.O.C.P. (L, q) has two solutions x, y, that is, $[L(x) + q] \wedge x = 0 = [L(y) + q] \wedge y$.

Then, since $L(y) + q \geq 0$, $y \geq 0$ we have $0 = [L(x) + q] \wedge x \geq [(L(x) + q) - (L(y) + q)] \wedge (x - y) = L(x - y) \wedge (x - y)$.

Symmetrically, $L(y - x) \wedge (y - x) \leq 0$.

Thus,

$$L(x - y) \wedge (x - y) \leq 0 \leq L(x - y) \vee (x - y).$$

Since L is $(\mathcal{P})$-operator we obtain $x = y$.

Conversely, suppose that L.O.C.P. (L, q) has at most one solution for each q.

Let $y: = L(x)$ and suppose that $y^+ \wedge x^+ = y^- \wedge x^- = 0$.

Let $q: = [L(x)]^+ - L(x^+) = y^+ - L(x^+) = y^- - L(x^-)$ (since $y = y^+ - y^-$ and $x = x^+ - x^-$).

Then

$$[L(x^+) + q] \wedge x^+ = y^+ \wedge x^+ = 0 = y^- \wedge x^- = [L(x^-) + q] \wedge x^-$$

and x^+ and x^- solve L.O.C.P. (L, q).

Thus $x = x^+ - x^- = 0$ and using implication (3) of Remark 5.1.2 we obtain that L is type $(\mathcal{P})$.

Definition 5.1.4

We say that $L \in L(E, F)$(E, F are vector lattices) is type (H^+) if

(6): $\quad L(x \wedge y) \geq [L(x)] \wedge [L(y)]$; for every $x, y \in E$.

Proposition 5.1.1

An operator $L \in L(E, F)$ is type (H^+) if and only if one of the following holds:

(7): $\quad x \wedge y = 0$ implies $[L(x)] \wedge [L(y)] \leq 0$,

(8): $\quad L(x \vee y) \leq [L(x)] \vee [L(y)]$, for all $x, y \in E$,

(9): $\quad L(|x|) \leq |L(x)|$, for all $x \in E$.

Proof

Since $L(0) = 0$ (7) is a consequence of (6). Conversely, for any $x, y \in E$, $(x - x \wedge y)$ and $(y - x \wedge y)$ are lattice orthogonal and (7) implies that

$$0 \geq [L(x - x \wedge y)] \wedge [L(y - x \wedge y)] = [L(x)] \wedge [L(y)] - L(x \wedge y)$$

and hence (6) is satisfied.

Replacing x and y by $-x$ and $-y$ we obtain that (8) is equivalent to (6) (and hence to (7)).

If in (8) we put $y: = -x$ we obtain (9).

The proposition is proved if we show that (9) implies (8).

Indeed, let $z := x - y$.

Since $x + y = x \vee y + x \wedge y$ we obtain $|z| = 2(x \vee y) - (x + y)$ and using (9) we get $2 L(x \vee y) - L(x + y) \leq |L(z)| = 2[(L(x)) \vee (L(y))] - (L(x) + L(y))$ and finally $L(x \vee y) \leq [L(x)] \vee [L(y)]$, that is, we have (8). □

<u>Proposition 5.1.2</u>

<u>Let $L \in L(E, E)$ be an arbitrary operator.</u>

<u>If every nonempty $F(L, q)$ has a least element then L is type (H^+).</u>

<u>Proof</u>

To prove this proposition we verify condition (7) of <u>Proposition 5.1.1</u>.

Let $x \quad y = 0$ and set $q := -[(L(x)) \wedge (L(y))]$.

Then x and y lie in $F(L, q)$.

By hypothesis a least element exists for $F(L, q)$, say z.

But then we have $0 = x \wedge y \geq z \geq 0$ and so $z = 0$.

Thus $L(0) + q \geq 0$ and $[L(x)] \wedge [L(y)] \leq 0$.

<u>Proposition 5.1.3</u>

<u>If $L \in L(E, E)$ is type (H^+) then any minimal point of $F(L, q)$ is the unique least element of $F(L, q)$.</u>

<u>Proof</u>

<u>Suppose that x_* is minimal for $F(L, q)$ and let $y \in F(L, q)$.</u>

Then by (8) of <u>Proposition 5.1.1</u> and the fact that $x \wedge y = -(-x) \vee (-y)$ we have

$$L(x_* \wedge y) \geq [L(x_*)] \wedge [L(y)] \geq -q \text{ and since } x_* \wedge y \geq 0, \text{ this point is}$$

feasible.

Since x_* is minimal and we have $x_* \geq x_* \wedge y$ we deduce $x_* = x_* \wedge y$ and finally $x_* \leq y$.

Thus x_* is the least element for the feasible set $F(L, q)$. □

<u>Remark 5.1.4</u>

A sufficient condition for $F(L, q)$ to possess minimal point is that $F(L, q)$ have a <u>weakly compact section</u>

$$D_* = F(L, q) \cap \{x \in E | x \leq x_*\}$$

for some x_* in $F(L, q)$.

In particular, this holds if order-intervals in E are weakly compact.

Indeed, it is sufficient to show that every chain $\{x_\alpha | \alpha \in I\}$ in D_* admits a lower bound in D_* and then to apply Zorn's Lemma. (We note that any minimal point of D_* is a minimal point for (L,q)).

Let $\{x_\alpha | \alpha \in I\}$ be a chain in D_*.

For every x_α we denote by S_α the section $\{x \in D_* | x \leq x_\alpha\}$.

Since $\{S_\alpha\}_{\alpha \in I}$ is a family of weakly closed subsets of D_* with the finite intersection property and D_* is weakly compact we have that $\bigcap_{\alpha I} S_\alpha$ is nonempty and any intersection point is then a minorant for the chain $\{x_\alpha | \alpha \in I\}$.

Definition 5.1.5

Let $L \in L(E, E)$ be an arbitrary operator.

We say that L is type $(\lambda(x))$ if there exists $\lambda: K \to R_+$ such that for each $x \in K$, $L(x) \leq \lambda(x)x$.

Remark 5.1.5

Every type (λI) operator is type $(\lambda(x))$.

Proposition 5.1.4

Every type $(\lambda(x))$ operator L is type (Z).

Proof

If $x \wedge y = 0$ and L is type $(\lambda(x))$ then $[L(x)] \wedge y \leq (\lambda(x)x) \wedge y \leq \lambda(x)(x \wedge y) = 0$ and L is type (Z). □

Proposition 5.1.5

Every type (Z) operator L is type (H^+).

Proof

If L is type (Z) and $x \wedge y = 0$ then

$$[L(x)]^+ \wedge y = (0 \wedge y) \vee [(L(x)) \wedge y] \leq 0.$$

But since $x \wedge y = 0$ we have $y \geq 0$ and because $(L(x))^+ \geq 0$ we obtain $[L(x)]^+ \wedge y \geq 0 \wedge y = 0$.

Finally we have that

(10): $x \wedge y = 0$ implies $[L(x)]^+ \wedge y = 0$.

If we put $z := [L(x)]^+$ and since $[L(x)]^+ \wedge y = 0$ we have $[L(y)] \wedge [L(x)] \le L(y) \wedge [L(x)]^+ \le 0$ (by (10) and by Proposition 5.1.1 we obtain that L is type (H^+). □

We note without proof the following result proved by Borwein and Dempster [A33].

Theorem 5.1.4

Let E be an order complete Banach lattice and let $L \in L(E, E)$ be continuous. If L is type (Z) then $L \leqslant \|L\| I$ (that is L is type (λI)). □

From this result and Theorem 5.1.2 we deduce the following result.

Theorem 5.1.5

Let E be an order complete Banach lattice and let $L \in L(E, E)$ be a continuous operator.

If L is a type (Z) operator then each L.O.C.P. (L, q) possesses a least element solution. □

5.2 The Generalized Order Complementarity Problem

In this section we consider the Order Complementarity Problem associated with a convex cone and a finite family of (not necessary linear) mapppings.

This problem seems to be quite important in Economics and in some applications in Mechanics and Engineering.

Let $(E, \| \ \|)$ be a Banach space ordered by a pointed, closed convex cone $K \subset E$. We denote by " $\le$" the ordering defined by K.

In this section we assume that the ordered vector space (E, K) is a vector lattice (Riesz space).

The Riesz space E is called Dedekind Complete if every nonempty subset of E which is bounded from above has a supremum.

Let (Ω, μ) be a measure space such that $\mu(\Omega)$ is finite and $\Omega \subset R^n$.

It is well known [Luxemburg W.A.J. and Zaanen A.C.: Riesz spaces, Vol. I, North-Holland (1971)] that $L_p(\Omega, \mu)$; $(1 \le p \le \infty)$ is a Dedekind Complete Riesz space.

Given m (nonlinear or linear) functions $f_1, f_2, \ldots, f_m : E \to E$, the Generalized Order Complementarity Problem associated with $\{f_i\}_{i=1,\ldots,m}$ and K is the following:

$$\text{G.O.C.P. } (\{f_i\}_1^m, K): \left\| \begin{array}{l} \text{find } x_o \in K \text{ such that} \\ f\,(x_o) \in K;\ i = 1, 2, \ldots, m \text{ and} \\ \wedge(x_o, f_1(x_o), \ldots, f_m(x_o)) = 0 \end{array} \right.$$

This problem contains as particular cases a problem studied by Cottle and Dantzig by Mangasarian and the problem studied by Oh in [Oh K.P.: The formulation of the Mixed Lubrication Problem as a Generalized Nonlinear Complementarity Problem. Transactions of ASME, Journal of Tribology. Vol. 108 (1986), 598-604] as the mathematical model of the mixed lubrication.

Proposition 5.2.1

The problem G.O.C.P. ($\{f_i\}_1^m$, K) is equivalent to the following fixed point problem:

$$(F.P.)_1: \quad \left\| \begin{array}{l} \text{find } x_o \in K \text{ such that} \\ F(x_o) = x_o, \text{ where } F(x) = V(0, x - f_1(x), \ldots, x - f_m(x)) \end{array} \right.$$

Proof

a) If $x_o \in K$ is a solution of the problem G.O.C.P. ($\{f_i\}_1^m$, K) then we have, $x_o - \Lambda (x_o, f_1(x_o), \ldots, f_m(x_o)) = x_o$, which implies $x_o + V(-x_o, -f_1(x_o), \ldots, -f_m(x_o)) = x_o$ and finally $V(0, x_o - f_1(x_o), \ldots, x_o - f_m(x_o)) = x_o$, that is x_o is a fixed point of F and $x_o \in K$.

b) If $x_o = F(x_o)$ then from the definition of F we deduce,

$$x_o = V(0, x_o - f_1(x_o), \ldots, x_o - f_m(x_o)) = x_o, \text{ that is}$$
$$V(-x_o, -f_1(x_o), \ldots, f_m(x_o)) = 0 \text{ which implies}$$
$$-\Lambda (x_o, f_1(x_o), \ldots, f_m(x_o)) = 0. \text{ Hence}$$
$$\Lambda (x_o, f_1(x_o), \ldots, f_m(x_o)) = 0.$$

Since $x_o = F(x_o)$ and $x_o \in K$, $f_i(x_o) \in K$, $i = 1, 2, \ldots, m$ and the proposition is proved. □

Remark 5.2.1

The problem G.O.C.P. ($\{f_i\}_1^m$, K) has a solution if and only if the mapping $F(x) = V(0, x - f_1(x), \ldots, x - f_m(x))$ has a fixed point in K.

Corollary

If for every $i = 1, 2, \ldots, m$, $f_i(x) = x - T_i(x)$ then the problem G.O.C.P. ($\{f_i\}_1^m$, K) is equivalent to the fixed point problem:

$$(F.P.)_2: \quad \left\| \begin{array}{l} \text{find } x_0 \in K \text{ such that,} \\ \mathcal{F}(x_0) = x_0, \text{ where} \\ \mathcal{F}(x) = V(0, T_1(x), \ldots, T_m(x)). \end{array} \right. \quad \square$$

Using the properties of the latticial operations "Λ", "V" we obtain the following result.

Proposition 5.2.2

The problem G.O.C.P. $(\{f_i\}_1^m, K)$ where $f_i(x) = x - T_i(x)$, $i = 1, 2, \ldots, m$ is equivalent to the following Order Complementarity Problem,

$$\text{O.C.P. } (H, K): \left\| \begin{array}{l} \text{find } x_0 \in K \text{ such that} \\ x_0 - H(x_0) \in K \text{ and} \\ x_0 \wedge (x_0 - H(x_o)) = 0, \end{array} \right.$$

where $H(x) = V(T_1(x), T_2(x), \ldots, T_m(x))$.

Proof

Since for every $i = 1, 2, \ldots, m$, we have $f_i(x) = x - T_i(x)$, we obtain (from the definition of G.O.C.P. $(\{f_i\}_1^m, K)$,

$$\text{G.O.C.P. } (\{f_i\}_1^m, K) \iff \left\{ \begin{array}{l} \text{find } x_0 \in K \\ x_0 - T_i(x_0) \in K;\ \forall\, i = 1, 2, \ldots, m, \text{ and} \\ x_0 \wedge \{\wedge(x_0 - T_1(x_0), \ldots, x_0 - T_m(x_o))\} = 0 \end{array} \right.$$

$$\iff \text{(O.C.P.)}: \quad \left\{ \begin{array}{l} \text{find } x_0 \in K \text{ such that} \\ T(x_0) = \wedge\, (x_0 - T_1(x_0), \ldots, x_0 - T_m(x_0)) \in K \text{ and} \\ x_0 \wedge T(x_o) = 0 \end{array} \right.$$

But, $T(x_0) = x_0 + \wedge\, (-T_1(x_0), \ldots, -T_m(x_0)) = x_0 - V\, (T_1(x_0), \ldots, T_m(x_0))$.

If we set,

$$H(x) = V(T_1(x), \ldots, T_m(x))$$

then the problem (O.C.P.) becomes

$$\text{(O.C.P.)}(H,K): \left\| \begin{array}{l} \text{find } x_0 \in K \text{ such that} \\ T(x_0) = x_0 - H(x_0) \in K \text{ and} \\ x_0 \wedge (x_0 - H(x_o)) = 0. \end{array} \right.$$

Corollary

If for every $i = 1, 2, \ldots, m$, $f_i(x) = x - T_i(x)$, then the problem G.O.C.P. $(\{f_i\}_1^m, K)$ is equivalent to the fixed point problem:

$$(F.P.)_3: \quad \left\| \begin{array}{l} \text{find } x_o \in K \text{ such that} \\ G(x_o) = x_o \\ \text{where } G(x) = V(0, H(x)) \text{ and} \\ H(x) = V(T_1(x), T_2(x), \ldots, T_m(x)). \ \square \end{array} \right.$$

A. The study of G.O.C.P. $(\{f_i\}_1^m, K)$ in the case $E = C(\Omega, R)$.

Let Ω be a compact topological space.

Consider the set, $C(\Omega, R) = \{x:\Omega \to R \mid x \text{ continuous}\}$.

It is well known that $C(\Omega, R)$ is a Banach space with respect to the vector structure:

α_1) $(x + y)(t) = x(t) + y(t)$; $\forall\ t \in \Omega$

α_2) $(ax)(t) = ax(t)$; $\forall\ a \in R$, $\forall\ t \in \Omega$ and the norm,

α_3) $\|x\| = \sup_{t \in \Omega} |x(t)|$

$C(\Omega, R)$ is an ordered Banach space with respect to the ordering defined by the pointed closed convex cone,

$$K = \{x \mid x \in C(\Omega, R),\ x(t) \geq 0;\ \forall\ t \in \Omega\}.$$

The ordered Banach space $(C(\Omega, R), K)$ is a Riesz space with:

$$(x \wedge y)(t) = \min\{x(t), y(t)\};\ \forall\ t \in \Omega,$$

$$(x \vee y)(t) = \max\{x(t), y(t)\};\ \forall\ t \in \Omega.$$

We recall that an operator (not necessary linear), on $C(\Omega, R)$ is completely continuous if it maps every bounded set in a relatively compact set and it is continuous.

We say that an operator $T:C(\Omega, R) \to C(\Omega, R)$ is isotone if $x \leq y \Longrightarrow T(x) \leq T(y)$, for every $x, y \in C(\Omega, R)$.

We use the following classical fixed point theorem:

Theorem 5.2.1 [Schauder-Tychonov]

If C is a closed convex set in a locally convex space and $f:C \to C$ a continuous mapping such that $f(C)$ is relatively compact, then there exists an element $x_o \in C$ such that $f(x_o) = x_o$. $\square$

Suppose we are given the mappings $T_i:C(\Omega, R) \to C(\Omega, R)$, $i = 1, 2, \ldots, m$.

Denote $H(x) = V(T_1(x), T_2(x), \ldots, T_m(x))$.

Theorem 5.2.2

Consider the G.O.C.P. $(\{f_i\}_1^m, K)$ where $f_i(x) = x - T_i(x)$, $i = 1, 2, \ldots, m$.

If the following assumptions are satisfied.

i) $H(x)$ is isotone,

ii) for every $i = 1, 2, \ldots, m$; T_i is completely continuous,

iii) the set $D = \{x \in K | H(x) \leq x\}$ is nonempty,

then the problem G.O.C.P. $(\{f_i\}_1^m, K)$ has a solution.

Proof

From Corollary of Proposition 5.2.2 the theorem is proved if we show that the mapping $G(x) = V(0, H(x))$ has a fixed point in K.

Indeed, from assumption iii) we have that D is nonempty, so let x_0 be an arbitrary element of D. We set,

$$S = \{x | x \in K \text{ and } x \leq x_0\}.$$

Since the cone K is normal in $C(\Omega, R)$, we observe that S is bounded.

So, S is a closed bounded convex set.

From assumption i) we deduce that $G(S) \subseteq S$.

Because in the space $C(\Omega, R)$ the latticial operation "V" is continuous, we know that H, and therefore G, are completely continuous operators.

So, the all assumptions of Theorem 5.2.1 are satisfied, and we obtain the existence of an element $x_* \in S \subset K$ such that $G(x_*) = x_*$. □

Remark 2

If every t_i $(i = 1, 2, \ldots, m)$ is isotone, then H is isotone, but H isotone does not imply that every T_i is isotone.

Theorem 5.2.3

If the all assumptions of Theorem 5.2.2 are satisfied, then the set D has a least element x_0. Moreover, x_0 is a solution of the problem G.O.C.P.$(\{f_i\}_1^m, K)$.

Proof

We consider the operator $G(x) - V(0, H(x))$ where $H(x) = V(T_1(x), T_2(x), \ldots, T_m(x))$.

By assumption iii), D is nonempty.

Choose an arbitrary element $x_0 \in D$.

We denote,

$$D_{x_o} = \{x \mid x \in K,\ x \leq x_o\} \text{ and}$$

$$S^* = \{x_* \in D_{x_o} \mid G(x_*) = x_*\}.$$

The set S^* is nonempty (by <u>Theorem 5.2.1</u>) and $S^* \subset D$.

Since G is continuous, S^* is closed.

Since $S^* \subset \overline{G(D_{x_o})}$ and G is completely continuous, we have that S^* is compact.

For every $t \in \Omega$ we denote,

$$S_t = \{x_* \in S^* \mid x_*(t) \leq y_*(t);\ \forall\, y_* \in S^*\}.$$

The set S_t is nonempty since, if we consider, for every $t \in \Omega$, the function

$$\Phi_t :\ S^* \longrightarrow R$$

$$x_* \longrightarrow \Phi_t(x_*) = x_*(t).$$

We have that Φ_t is continuous and because S^* is compact, we have that S_t is the set of global minima of Φ_t on S^* and so S_t is nonempty.

Moreover, S_t is closed. Let $I = \{t^1, t^2, \ldots, t^n\} \subset \Omega$ be an arbitrary finite subset of Ω.

We will show that $\bigcap_{t \in I} S_t \neq \phi$.

Indeed, for every $t \in I$, we choose $x_t^* \in S_t$.

We denote, $x^{oo} = \Lambda\, x_t^*$. Using assumption i) and the definition of S^*, we deduce that $x^{oo} \in D_{x_o}$.

Applying <u>Theorem 5.2.1</u> we find that G(x) has a fixed point z_* in $\{x \mid x \in K,\ x \leq x^{oo}\}$. So $z_* \in S^*$ and $z_* \leq x^{oo}$ imply that $z_* \in \bigcap_{t \in I} S_t$.

But because S^* is compact, $\bigcap_{t \in \Omega} S_t \neq \phi$.

Let $x_\bullet$ be an element of $\bigcap_{t \in \Omega} S_t$. We have,

1) $x_\bullet \in S^*$,

2) $x_\bullet \leq x_*;\ \forall\, x_* \in S^*$.

Let $x \in D$ be an arbitrary element.

We set, $\hat{x} = x \wedge x \ \in D \ \cap D_o$.

Using again Theorem 5.2.1 for G and $\{x \in K | x \le \hat{x}\}$ we obtain an element $x_* \in S^*$ such that $x_* \le \hat{x} \le x$, whence $x_\bullet \le x$ for every $x \in D$.

To finish the proof we observe that $x_\bullet$ is a solution of G.O.C.P. $(\{f_i\}_1^m, K)$ since $x_\bullet$ is a fixed point of G. □

We study now the uniqueness of the solution of the problem G.O.C.P. $(\{f_i\}_1^m, K)$.

Theorem 5.2.4

Consider the G.O.C.P. $(\{f_i\}_1^m, K)$ where $f_i(x) = x - T_i(x)$; $i = 1, 2, \ldots, m$.

Assume:

i) $H(x)$ is isotone,

ii) for every $i = 1, 2, \ldots, m$; T_i is completely continuous,

iii) the set $D = \{x \in K | H(x) \le x\}$ is nonempty,

iv) $y - x \in K \setminus \{0\} \Longrightarrow x - H(x) \ge y - H(y)$,

then the problem G.O.O.P. $(\{f_i\}_1^m, K)$ has a solution and this solution is unique.

Proof

From Theorem 5.2.2 we have that the problem G.O.C.P. $(\{f_i\}_1^m$, K has a solution.

Now we must show that this solution is unique.

Suppose that x_* and z_* are two solutions of G.O.C.P. $(\{f_i\}_1^m, K)$ such that $x_* \ne z_*$.

We set, $U = \{t \in \Omega | x_*(t) < z_*(t)\}$.

We can suppose $U \ne \phi$.

So, we have $z_*(t) > 0$, $\forall\, t \in U$.

Since x_* is a solution of G.O.C.P. $(\{f_i\}_1^m, K)$ and the problem G.O.C.P. $(\{f_i\}_1^m, K)$ is equivalent to the problem O.C.P. (H, K) (see the proof of Proposition 5.2.2) we must have,

(P_1): $\quad (z_* - H(z_*))(t) = 0$; for every $t \in U$.

Let $y := x_* \vee z_*$.

We remark that H is isotone (from assumption i)) and from assumption iv) we have,

$$(y - H(y))(t) > (x_* - H(x_*))(t) \ge 0, \text{ for some } t \in U.$$

Because H is isotone we get,

(P_2): $(z_* - H(z_*))(t) \geq (y - H(y))(t); \; \forall \, t \in U.$

Hence from (P_1) and (P_2) we obtain, $(z_* - H(z_*))(t) > 0$, for some $t \in U$ which contradicts (P_1). So $x_* = z_*$. □

Remark 5.2.3

1) Assumption iv) is satisfied if H is a strict contraction with respect to the norm of $C(\Omega, R)$, that is, if $x \neq y \Longrightarrow \|H(x) - H(y)\| < \|x - y\|$.

2) In the proofs of Theorems 5.2.2, 5.2.3, 5.2.4 we generalize the principal results of [100], using similar reasoning.

B. The study of G.O.C.P. $(\{f_i\}_1^m, K)$ in the case $E = L_p(W, m)$, $1 \leq p \leq \infty$.

Let $L_p(\Omega)$ be the Banach space of all real-valued Lebesgue measurable functions $x:\Omega \to R$, where $\Omega \subset R^n$ is such that $\mu(\Omega)$ is finite (μ is the Lebesgue measure).

The norm is $\|x\| = [S_\Omega |x(t)|^p]^{1/p}$; $1 \leq p \leq \infty$.

Let K $L_p(\Omega)$ be the cone of functions which are nonnegative almost everywhere.

$L_p(\Omega)$ is an ordered Banach space with the ordering defined by K, that is, for $x, y \in L_p(\Omega)$, $x \leq y \iff x(t) \leq y(t)$, a.e.

It may also be noted that functions which are almost everywhere equal are considered identical.

It is known [Luxemburg W.A.J. and Zaanen A.C.: Riesz spaces, Vol. I, North-Holland (1971)] that $L_p(\Omega)$ is a Dedekind complete Riesz space.

We use now the following classical fixed point theorem.

Theorem 5.2.5 [Tarski-Knaster-Kantorovich-Birkhoff]

Let T be an isotone map of a Dedekind complete Riesz space E into itself.

If there exist vectors u and v in E such that $u \leq T(u)$ and $T(v) \leq v$, then the set of fixed point $u \leq x_* \leq v$ is not empty and possesses a minimum and a maximum element.

For this theorem see [Tarski A.: The lattice theoretical fixed point theorem and its applications. Pacific J. Nath. Vol. 5 (1955), 285-309.], [A130]

We suppose now that m mappings T_i, $i = 1, \ldots, m$ are given where, $T_i : L_p(\Omega) \to L_p(\Omega)$.

Theorem 5.2.6

Consider the problem G.O.C.P.$(\{f_i\}_1^m, K)$ where $f_i(x) = x - T_i(x)$, $i = 1, 2, \ldots, m$ and denote $H(x) = V(T_1(x), T_2(x), \ldots, T_m(x))$; $x \in L_p(\Omega)$.

<u>If the following assumptions are satisfied:</u>

i). <u>H is isotone,</u>

ii). <u>the set $D = \{x \in K | H(x) \le x\}$ is non-empty,</u>

<u>then the problem G.O.C.P. ($\{f_i\}_1^m$, K) has a solution which is also the least element of D.</u>

<u>Proof</u>

From the <u>Corollary</u> of <u>Proposition 5.2.2.</u> it is sufficient to show that the mapping $G(x) = V(0, H(x))$ has a fixed point in K.

Indeed, since $D \ne \Phi$, then there exists $x_o \in D$. We denote,

$$D_{x_o} = \{x | x \in K, x \le x_o\}.$$

For every $x \in D_{x_o}$ we have,

$$0 \le G(0) \le G(x) \le G(x_o) \le x_o$$

We observe now that the all assumptions of <u>Theorem 5.2.5.</u> are satisfied and hence G has a fixed point $x_\bullet$ which is the minimum element of

$$S^* = \{x_* \in D_{x_o} | G(x_*) = x_*\}.$$

Let $x \in D$ be an arbitrary element.

We set, $\hat{x} := x \wedge x_o \in D \cap D_{x_o}$

Using again <u>Theorem 5.2.5.</u> for G and the set $\{x \in K | x \le \hat{x}\}$ we obtain an element $x_{**} \in S^*$ such that $x_{**} \le \hat{x} \le x$, which implies $x_\bullet \le x_{**} \le x$ and we have that $x_\bullet$ is the least element of D.

<u>Remark 5.2.4</u>

In this case the compactness is not used.

The next result is the uniqueness theorem.

<u>Theorem 5.2.7</u>

<u>If the mapping $H(x) = V(T_1(x), \ldots, T_m(x))$ satisfies the following assumptions:</u>

i) <u>H is isotone,</u>

ii) <u>the set $D = \{x \in K | H(x) \le x\}$ is nonempty,</u>

iii) <u>for every $x, y \in H$ such that $C(x,y) = \{t \in \Omega | x(t) < y(t)\}$ is nonempty, then there exists a subset $D(x,y)$ of $C(x,y)$ of positive measure such that, $H(y)(t) - H(x)(t) < y(t) - x(t)$, for all $t \in D(x,y)$,</u>

<u>then the problem G.O.C.P. ($\{f_i\}_1^m$, K) has a solution and this solution is unique.</u>

Proof

The proof is similar to the proof of Theorem 5.2.4. □

C. The general case

Let $E\ (\tau)$ be a locally convex space ordered by a closed convex cone $K \subset E$.

We use the concept of normal cone [Peressini A.L.: Ordered topological vector spaces. Harper & Row (1967)].

Definition 5.2.1

K is called completely regular [resp. regular] if every monotone increasing and topologically bounded [resp. ordered bounded] sequence of elements of K is τ-convergent.

Definition 5.2.2

We say that $T:A \to E$, $A \subset E$ is a (sm)- compact operator on A if and only if, all sequences of the form,

$T(x_1) \geq T(x_2) \geq \ldots \geq T(x_m) \geq \ldots$; $\forall m \in N$, $\forall x_m \in A$ contains a τ convergent subsequence.

Examples

1°) If $\overline{T(A)}$ is a sequentially compact set, then $T:A \to E$ is a (sm)-compact operator.

2°) If $T:A \to E$ is such that $T(A)$ is a bounded set and K is a completely regular cone, or $T(A)$ is an ordered bounded set and K is a regular cone then T is (sm)-compact.

Let $\Psi:E \to E$ be a mapping.

Definition 5.2.3.

We say that $T:A \to E$ is Ψ-isotone if:

i) there exists $(I + \Psi)^{-1}$ and it is isotone,

ii) $T + \Psi$ is isotone on A.

3°) If T is isotone then T is Ψ-isotone when $\Psi(x) = 0$ for every $x \in E$.

4°) $f(x) = \sin x$, $x \in R$ is not isotone but there exists a constant $a > 0$ such that f is Ψ-isotone where $\Psi(x) = a$; $\forall x \in R$.

5°) Let $T:R^m_+ \to R^m$ be a differentiable function and consider $K = R^m_+$.

If there exists a (M)-matrix $A = (a_{ij})$ such that $T'(x) + A$ is a positive matrix for all $x \in R^m_+$, then T is A-isotone.

In [ISAC G.: Un thèorème de point fixe. Application à la comparaison des equations differentielles dans les Espaces de Banach ordonnés. Libertas Math. Vol. 1, (1981), 75-89] the following fixed point theorem was proved.

Theorem 5.2.8 [ISAC]

Let $E(\tau)$ be a metrizable locally convex space ordered by a normal closed convex cone $K \subset E$.

Let $A \subset E$ be a closed subset and let $T:A \to A$ be a (sm)-compact isotone operator (not necessary continuous).

If there exists $x_0 \in A$ such that $T(x_0) \leq x_0$, then T has a fixed point in A. □

Suppose given the mappings $T_i:E \to E$, $i = 1, 2, \ldots, m$, and consider the problem

$G.O.C.P.(\{f_i\}_1^m, K)$ where $f_i(x) = x - T_i(x)$, $i = 1, 2, \ldots, m$.

Define $H(x) = V(T_1(x), \ldots T_m(x))$ and $G(x) = V(0, H(x))$.

Theorem 5.2.9

Let $(E(\tau), K)$ be an ordered locally convex space. Suppose that E is a Riesz space and K is a regular closed convex cone.

If the following assumptions are satisfied:

1°) G is Ψ-isotone,

2°) $(I + \Psi)^{-1}(G + \Psi)(K) \subset K$,

3°) Ψ, G and $(I + \Psi)^{-1}$ are continuous,

4°) $D = \{x \in K | H(x) \leq x\}$ is nonempty,

then the sequence $\{x_n\}_{n \in N}$ defined by:

$$x_0 = 0$$

$$x_{n+1} + \Psi(x_{n+1}) = G(x_n) + \Psi(x_n); \ \forall n \in N,$$

is convergent and if $x_* = \lim_{n \to \infty} x_n$ then x_* is a solution of the problem $G.O.C.P. (\{f_i\}_1^m, K)$.

Moreover, x_* is the least element of D.

Proof

Since $D \neq \Phi$ there exists an element y in D.

Because $H(y) \leq y$ we have $G(y) \leq y$ and from assumption 1°) we deduce,

$G(y) + \Psi(y) \leq y + \Psi(y)$ which implies

$(I + \Psi)^{-1}[G(y) + \Psi(y)] \in y$ and since

$(I + \Psi)^{-1}(G + \Psi)(0) \in K$ we obtain

$0 \le x_1 \le x_2 \le \dots \le x_m \le \dots \le y.$

The assumptions that K is regular and closed imply that there exists

$x_* = \lim_{n\to\infty} x_n$ and $x_* \in K$.

From the continuity and the definition of $\{x_n\}_{n\in N}$ we have,

$x_* = (I+\Psi)^{-1}(G+\Psi)(x_*)$, or equivalently $G(x_*) = x_*$. So x_* is a solution of G.O.C.P. $(\{f_i\}_1^m, K)$.

Obviously, $x_* \in K$ and we observe that $x_* \le y$, for every $y \in D$, that is x_* is the least element of D. □

The next results are very interesting because the continuity is not used.

Theorem 5.2.10

Let E(τ) be a metrizable locally convex space ordered by a normal closed convex cone K ⊂ E. Suppose that E is a Riesz space.

If the following assumptions are satisfied:

1°) *G is Ψ-isotone,*

2°) $(I + \Psi)^{1}(G + \Psi)(K) \subseteq K$,

3°) $(I + \Psi)^{-1}(G + \Psi)$ *is (sm)-compact,*

4°) $D = \{x \in K \mid H(x) \le x\}$ *is nonempty,*

then the problem G.O.C.P.$(\{f_i\}_1^m, K)$ has a solution.

Proof

Let y be an element of D. We have $H(y) \le y$ and hence $G(y) \le y$.

Consider now the mapping,

$T(x) = (I + \Psi)^{-1}(G + \Psi)(x)$

and the set $A = \{x \mid 0 \le x \le y\}$ which is closed since K is a closed convex cone.

Since G is Ψ-isotone we have $T(y) \le y$, $T(A) \subseteq A$ and T is (sm)-compact.

Apply Theorem 5.2.8 with respect to T and A we obtain a fixed point x_* for T and A and x_* is also a fixed point for G in K. This fixed point of G is a solution of the problem G.O.C.P.$(\{f_i\}_1^m, K)$. □

Thoerem 5.2.11

Let E(τ) be a metrizable locally convex space ordered by a normal closed convex cone K ⊂ E.

Suppose that E is a Riesz space.

Assume that:

1°) G is Ψ-isotone,

2°) $(I + \Psi)^{-1}(G + \Psi)(K) \subseteq K$,

3°) $(I + \Psi)^{-1}(G + \Psi)$ is (sm)-compact on intervals,

4°) there exists $x_o \in D = \{x \in K | H(x) \leq x\}$ such that $\{x \in K | x \leq x_o$ and $G(x) = x\}$ is nonempty and compact.

Then D has a least element x_{**} which is also a least element of the set $M = \{x \in K | G(x) = x\}$ and a solution of the problem G.O.C.P.($\{f_i\}_{i=1}^m$, K).

Proof

Let $D_* = \{x \in K | x \leq x_o$ and $G(x) = x\}$ and for each $x_* \in D_*$ consider the nonempty closed set $D_{x_*} = [0, x_*] \cap D_*$.

Let $I = \{x_*^1, x_*^2, \ldots, x_*^n\}$ be an arbitrary finite subset of D_*.

We have

(i_1): $\bigcap_{r=1}^{n} D_{x_*}^r \neq \Phi$ where $D_{x_*}^r = [0, x_*^r] \cap D_*$.

Indeed, for each $r = 1, 2, \ldots, n$ consider $\bar{x}_*^r \in D_{x_*}^r$ and denote $x^{oo} = \bigcap_{r=1}^{n} \bar{x}_*^r$.

Since $x^{oo} \leq \bar{x}_*^r$, $\forall r = 1, 2, \ldots, n$ we have

$(I + \Psi)^{-1}(G + \Psi)(x^{oo}) \leq (I + \Psi)^{-1}(G + \Psi)(\bar{x}_*^r) = x_*^r$, $\forall r = 1, 2, \ldots, n$; which implies

$(I + \Psi)^{-1}(G + \Psi)(x^{oo}) \leq x^{oo}$.

Applying Theorem 5.2.8 to G and $A = [0, x^{oo}]$ we obtain an element $z_* \in [0, x^{oo}]$ such that $G(z_*) = z_*$.

Hence, $z_* \leq x^{oo} \leq \bar{x}_*^r \leq x_o$; $\forall r = 1, 2, \ldots, n$; $z_* \in D_*$ and $z_* \in \bigcap_{r=1}^{n} D_{x_*}^r$, which proves (i_1).

Then, since D_* is a compact set we have

(i_2): $\bigcap_{x_* \in D_*} D_{x_*} \neq \phi$.

If $x_{**} \in \bigcap_{x_* \in D_*} D_{x_*}$, then $x_{**} \in D_*$ and $x_{**} \leq x_*$ for each $x_* \in D_*$. So x_{**} is the least element of D_*.

But x_{**} is also the least element of D.

Indeed, if x_{**} is not the least element of D there exists $\hat{x} \in D$ such that,

(i_3): or, $\hat{x} < x_{**}$,

(i_4): or, $\hat{x}$ and x_{**} are not comparable.

Applying Theorem 5.2.8 to G and $A = [0, \overset{*}{x} = \hat{x} \wedge x_o]$ we obtain an element $\tilde{x} \in A$ such that $G(\tilde{x}) = \tilde{x}$, $0 \leq \tilde{x} \leq x_o$ and $\tilde{x} \in D_*$.

Hence, we have $x_{**} \leq \tilde{x} \leq \hat{\hat{x}} = \hat{x} \wedge x_o$ which contradicts i_3) and i_4).

Consequently, x_{**} is the least element of D. Obviously since $M \subset D$ we have that x_{**} is the least element of M.

The proof is now finished since from the Corollary of Proposition 5.2.2. we have that x_{**} is a solution of the problem G.O.C.P.($\{f_i\}_1^m$, K). □

Remark 5.2.5

If we have the problem G.O.C.P.($\{f_i\}_1^m$, K) where $E = (H, <,>)$ is a Hilbert space, $f_i(x) = x - T_i(x)$, $i = 1, 2, \ldots, m$ and $K = K^*$, then if the inner product $<,>$ is K-local, that is $< x,y > = 0$, whenever $x, y \in K$ and $x \wedge y = 0$ we can show that every solution of the problem G.O.C.P.($\{f_i\}_1^m$, K) is a solution of the classical complementarity Problem:

$$\text{C.P.}(H,K): \quad \left\| \begin{array}{l} \text{find } x_o \in K \text{ such that,} \\ x_o - H(x_o) \in K = K^* \text{ and} \\ < x_o, x_o - H(x_o) > = 0. \end{array} \right.$$

Applications

The Generalized Order Complementarity Problem had its first formal applications in the work of K.P. Oh [The formulation of the mixed lubrication problem as a Generalized Complementarity Problem. Transaction of ASME, Journal of Tribology, vol. 108, (1986), 598-604.]

Oh considered the case of mixed lubrication in the context of a journal bearing with elastic support.

In this case, there are three distinct functions which cause the decomposition of the spatial area into three disjoint regions. Innermost is the region of solid-to-solid contact, followed by an elasto-hydrodynamic lubrication region and finally, the cavitation region in which the pressure returns to the ambiant value.

For the innermost (solid-to-solid) region, Oh considered the linear elastic model,

$$f_1(X) = H_o + a + L(X) = 0$$

where X is the contact pressure, $f_1(X)$ measures the gap between the bearing surfaces H_o represents surface separation when the load is zero, L is a linear compliance operator and a defines whole-body approach.

Note that in the next two regions we are about to define, $f_1(X) > 0$.

This natural identity serves as a strong geometric motivation for the Generalized Order Complementarity Problem.

In the elasto-hydrodynamic region the basic premise is that the load bearing surfaces are separated by a lubricant film.

The lubricant pressure is governed by the Reynold's differential equation

$$f_2(X) = -R(X) + U\left(\frac{\partial H}{\partial x}\right) + V\left(\frac{\partial H}{\partial t}\right) = 0$$

where R is the second order nonlinear operator $\nabla .\ e^{GX}H^3(x,y,t)\nabla X$, x is the spatial variable in the direction of rotation, y is the transverse spatial variable, H is film thickness, U is the entrainment velocity, G is a piezo-viscous coefficient and V is a positive number which is problem dependent.

By $f_2(X) = 0$ in this region, we are saying that Reynold's equation must be satisfied in the fluid lubricant.

In the outermost region, cavitation occurs and neither the linear elastic model nor the Reynold's lubrication model apply. What does apply is that the pressure function $X = X(x,y) = 0$.

So we take $f_3(X) = X$ to complete this Generalized Order Complementarity Problem.

In addition to the above formulation, it is necessary that the bodies in contact must be in dynamic or static equilibrium at all times.

For this problem we get $\int_\Omega X \, d\,\Omega = F_A$, the integral of the pressure is the applied load.

Although boundary conditions are not stressed on Oh's paper, the condition that the pressure function $X = X(x,y)$ be continuous across the boundaries of the regions is quite reasonable.

Also we wish to enforce the condition that no flow is occuring across the outermost boundary, so $\frac{\partial X}{\partial n} = 0$ should hold on that boundary.

Using the G.O.C.P. we can study other lubrication problems.

For example, a model of an extremely heavily loaded bearing my develop yet another set of regions in which boundary lubrication takes place. In boundary lubrication, temperature and chemical effects on the molecular level need to be modeled as well as surface asperities.

Little work has been done on modeling boundary lubrication [see Chapter 2 models 2.4.3. B], but G.O.C.P. opens the door for research in a systematic mathematical framework.

Many practical applications exist for mixed lubrication models.

The process of shaving with razor and lather which is an almost daily experience for many is certainly one process which exhibits the mixed lubrication phenomena including the solid-to-solid contact region.

Every time one uses the windshield wipers on an automobile, the windshield is wiped cleaned well only if both lubrication and solid-to-solid contact occurs.

There are many other daily life experiences which one may understand much better if one knows about the G.O.C.P.

Also, the Generalized Order Complementarity Problem has interesting applications in economics.

Note

The results presented in section 5.1 was obtained by Borwein and Dempster [33] and the results presented in section 5.2 by Isac and Kostreva [G. ISAC and M. Kostreva: The generalized order complementarity problem. To appear J. Opt. Theory and Appl. 1991].

Chapter 6

The Implicit Complementarity Problem

In this chapter we consider the Implicit complementarity Problem which was first considered by Bensoussan and Lions [A19] Bensoussan, Gourset and Lions [A18] and studied by Capuzzo-Dolcetta and Mosco [A37] Mosco [222], Pang [A241 - A242] Chan and Pang [A38] Noor [C77], Noor and Zarae [C78] and recently by Isac [A130-131] [C26], [C29].

As remarked in [19], [18] and [37], the Implicit Complementarity Problem has interesting applications in Stochastic Optimal Control.

The fixed point theory and the coincidence equations on cones theory play an important role in the study of the Implicit Complementarity Problem and conversely, the Implicit Complementarity Problem may be used to obtain new coincidence theorems on convex cones.

We give now the definition of principal Implicit Complementarity Problems studied by the cited authors.

Let $\langle E,E^* \rangle$ be a dual system of locally convex spaces and suppose defined a closed convex cone $K \subset E$.

Given the mappings $f:K \to E^*$ and $g:D \to E$, where D is a subset of E, the Implicit complementarity Problem associated to f, g and K is:

$$\text{I.C.P.}(f,g,K): \quad \left\| \begin{array}{l} \text{find } x_o \in D \text{ such that} \\ g(x_o) \in K, \quad f(x_o) \in K^* \text{ and} \\ \langle g(x_o), f(x_o) \rangle = 0 \end{array} \right.$$

When $g(x) = x$, for every $x \in D = K$ we have the problem C.P. (f,K).

An other Implicit complementarity Problem more general than the problem I.C.P.(f,g,K) is the following.

Let E and F be locally convex spaces and let $\langle \, , \, \rangle$ be a bilinear form on E x F. Suppose defined two closed convex cones $K_1 \subset E$ and $K_2 \subset F$.

Given two mappings $f:E \to E$ and $g:E \to F$, the Implicit Complementarity Problem associated to f, g, K_1 and K_2 is:

$$\text{I.C.P.}(f,g,K_1,K_2): \quad \left\| \begin{array}{l} \text{find } x_o \in E \text{ so that} \\ f(x_o) \in K_1, \; g(x_o) \in K_2 \text{ and} \\ \langle f(x_o), g(x_o) \rangle = 0 \end{array} \right.$$

This problem with $E = F = H$, where H is a Hilbert space, $K_1 = K_2 \subset H$, $f(x) = M(x) - x$ and $g(x) = b - A(x)$ was considered in [A19], [A18], [A37], [A222]

The same problem was studied in R^n by Pang [A241], [A242]

Finally, we consider the Implicit Complementarity Problem associated to a cone-valued mapping.

Given a dual system $< E,E^* >$ of locally convex spaces we suppose defined a cone-valued mapping $K:E \to E$, that is, for every $x \in E$, $K(x)$ is a closed convex cone in E.

If $f:E \to E^*$ and $g:E \to E$ are given the Implicit Complementarity Problem associated to f, g and $K(x)$ is:

$$\text{I.C.P.}(f,g,K(x)): \quad \left\| \begin{array}{l} \text{find } x_o \in g(x_o) + K(x_o) \\ \text{such that } f(x_o) \in [K(x)]^* \text{ and} \\ < x_o - g(x_o), f(x_o)> = 0 \end{array} \right.$$

We note that the last problem is not so much studied. Pang, Chan and Pang and also Noor obtained some results on this problem in R^n.

6.1 The Implicit complementarity Problem and the fixed point theory

We consider in this section the Implicit Complementarity Problem in a Hilbert space.

Les $(H,< , >)$ be a Hilbert space and let $K \subset H$ be a closed convex cone.

If $D \subseteq H$ is a subset and $f,g:D \to H$ are two mappings, we consider the following Implicit Complementarity Problem:

$$\text{I.C.P.}(f,g,K): \quad \left\| \begin{array}{l} \text{find } x^* \in D \text{ such that} \\ g(x^*) \in K,\ f(x^*) \in K^* \text{ and} \\ < g(x^*), f(x^*) > = 0. \end{array} \right.$$

If we denote by $P_K(x)$ the projection onto K, that is the operator defined by:

$$\| x - P_K(x) \| = \min_{y\in K} \| x-y \| \ ; \ \forall x \in H$$

then we know that P_K is well defined and it is characterized by the following classical result.

Proposition 6.1.1.

For every element $x \in H$, $P_K(x)$ is characterized by the following properties:

1°) $< P_K(x) - x, y > \geq 0 \ ; \ \forall y \in K$,

2°) $< P_K(x) - x, P_K(x) > = 0$

The principal result of this section is based on the following fixed point theorem.

First, we recall that a metric space (X,ρ) is said to be metrically convex, if for each $x, y \in X$, $(x \neq y)$ there is a $z \neq x, y$ for which $\rho(x,y) = \rho(x,z) + \rho(z,y)$

We write $R_\rho = \{\rho(x,y) | x, y \in X\}$.

Theorem 6.1.1. [Boyd and Wong] [C75]

Let (X,ρ) be a complete metrically convex metric space. If for the mapping $T:X \to X$ there is a mapping $\Phi:\bar{R}_\rho \to R_+$ satisfying,

1°). $\rho(T(x), T(y)) \leq \Phi(\rho(x,y))$; $\forall x, y \in X$,

2°). $\Phi(t) < t$, for all $t \in \bar{R}_\rho \{0\}$

then T has a unique fixed point x_o and $T^n(x) \to x_o$ for each $x \in X$.

Definition 6.1.1

Given a subset $D \subseteq H$, we consider the mappings $f, g:D \to H$; $\Phi, \Psi:R_+ \to R_+$ and we say that:

i°). f is a Φ-Lipschitz mapping with respect to g if,

$\| f(x) - f(y) \| \leq \| g(x) - g(y) \| \Phi(\| g(x) - g(y) \|)$; $\forall x, y \in D$,

ii°). f is a Ψ-strongly monotone mapping with respect to g if

$< f(x) - f(y), g(x) - g(y) > \geq \| g(x) - g(y) \|^2 \Psi(\| g(x) - g(y) \|)$; $\forall x, y \in D$

Remark 6.1.1

1°) If in Definition 6.1.1, $g(x) = x$; $\forall x \in D$ then we say that f is a Φ-Lipschitz mapping (respectively, f is a Ψ-strongly monotone mapping).

2°) If Φ and Ψ both are strictly positive constants, we obtain from Definition 6.1.1 that f is a Lipschitz (respectively strongly monotone) mapping.

Theorem 6.1.2

Let $(H, <,>)$ be a Hilbert space and let $K \subset H$ be a closed convex cone. If, for a subset $D \subseteq H$, the mappings $f, g:D \to H$ satisfy the following assumptions:

1°) f is a Φ-Lipschitz mapping with respect to g,

2°) f is a Ψ-strongly monotone with respect to g,

3°) there exists a real number $\tau > 0$ such that,

$\tau \Phi^2(t) < 2 \Psi(t) < \frac{1}{\tau} + \tau \Phi^2(t)$; $\forall t \in R_+$

4°) $K \subseteq g(D)$,

then the problem I.C.P.(f,g,K) has a solution.

Moreover, if g is one to one, then the problem I.C.P.(f,g,K) has a unique solution.

Proof

Using assumption 4°), we consider the mapping $h:K \to H$ (which is not unique) defined by, $h(u) := f(x)$, where x is an arbitrary element of $g^{-1}(u)$ and $u \in K$.

From this definition we observe that h has the following properties:

5°) $\| h(u) - h(v) \| \leq \| u - v \| \Phi(\| u-v \|)$; $\forall u, v \in K$,

6°) $\langle h(u) - h(v), u - v \rangle \geq \| u-v \|^2 \Psi(\| u-v \|$; $\forall u, v \in K$.

Now, we observe that the problem I.C.P.(f,g,K) is equivalent to the following explicit complementarity problem:

$$\text{E.C.P.}(h,K): \left\| \begin{array}{l} \text{find } u_* \in K \text{ such that} \\ h(u_*) \in K^* \text{ and } \langle u_*, h(u_*)\rangle = 0 \end{array} \right.$$

But, from Proposition 6.1.1. we deduce that the problem E.C.P.(h,K) has a solution if and only if the mapping $T:K \to K$ defined by $T(u) = P_K(u - \tau h(u))$; $\forall u \in K$ has a fixed point (where τ is the real number used in assumption 3°).

So, the problem is to show that T has a fixed point.

Indeed, we have

$$\| T(u) - T(v) \|^2 = \| P_K(u - \tau h(u)) - P_K(v - \tau h(v)) \|^2 \leq$$

$$\leq \| (u - \tau h(u)) - (v - \tau h(v)) \|^2 =$$

$$= \| (u - v) - \tau(h(u) - h(v)) \|^2 =$$

$$= \| u - v \|^2 - 2 \tau \langle u - v, h(u) - h(v) \rangle + \tau^2 \| h(u) - h(v) \|^2$$

$$\| u - v \|^2 - 2 \tau \| u - v \|^2 \Psi(\| u - v \|) + \tau^2 \| u - v \|^2 \Phi^2(\| u - v \|) =$$

$$= \| u - v \|^2 [1 - 2 \tau \Psi(\| u - v \|) + \tau^2 \Phi^2(\| u - v \|)],$$

which implies

$$\| T(u) - T(v) \| \leq \| u - v \| [1 - 2 \tau \Psi(\| u - v \|) + \tau^2 \Phi^2(\| u - v \|)]^{\frac{1}{2}}; \ \forall u, v \in D.$$

If we define $(t) = t[1 - 2 \tau \Psi(t) + \tau^2 \Phi^2(t)]^{\frac{1}{2}}$; $\forall t \in R_+$ we observe, using assumption 3°) and the fact that a Hilbert space is a complete metrically convex metric space (and the same property has K), that all assumptions of Theorem 6.1.1. are satisfied. Also $\overline{R}_\rho = R_+$, where $\rho(x,y) = \|x-y\|$: $\forall x\ y \in H$.

Hence, T has a unique fixed point u_* and for every $u \in K$, $u_* = \lim_{n \to \infty} T^n(u)$

Obviously, if g is one to one then the problem I.C.P.(f,g,K) has a unique solution. □

Remark 6.1.2

From the proof of Theorem 6.1.2 we obtain that a solution of the problem I.C.P.(f,g,K) is a solution of the equation, $g(x) = u_*$; $x \in D$, where u_* is obtained by successive approximations using the operator T.

Corollary 1

Let $(H, <,>)$ be a Hilbert space and let $K \subset H$ be a closed convex cone. If for a subset $D \subset H$ the mappings, $f,g:D \to H$ satisfy the following assumptions:

1°) f is k-Lipschitz with respect to g,

2°) f is c-strongly monotone with respect to g,

3°) $K \subseteq g(D)$,

then the problem I.C.P.(f,g,K) has a solution and this solution is unique if g is one to one.

Proof

Replacing the constant c by a smaller constant $c_1 (0 < c_1 < c)$ and noting that f is still c_1-strongly monotone with respect to g, we may find a real number $\tau > 0$ such that $\tau k^2 < 2c < \frac{1}{\tau} + \tau k^2$ and we can apply Theorem 6.1.2.

Indeed, as in the proof of Theorem 6.1.2 we consider the mapping $T(u) = P_K(u - \tau h(u))$, where $0 < \tau < \frac{2c}{k^2}$ and replacing c by $c_1 (0 < c_1 < c)$ such that $\tau k^2 < 2c_1 < \min(\frac{1}{\tau} + \tau k^2, 2c)$, we obtain that assumption 3°) of Theorem 6.1.2 is satisfied with f considered c_1-strongly monotone with respect to g. □

Corollary 2

If $(H, <,>)$ is a Hilbert space, $K \subset H$ a closed convex cone and $f:K \to H$ satisfies the following assumptions:

1°) f is k-Lipschitz,

2°) f is c-strongly monotone,

then the problem C.P. (f,K) has a solution and this solution is unique. □

Carollary 3

Let $(H, <,>)$ be a Hilbert space and let $K \subset H$ be a closed convex cone. If for a subset $D \subset H$ and $f,g:D \to H$ the following assumptions are satisfied:

1°) f is a Ψ-strongly monotone mapping with respect to g,

2°) g is an expansive mapping, that is, $(\exists \lambda \geq 1)(\forall x,y \in D)(\| g(x)-g(y) \| \geq \lambda \| x-y \|)$,

3°) $\| f(x)-f(y) \| \le \| x-y \| \Phi (\| g(x)-g(y) \|$; $\forall x,y \in D$,

4°) there exists a real number $\tau > 0$ such that,

$$\tau\Phi^2(t) < 2\,\Psi(t) < \frac{1}{\tau} + \tau\,\Phi^2(t);\ \forall t \in R_+$$

5°) $K \subseteq g(d)$,

then the problem I.C.P.(f,g,K) has a solution and this solution is unique. □

A mapping $h:D \to H$ is said to be accretive if and only if,

$\| x-y \| \le \| (x-y) + \lambda(h(x) - h(y)) \|$; for all x,y D and all $\lambda \ge 0$.

Also, $U:D \to H$ is said to be pseudo-contractive if and only if, for all $x,y \in D$ and all $\lambda > 0$ we have,

$\| x-y \| \le \| (1+\lambda)(x-y) - (U(x) - U(y)) \|$.

A classical result proved by Kato and Browder is the following.

If $g = Id - U$, where $U:H \to H$, then the mapping U is pseudo-contractive if and only if, g is accretive. (Id is the identity mapping).

Corollary 4

Let $(H, <, >)$ be a Hilbert space and let $K \subset H$ be a closed convex cone.

If for a subset $D \subset H$ and $f,g:D \to H$ the following assumptions are satisfied:

1°) f is Ψ-strongly monotone with respect to g

2°) $g - \rho Id$ is accretive for some $\rho > 0$ on D,

3°) $\| f(x) - f(y) \| \le \| x-y \| \Phi (\| g(x) - g(y) \|)$; $\forall x,y \in D$,

4°) $K \subseteq g(D)$,

5°) there exists a real number $\tau > 0$ such that,

$$\tau\,\Phi^2(\tau) < 2\rho^2\,\Phi(t) < \frac{\rho^2}{\tau} + \tau\,\Phi^2(t)\ ;\ \forall t \in R_+,$$

then the problem I.C.P.(f,g,K) has a solution.

Proof

Since we can show that in this case we have, $\| x-y \| \le \rho^{-1} \| g(x) - g(y) \|$; $\forall x, y \in D$, we deduce (using assumption 3°) that

$$\| f(x) - f(y) \| \leqslant \rho^{-1} \| g(x) - g(y) \| \Phi (\| g(x) - g(y) \|)$$

and we can apply Theorem 6.1.2. □

Given $f,g : D \to H$ we say that f is α-monotone with respect to g, if there exists a strictly increasing function $\alpha:[0, +\infty) \to [0, +\infty)$ with $\alpha(0) = 0$ and $\lim_{t\to+\infty} \alpha(t) = +\infty$ such that,

$$< f(x) - y(y), g(x) - g(y) > \ge \| g(x) - g(y) \| \alpha(\| g(x) - g(y) \|)\ ;\ \forall x, y \in D$$

Proposition 6.1.2

Let $(H, <,>)$ be a Hilbert space and $K \subset H$ a closed convex cone.

If for a subset $D \subset H$ the mappings $f,g : D \to H$ satisfy the following assumptions:

1°) f is a Φ-Lipschitz mapping with respect to g and $\lim_{r\to 0} \Phi(r) \neq \infty$,

2°) f is α-monotone with respect to g,

3°) $K \subseteq g(D)$,

then the problem I.C.P.(f,g,K) has a solution.

Proof

We consider the problem E.C.P.(h,K) where the mapping $h:K \to K$ is defined as in the proof of Theorem 6.1.2. and we observe that all the assumptions of Corollary of Luna's Theorem [Theorem 4.3.8] are satisfied. □

If in the problem I.C.P.(f,g,K), where K is again a closed convex cone in a Hilbert space $(H, <,>)$, the mapping g has the form, $g(x) = x - h(x)$; $\forall x \in H$, then we have the following result.

In this case $D = H$ and $f:H \to H$.

Theorem 6.1.3

Let α, β, γ be positive real constants.

If f is α-strongly monotone and β-Lipschitz, h is γ-Lipschitz and $\rho > 0$ is a real constant such that $0 < 2\gamma = \sqrt{1-2\alpha\rho+\rho^2\beta^2} < 1$, then for every $0 < \lambda \leq 1$ the sequence $\{x_n\}_{n\in N}$ defined by:

$$x_{n+1} = \lambda h(x_n) + \lambda P_K [x_n - \rho f(x_n) - h(x_n)] + (1-\lambda)x_n; \ n = 0, 1, 2, \ldots$$

and x_0 an arbitrary element in K, is convergent to a solution of the problem I.C.P.(f, x-h(x), K).

Proof

For an arbitrary λ such that $0 < \lambda \leq 1$ we consider the mapping

$T(x) = \lambda h(x) + \lambda P_K[x - \rho f(x) - h(x)] + (1-\lambda)x$, where $x \in H$.

Using Proposition 6.1.1. we can show that a fixed point of T is a solution of the problem I.C.P.(f, x-h(x), K).

The Theorem is proved if we show that T is a contraction.

Indeed, for every $x, y \in H$ we have,

$$||T(x) - T(y)|| = ||\lambda h(x) + \lambda P_K[x-\rho f(x) - h(x)] + (1-\lambda) x -$$
$$- \lambda h(y) - \lambda P_K[y - \rho f(y) - h(y)] - (1-\lambda)y|| \leq$$

$\leq \lambda\| h(x) - h(y) \| + \lambda\| P_K[x-\rho f(x) - h(x)] - P_K[y-\rho f(y) - h(y)] \|$

$+ (1-\lambda)\| x-y\| \leq$

$\leq \lambda\|h(x) - h(y)\| + \lambda\|(x - \rho f(x) - h(x)) - (y - \rho f(y) - h(y))\| + (1-\lambda)\|x - y\| \leq$

$2\lambda\|h(x) - h(y)\| + \lambda\|(x-y) - \rho(f(x) - f(y))\| + (1-\lambda)\| x-y\|.$

Since,

$\|(x-y) - \rho(f(x) - f(y))\|^2 = \|x-y\|^2 - 2 \langle x-y, \rho(f(x) - f(y))\rangle +$

$+ \rho^2\|f(x) - f(y)\|^2 \leq [1 - 2\alpha\rho^2 + \rho^2\beta^2] \|x-y\|^2$,

from assumptions we obtain,

$$\|T(x) - T(y)\| \leq [2\lambda\gamma + (1-\lambda) + \lambda \sqrt{1-2\alpha\rho+\rho^2\beta^2}\ ] \|x-y\|$$

Now, if we denote

$\delta = 2\lambda\gamma + (1-\lambda) + \lambda \sqrt{1-2\alpha\rho + \rho^2\beta^2}$, we have,

$\|T(x) - T(y)\| \leq \delta\|x-y\|$; $\forall x, y \in H$ where $0 < \delta < 1$ and the sequence $\{x_n\}_{n\in N}$ is the sequence defined by the Banach's contraction theorem, which implies that $\{x_n\}_{n\in N}$ is convergent to a fixed point of T. □

Remark 6.1.3

The condition $0 < \delta < 1$, where $\delta = 2\lambda\gamma + (1-\lambda) + \lambda \sqrt{1-2\alpha\rho + \rho^2\beta^2}$ is obtained if

$$0 < \lambda \leq 1,\ 0 \leq \gamma \leq \frac{1}{2},\ \alpha > 2\beta \sqrt{\gamma - \gamma^2} \text{ and } \left|\rho - \frac{\alpha}{\beta^2}\right| < \frac{\sqrt{\alpha^2 - 4\beta^2(\gamma-\gamma^2)}}{\beta^2}.$$

6.2 The Implicit Complementarity Problem and a special variational inequality.

We study in this section the Implicit Complementarity Problem using a variational inequality having a special form.

Let $\langle E,E^*\rangle$ be a dual system of locally convex spaces. Suppose that $K \subset E$ is a closed convex cone.

Given two mappings, $f,g:K \to E^*$ we consider the following Implicit Complementarity Problem associated to f,g and K:

I.C.P.(f,g,K): find $x_o \in K$ such that $g(x_o) \in K$, $f(x_o) \in K^*$ and $\langle g(x_o), f(x_o) \rangle = 0$.

To study this problem we consider also the following special variational inequality:

S.V.I.(f,g,K): find $x_o \in K$ such that $g(x_o) \in K$ and $\langle x - g(x_o), f(x_o) \rangle \geq 0$; $\forall x \in K$.

Proposition 6.2.1.

The problem I.C.P.(f,g,K) and S.V.I(f.g.K) are equivalent.

Proof

If $x_o \in K$ is a solution of the problem S.V.I.(f,g,K) then we have $g(x_o) \in K$ and

(1): $\langle x - g(x_o), f(x_o) \rangle \geq 0$; for all $x \in K$.

If $u \in K$ is an arbitrary element and if we set $x = u + g(x_o)$ in (1) then we have $\langle u, f(x_o) \rangle \geq 0$; for all $u \in K$, that is $f(x_o) \in K^*$.

If we put $x = 0$ in (1) then we get $\langle g(x_o), f(x_o) \rangle \leq 0$ and if we set $x = 2g(x_o)$ again in (1) we deduce $\langle g(x_o), f(x_o) \rangle \geq 0$, which implies $\langle g(x_o), f(x_o) \rangle = 0$, that is, x_o is a solution of the problem I.C.P.(f,g,K).

Conversely, if x_o is a solution of the problem I.C.P.(f,g,K) then we have

$$(2): \quad \left\| \begin{array}{l} g(x_o) \in K, \\ \langle x, f(x_o) \rangle \geq 0 \ ; \ \forall x \in K \text{ and} \\ \langle g(x_o), f(x_o) \rangle = 0 \end{array} \right.$$

which imply $g(x_o) \in K$ and $\langle x - g(x_o), f(x_o) \rangle \geq 0$; for all $x \in K$. □

We will use the following result.

Theorem 6.2.1. [Ky Fan][C76]

Let X be a nonempty compact convex set in a Hausdorff topological vector space. Let A be a subset of X x X having the following properties:

1°) for every $x \in X$, $(x,x) \in A$,

2°) for each fixed $x \in X$ the set $A_x = \{y \in X \mid (x,y) \in A\}$ is closed in X,

3°) for each fixed $y \in X$ the set $A_y = \{x \in X \mid (x,y) \notin A\}$ is convex.

Then there exists a point $y_o \in X$ such that $X \times \{y_o\} \subset A$. □

Let $D \subset K$ be a nonempty compact convex set. Associated to the problem S.V.I.(f,g,K) we consider also the following problem:

$$\text{S.V.I.}(f,g,D): \quad \left\| \begin{array}{l} \text{find } x_o \in D \text{ such that} \\ \langle x - g(x_o), f(x_o) \rangle \geq 0 \ ; \ \forall x \in D \end{array} \right.$$

The bilinear form $\langle \, , \, \rangle$ is supposed now to be continuous.

Proposition 6.2.2.

If f and g are continuous and $\langle g(x), f(x) \rangle \leq \langle x, f(x) \rangle$ for every $x \in D$ then the problem S.V.I.(f,g,D) has a solution.

Proof

We apply Theorem 6.2.1 for $X = D$ and $A + \{(x,y) \in D \times D \mid < g(y) - x, f(y) > \leq 0\}$

The proposition will be proved if we show that assumptions 1°), 2°) and 3°) of Theorem 6.2.1 are satisfied.

Indeed, for every $x \in D$, $(x,x) \in A$ if and only if (by the definition of A) $< g(x) - x, f(x) > \leq 0$, but from assumptions this inequality is satisfied.

From the continuity of f and g the set $A_x = \{y \in D \mid < g(y) - x, f(y) > \leq 0\}$ is closed for every $x \in D$.

To finish we show that for every $y \in D$ the set $B_y = \{x \in D \mid (x,y) \notin A\} =$ $= \{x \in D \mid < g(y) - x, f(y) > > 0\}$ is convex.

Indeed if $x_1, x_2 \in B_y$ and $\alpha, \beta \in R_+$ are such that $\alpha + \beta = 1$, then we have

$< \alpha g(y) - \alpha x_1, f(y) > > 0$ and

$< \beta g(y) - \beta x_2, f(y) > > 0$

which imply, $< g(y) - (\alpha x_1 + \beta x_2), f(y) > > 0$, that is $\alpha x_1 + \beta x_2 \in B_y$.

Now, from Theorem 6.2.1 there exists $x_0 \in D$ such that $D \times \{x_0\} \subset A$, which implies that $< x - g(x_0), f(x_0) > \geq 0$ for every $x \in D$ and the proof is finished. □

The next result is a variant of Proposition 6.2.2. without continuity but in Banach spaces.

If $< E,E^* >$ is a dual system of Banach spaces, $K \subset E$ a closed convex cone and $D \subset K$ a weakly compact nonempty convex set, then we have the following result.

Proposition 6.2.3

If $f,g:K \to E^*$ satisfy the following assumptions:

1°) $< g(x), f(x) > \geq < x, f(x) >$, for every $x \in D$,

2°) for each sequence $\{y_n\}_{n\in N} \subset D$, weakly convergent to y_* we have,

$\liminf_{n\to\infty} < x, f(y_n) > \leq < x, f(y_*)>$, for every $x \in D$,

3°) $x \to < g(x), f(x) >$ is sequentially weakly lower semicontinuous on D,

then the problem S.V.I.(f,g,D) has a solution.

Proof

The proof is similar to the proof of Proposition 6.2.2., but using the weak topology.

Indeed, we consider $X = D$, $A = \{(x,y) \in D \times D \mid < g(y) - x, f(y) > \leq 0\}$ and we verify the assumptions of Theorem 6.2.1.

As in the proof of Proposition 6.2.2. we have that $(x,x) \in A$ for every $x \in D$ and $B_y = \{x \in D \mid (x,y) \notin A\}$ is convex for every $y \in D$.

The proposition is proved if we show that for every $x \in D$ the set $A_x = \{y \in D \mid \langle g(y) - x, f(y) \rangle \leq 0\}$ is weakly closed.

Indeed, let $x \in D$ be fixed and pick $\{y_n\}_{n\in N} \subset A_x$.

Since $A_x \subset D$ and D is weakly compact we may extract a subsequence $\{y_{n_k}\}_{k\in N}$ of $\{y_n\}_{n\in N}$ weakly convergent to an element $y_* \in D$.

Because $y_{n_k} \in A_x$ for every $k \in N$ we have, $\langle g(y_{n_k}), f(y_{n_k}) \rangle \leq \langle x, f(y_{n_k})\rangle$.

From assumptions 2°) and 3°) we have $\langle g(y_*), f(y_*) \rangle \leq \liminf_{k\to\infty} \langle g(y_{n_k}), f(y_{n_k})\rangle \leq \liminf_{k\to\infty} \langle x, f(y_{n_k}) \rangle \leq \langle x, f(y_*)\rangle$.

So, we have $\langle g(y_*) - x, f(y_*) \rangle \leq 0$, that is, $y_* \in A_x$ and hence A_x is weakly sequentially compact.

But sence A_x is weakly sequentially compact it is weakly countable compact and by Eberlein's Theorem it is weakly compact and hence weakly closed. □

Theorem 6.2.2

Let $K \subset E$ be a locally compact convex cone, $S:K \to K$ a continuous mapping, $T_1, T_2:K \to E^*$ mappings such that $T_1 - T_2$ is continuous and $f:K \to K$ a positive homogeneous mapping of order $p_2 > 0$

If the following assumptions are satisfied

1°) the mapping $x \to \langle f(x), T_1(x) \rangle$ is lower semicontinuous in K and $\langle f(x), T_1(x)\rangle > 0$, for every $x \in K$ with $\|x\| = 1$,

2°) there exist $c_1 > 0$, $r_1 > 0$ and $p_1 > 0$ such that $T_1(\lambda x) \leq_{K^*} c_1 \lambda^{p_1} T_1(x)$, for every $x \in K$ with $\|x\| \geq r_1$ and $0 < \lambda \leq 1$,

3°) there exist $c_2 > 0$ and $r_2 > 0$ such that $\langle f(x), T_1(x) \rangle \leq c_2 \langle S(x), T_1(x)\rangle$ for every $x \in K$ with $\|x\| \geq r_2$,

4°) $\limsup_{\|x\| \to\infty} \dfrac{\langle S(x), T_2(x)\rangle}{\|x\|^{p_1+p_2}} \leq 0$

5°) $\langle S(x), T_1(x) - T_2(x)\rangle \leq \langle x, T_1(x) - T_2(x)\rangle$, for every $x \in K$,

then the problem I.C.P.$(T_1 - T_2, S, K)$ has a solution.

Proof

Since K is locally compact we have that for every $n \in N$ the set $D_n = \{x \in K \mid \|x\| \leq n\}$ is convex compact and $o \in D_n$.

Moreover, $K = \bigcup_{n=1}^{\infty} D_n$ and from Propositon 6.2 we have that the problem

I.C.P. (T_1-T_2, S, K) is equivalent to the problem S.V.I.(T_1-T_2, S, K).

Since the all assumptions of Proposition 6.2.2. are satisfied for every problem S.V.I.(T_1-T_2, S, D_n), where $n \in N$, we deduce that for every $n \in N$ there exists $x_n^* \in D_n (\subset K)$ such that,

(3) $< y - S(x_n^*), T_1(x_n^*) - T_2(x_n^*) > \geq 0 \; ; \forall y \in D_n$.

The sequence $\{x^*\}_{n\in N}$ is bounded.

Indeed, if we suppose that $\{x_n^*\}_{n\in N}$ is not bounded then we can consider (eventually considering a subsequence) that $\lim_{n\to\infty} \|x_n^*\| = +\infty$ and $\|x_n^*\| \neq 0$ for every $n \in N$.

We consider the sequence $\{y_n\}_{n\in N}$ defined by, $y_n = \frac{x_n^*}{\|x_n^*\|}$ for every $n \in N$.

Since D_1 is compact the sequence $\{y_n\}_{n\in N}$ has a subsequence $\{y_{n_k}\}_{k\in N}$ convergent to an element y_* such that $\|y_*\| = 1$

From assumption 1°) we have that $< f(y_*), T_1(y_*) > > 0$.

Since $0 \in D_n$ and $x_n^* \in D_n$ for every $n \in N$, we obtain from inequality (3) that

(4) $< S(x_n^*), T_1(x_n^*) - T_2(x_n^*) > \leq 0$, for every $n \in N$, or

(5) $< S(x_n^*), T_1(x_n^*) > \leq < S(x_n^*), T_2(x_n^*) >$, for every $n \in N$

Using assumptions 2°), 3°) and formula (5) we deduce for every x_n^* with

$$\|x_n^*\| > \max(1, r_1, r_2),\ < f(\frac{x_n^*}{\|x_n^*\|}), T_1(\frac{x_n^*}{\|x_n^*\|}) > \leq < f(\frac{x_n^*}{\|x_n^*\|}), \frac{c_1}{\|x_n^*\|^{p_1}} T_1(x_n^*) > =$$

$$= \frac{c_1}{\|x_n^*\|^{p_1+p_2}} < f(x_n^*), T_n(x_n^*) > \leq \frac{c_1 c_2}{\|x_n^*\|^{p_1+p_2}} < S(x_n^*), T_1(x_n^*) > \leq$$

$$\leq \frac{c_1 c_2}{\|x_n^*\|^{p_1+p_2}} < S(x_n^*), T_2(x_n^*) > .$$

Using the last formula and assumptions 1°) and 4°) we have

$$0 < < f(y_*), T_1(y_*) > \leq \liminf_{n\to\infty} < f(\frac{x_n^*}{\|x_n^*\|}) , T_1(\frac{x_n^*}{\|x_n^*\|}) > \leq$$

$$\leq c_1 c_2 \limsup_{n\to\infty} \frac{< S(x_n^*), T_2(x_n^*) >}{\|x_n^*\|^{p_1+p_2}} \leq 0, \text{ which is impossible.}$$

So, we have that $\{x_n^*\}_{n\in N}$ is bounded and because K is locally compact $\{x_n^*\}_{n\in N}$ has a convergent subsequence.

Considering eventually a subsequence we can suppose that $\{x_n^*\}_{n\in N}$, is convergent.

If we denote $x_* = \lim x_n^*$, then we have that $x_* \in K$.

The proof is finished if we show that x_* is a solution of the problem

I.C.P. $(T_1 - T_2, S, K)$.

Indeed, let $y \in K$ be an arbitrary element.

Then there exists $m \in N$ such $y \in D_m$.

Hence $y \in D$ for every $n \geq m$, which implies,

(6) $< y - S(x_n^*), T_1(x_n^*) - T_2(x_n^*) > \geq 0$; $\forall n \geq m$

Since S and $T_1 - T_2$ are continuous, computing the limit in (6°) we get finally

$< y - S(x_*), T_1(x_*) - T_2(x_*) > \geq 0$; $\forall y \in K$

and using again Proposition 6.2.1. we have that x_* is a solution of the problem I.C.P.(T_1-T_2, S, K) and the proof is finished. □

Remarks 6.2.1.

i) Assumption 2°) in Theorem 6.2.2. is satisfied in particular if T_1 is a convex operator with respect to the order " $\leq_{K^*}$" and $T_1(0) \leq_{K^*} 0$.

(ii) Since the problem I.C.P.(T_1-T_2, S, K) is equivalent to the problem I.C.P. $(T_1-T_2, \frac{1}{\alpha} S, K)$ for every $\alpha \in R_+\setminus\{0\}$, we remark that assumption 5°) in Theorem 6.2.2. can be: "there exists $\alpha \in R_+\setminus\{0\}$ such that

$< S(x), T_1(x) - T_2(x) > \leq \alpha < x, T_1(x) - T_2(x) >$; for every $x \in K$.

By a similar proof as for Theorem 6.2.2. we can prove also the following result.

Theorem 6.2.3.

Let $K \subset E$ be a locally compact cone, $S:K \to K$ a continuous mapping, $T_1, T_2:K \to E^*$ mappings such that $T_1 - T_2$ is continuous and $f:K \to K$ a positive homogeneous mapping of order $p_2 > 0$.

If the following assumptions are satisfied:

1°) the mapping $x \to \langle f(x), T_2(x)\rangle$ is upper semicontinuous in K and $f(x), T_2(x) >< 0$ for every $x \in K$ with $\|x\| = 1$,

2°) ther exist $c_1 > 0$, $r_1 > 0$ and $p_1 > 0$ such that $c_1\lambda^{p_1} T_2(x) \leq_{K^*} T_2(\lambda x)$; for every $x \in K$ with $\|x\| \geq r_1$ and $0 < \lambda \leq 1$,

3°) there exist $c_2 > 0$ and $r_2 > 0$ such that $\langle f(x), T_2(x)\rangle \geq c_2 < S(x), T_2(x) >$, for every $x \in K$ with $\|x\| \geq r_2$,

4°) $\liminf_{\|x\| \to \infty} \dfrac{\langle S(x), T_2(x)\rangle}{\|x\|^{p_1 + p_2}} \geq 0.$

5°) there exists $c_3 > 0$ such that $< S(x), T_1(x) - T_2(x) > \leq c_3 < x, T_1(x) - T_2(x) >$, for every $x \in K$

then the problem I.C.P.(T_1-T_2, S, K) has a solution. □

If E and F are two Banach spaces and $D \subset E$ a subset, then we say that a mapping $f:D \to F$ is strongly continuous at the point $x_* \in D$, if and only if for every sequence $\{x_n\}_{n\in N} \subset D$ weakly convergent to x_* we have that $\{f(x_n)\}_{n\in N}$ is strongly convergent to $f(x^*)$.

Obviously f is strongly continuous on D if it is strongly continuous at every point of D.

More interesting as Theorem 6.2.2. for applications in a reflexive Banach spaces is the following result.

Theorem 6.2.4.

Let $K \subset E$ be a weakly locally compact convex cone, $S:K \to K$ a strongly continuous mapping, $T_1, T_2:K \to E^*$ mappings such that $T_1 - T_2$ is strongly continuous and $f:K \to K$ a positive homogeneous mapping or order $p_2 > 0$.

If the following assumptions are satisfied:

1°) the mapping $x \to < f(x), T_1(x)>$ is weakly lower semicontinuous in K and $< f(x), T_1(x) >> 0$, for every $x \in K\backslash\{0\}$,

2°) there exist $c_1 > 0$ and $p_1 \geq 0$ such that $T_1(\lambda x) \leq_{K^*} c_1 \lambda^{p_1} T_1(x)$, for every $x \in K\backslash\{0\}$ and $0 < \lambda \leq 1$,

3°) there exist $c_2 > 0$ and $r > 0$ such that $<f(x), T_1(x)> \leq c_2 < S(x), T_1(x)>$, for every $x \in K$ with $\|x\| \geq r$,

4°) $\limsup_{\|x\| \to \infty} \dfrac{<S(x), T_2(x)>}{\|x\|^{p_1+p_2}} \leq 0$,

5°) $< S(x), T_1(x) - T_1(x) > \geq < x, T_1(x) - T_1(x)>$, for every $x \in K$,

then the problem I.C.P.($T_1 - T_2$, S, K) has a solution.

Proof

The proof follows the principal ideas of the proof of Theorem 6.2.2. but with some specific details.

Since K is weakly locally compact there exists a continuous linear functional $\Phi:E \to R$ such that

(7) $\|x\| \leq \Phi(x)$, for every $x \in K$ and the set $B = \{x \in K | \Phi(x) = 1\}$ is a base for K, that is, for every $x \in K\backslash\{0\}$ there exist a unique $b \in B$ and a unique $\lambda \in R_+\backslash\{0\}$ such that $x = \lambda b$.

that is, for every $x \in K\backslash\{0\}$ there exist a unique $b \in B$ and a unique $\lambda \in R_+\backslash\{0\}$ such that $x = \lambda b$.

Moreover, in this case B is a weakly compact set.

For every $n \in N$ the set $D_n = \{x \in K | \Phi(x) \le n\}$ is convex weakly compact and $0 \in D_n$.

We observe that $K = \bigcup_{n=1}^{\infty} D_n$ and for every $n \in N$ the problem S.V.I.$(T_1 - T_2, S, D_n)$ is solvable since the all assumptions of Theorem 6.2.1. are satisfied with the detail that in this case we deduce (using the strongly continuity of S and $T_1 - T_2$ and Eberlein's Theorem) that A_x for $x \in D_n$ is weakly compact and hence weakly closed

So, we have that for every $n \in N$ there exists $x_n^* \in D_n$ ($\subset K$) such that

(8) $< y - S(x_n^*), T_1(x_n^*) - T_2(x_n^*) > \ \ge 0$; $\forall y \in D_n$

The sequence $\{x_n^*\}_{n \in N}$ is bounded.

Indeed, if $\{x_n^*\}_{n \in N}$ is not bounded we can suppose that $\lim_{n\to\infty} \|x_n^*\| = +\infty$ and $\|x_n^*\| \ne 0$, for every $n \in N$ and from (7) we have that $\lim_{n\to\infty} \Phi(x_n^*) = +\infty$.

We consider the sequence $\{y_n\}_{n\in N}$ defined by $y_n = \dfrac{x_n^*}{\Phi(x_n^*)}$, for every $n \in N$.

For every $n \in N$, $y_n \in B$ and since B is weakly compact the sequence $\{y_n\}_{n\in N}$ has a subsequence $\{y_{n_k}\}_{k\in N}$ weakly convergent to an element $y_* \in B$ (and hence $y_* \ne 0$).

From assumption 1°) we have that $<f(y_*), T_1(y_*) > \ > 0$.

Since $0 \in D_n$ and $x_n^* \in D$ for every $n \in N$ we obtain from (8):

(9) $< S(x_n^*), T_1(x_n^*) > \ \le \ < S(x_n^*), T_2(x_n^*) >$; $\forall n \in N$

Using assumptions 2°), 3°) and formula (9) we deduce for every x_n^* with

$$\|x_n^*\| \le \max(1,r), \ < f(\frac{x_n^*}{\Phi(x_n^*)}), T_1(\frac{x_n^*}{\Phi(x_n^*)}) > \ \le \ < f(\frac{x_n^*}{\Phi(x_n^*)}), \frac{c_1}{[\Phi(x_n^*)]^{p_1}} T_1(x_n^*) > =$$

$$= \frac{c_1}{[\Phi(x_n^*)]^{p_1+p_2}} < f(x_n^*), T_1(x_n^*) > \ \le$$

$$\le \frac{c_1 c_2}{[\Phi(x_n^*)]^{p_1+p_2}} < S(x_n^*), T_1(x_n^*) > \ \le \frac{c_1 c_2}{\|x_n^*\|^{p_1+p_2}} < S(x_n^*), T_2(x_n^*) >.$$

From the last formula and assumptions 1°) and 4°) we get,

$$0 < \ < f(y_*), T_1(y_*) > \ \le \liminf_{n\to\infty} < f(\frac{x_n^*}{\Phi(x_n^*)}), T_1(\frac{n_n^*}{\Phi(x_n^*)}) > \ \le$$

$$\le c_1 c_2 \limsup_{n\to\infty} \frac{< S(x_n^*), T_2(x_n^*) >}{\|x_n^*\|^{p_1+p_2}} \le 0,$$

which is impossible. Hence $\{x_n^*\}_{n\in N}$ is bounded and it has a weakly convergent subsequence. Suppose that the limit of this subsequence is $x_* \in K$.

Now, using the strong continuity we can show as in the proof of Theorem 6.2.2. that x_* is a solution of the problem I.C.P. $(T_1 - T_2, S, K)$. □

The next result is a variant of Theorem 6.2.2. for a closed convex cone K in a Hilbert space where K is not necessary locally compact but approximable by a countable family of locally compact cones, that is for Galerkin cones as defined in [chapter 4, 2].

Let $(H, <,>)$ be a Hilbert space and let $K(K_n)_{n\in N}$ be a Galerkin cone.

We can prove that for every $n \in N$ there exists a projection onto K_n, denoted by P_n such that for every $x \in K$, $\lim_{n\to\infty} P_n(x) = x$. [We recall that if $C \subset H$ is a closed convex set, a projection onto C is a continuous operator (not necessary linear) $P:H \to H$ such that $P(H) = C$ and $P(x) = x$, for every $x \in C$].

Definition 6.2.1

We say that $S:K \to K$ is subordinate to the approximation $(K_n)_{n\in N}$ of K, if there exists $n_0 \in N$ such that for every $n \geq n_0$ $S(K_n) \subseteq K_n$.

Examples

1°) $S(x) = x$, for every $x \in K$ is subordinate to every Galerkin approximation of K.

2°) Suppose that $S \in L(H)$ and $S(K) \subseteq K$. We say that S is of finite rank if there exist $y_1, y_2, \ldots, y_n; x_1^*, x_2^*, \ldots, x_n^* \in H$ such that for all $x \in H$,

$$S(x) = \sum_{i=1}^{n} < x_i^*, x > \cdot y_i.$$

In this case if $(K_n)_{n\in N}$ is a finite dimensional Galerkin approximation of K then S is subordinate to $(K_n)_{n\in N}$.

3°) If $(H, <,>)$ is a Hilbert space of real functions defined on a measurable set $G \subset R^n$ and $S:H \to H$ is a Hammerstein operator of the form

$S[x(t)] = \int_G [\sum_{j=1}^{n} \lambda_j e_j(t)\, g_j(s)]\, f(s, x(s)) \mid ds$ such that $S(K) \subseteq K$ (a closed convex cone in H) and e_j; g_j, $j = 1, 2, \ldots, m$ are given functions in H, then S is subordinate to every finite dimensional approximation of K

If $T_1:K \to H$ is an operator, positive homogeneous of order $p > 0$ then in this case we denote, $\omega_K(T_1) = \{< x, T_1(x) > \mid x \in K, \|x\| = 1\}$ and we say that T_1 is K-range bounded if $\omega_K(T_1)$ is a bounded subset of R.

If T_1 is K-range bounded then $M_K(T_1) = \sup \omega_K(T_1)$ and $m_K(T_1) = \inf \omega_K(T_1)$ are finite real numbers and for every $x \in K\setminus\{0\}$ we have,

$$m_K(T_1)\|x\|^{p+1} \le \langle x, T_1(x)\rangle \le M_K(T_1)\|x\|^{p+1}.$$

Theorem 6.2.5

Let $K(K_n)_{n\in N}$ be a Galerkin cone in Hilbert space $(H, \langle\, ,\, \rangle)$.

Suppose that $S:K \to K$ is a strongly continuous mapping subordinate to the approximation $(K_n)_{n\in N}$, $T_1, T_2:K \to H$ mappings such that $T_1 - T_2$ is strongly continuous.

If the following assumptions are satisfied:

1°) the mapping $x \to \langle x, T_1(x)\rangle$ is lower semicontinuous in K,

2°) T_1 is positive homogeneous of order $p > 0$

3°) there exist $c_1 > 0$ and $r > 0$ such that

$\langle x, T_1(x)\rangle \le c_1 \langle S(x), T_1(x)\rangle$, for every $x \in K$ with $\|x\| \ge r$,

4°) $\lim_{\|x\|\to\infty} \dfrac{\langle S(x), T_2(x)\rangle}{\|x\|^{p+1}} = 0$

5°) there exists $c_2 > 0$ such that $c_2 \langle S(x), T_1(x) - T_2(x)\rangle \le \langle x, T_1(x) - T_2(x)\rangle$ for every $x \in K$,

6°) T_1 is K-range bounded and $m_K(T_1) > 0$,

then the problem I.C.P. $(T_1 - T_2, S, K)$ has a solution.

Proof

We know that the problem I.C.P. $(T_1 - T_2, S, K)$ is equivalent to the problem S.V.I. $(T_1 - T_2, S, K)$.

Since S is subordinate to the approximation $(K_n)_{n\in N}$ we have that for every $n \ge n_o$ every problem S.V.I. $(T_1 - T_2, S, K_n)$ is well defined and the all assumptions of Theorem 6.2.2 are satisfied (where $f(x) = x$, for every $x \in K_n$).

Hence, for every $n \ge n_o$ the problem S.V.I. $(T_1 - T_2, S, K_n)$ has a solution x_n^*.

We prove now that the sequence $\{x_n^*\}_{n\ge n_o}$ is bounded.

Indeed, if we suppose that $\{x_n^*\}_{n\ge n_o}$ is not bounded then we can suppose that $\lim_{n\to\infty} \|x_n^*\| = +\infty$ and $\|x_n^*\| \ne 0$, for every n.

Using assumptions 3°) and 6°) we have for every $n \ge n_o$ such that $\|x_n^*\| \ge r$ the following inequalities:

(10) $< S(x_n^*), T_1(x_n^*) - T_2(x_n^*) > = < S(x_n^*), T_1(x_n^*) > -$

$- < S(x_n^*), T_2(x_n^*) > \geq \frac{1}{c_1} < x_n^*, T_1(x_n^*) > -$

$< - \quad S(x_n^*), T_2(x_n^*) > \geq \frac{1}{c_1} m_K(T_1) \| x_n^* \|^{p+1} - < S(x_n^*), T_2(x_n^*) >.$

But from assumptions 4°) and 6°) there exists $m \in N$ such that for every $n \geq m$ with $\|x_n^*\| \geq r$ we have,

$$\frac{<S(x_n^*), T_2(x_n^*)>}{\|x_n^*\|^{p+1}} < \frac{1}{c_1} m_K(T_1)$$

which implies using (10) that for every n sufficiently big,

(11) $< S(x_n^*), T_1(x_n^*) - T_2(x_n^*) > > 0.$

Now, we remark that formula (11) is impossible because for every $n \geq n_o$ we have,

$$< S(x_n^*), T_1(x_n^*) - T_2(x_n^*) > = 0.$$

So, we have finally that $\{x_n^*\}_{n \geq n_o}$ is bounded and since H is reflexive it has a subsequence $\{x_{n_k}^*\}_{k \in N}$ weakly convergent to an element $x^* \in K$.

The proof is finished if we prove that x^* is a solution of the problem S.V.I. $(T_1 - T_2, S, K)$.

Let $x \in K$ be an arbitrary element.

For every $n \geq n_o$ we have, $<P_n(x) - S(x_n^*), T_1(x_n^*) - T_2(x_n^*) > \geq 0$, and computing the limit when $n \to \infty$ we obtain (since $\lim_{n\to\infty} P_n(x) = x$ and the the operators S and $T_1 - T_2$ are strongly continuous), $< x - S(x^*), T_1(x^*) - T_2(x^*)> \geq 0$, for every $x \in K$, that is x^* is a solution of the problem I.C.P. $(T_1 - T_2, S, K)$. □

The following result is a generalization in the context of the <u>Implicit Complementarity Problem of Karamardian's Theorem</u>, used in Chap. 4, § 3.

Let $< E, E^* >$ be a dual system of locally convex spaces and let $K \subset E$ be a closed convex cone.

<u>Theorem 6.2.6</u>

<u>Let $S:K \to E$ and $T:K \to E^*$ be two mappings.</u>

<u>If there exists a real-valued function $h:K \times K \to R$ such that:</u>

1°) <u>$< S(y) - x, T(y) > \leq h(x,y)$, for every $(x, y) \in K \times K$,</u>

2°) <u>the mapping $y \to <S(y) - x, T(y) >$ is lower semicontinous for every $x \in K$,</u>

3°) <u>the set $\{x \in K \mid h(x, y) > 0\}$ is convex for every $y \in K$,</u>

4°) <u>$h(x, x) \leq 0$, for all $x \in K$,</u>

5°) there exists a non-empty compact subset $D \subset K$ such that for every $y \in K\backslash D$, there exists a point $x \in D$ with $< S(y) - x, T(y) > > 0$,

then the problem I.C.P.(T, S, K) has a solution $x_* \in D$.

Proof

For each element $x \in K$ we denote,

$$D(x) = \{y \in D \mid < S(y) - x, T(y) > \leq 0\}$$

and from assumption 2°) we have that $D(x)$ is closed in D.

If we prove that $\bigcap_{x \in K} D(x) \neq \Phi$ then our theorem is proved since every element $x_* \in \bigcap_{x \in K} D(x)$ is a solution of the problem S.V.I. (T,S,K).

Since D is compact it is sufficient to show that the family $\{D(x)\}_{x \in K}$ has the finite intersection property.

Indeed, let $x_1, x_2, \ldots, x_m \in K$ be given.

We put, $A = \text{conv}(D \cup \{x_1, x_2, \ldots, x_m\})$ and we have that A is a compact convex subset of K.

We consider the following point-to-set mappings:

$$f_1(x) = \{y \in A \mid < S(y) - x, T(y) > \leq 0\},$$

$$f_2(x) = \{y \in A \mid h(x,y) \leq 0\}$$

defined for every $x \in K$.

From assumptions 1°), 4°) and 2°) we obtain that $f_1(x)$ is non-empty and compact (as closed subset of A).

We prove now that f_2 is a KKM-map.

Indeed, if we suppose that there exist $v_1, v_2, \ldots, v_n \in A$ and $\lambda \geq 0$; $i = 1, 2, \ldots, n$ with $\sum_{i=1}^{n} \lambda_i = 1$ such that, $\sum_{i=1}^{n} \lambda_i v_i \notin \bigcup_{j=1}^{n} f_2(v_j)$, then we have that $h(v_j, \sum_{i=1}^{n} \lambda_i v_i)$ 0, for $1 \leqslant j \leqslant n$ and from assumption 3°) $h(\sum_{i=1}^{n} \lambda_i v_i, \sum_{i=1}^{n} \lambda_i v_i) > 0$ which is in contradiction with assumption 4°).

Since from assumption 1°) we have $f_2(x) \subset f_1(x)$, for every $x \in K$, we obtain that f_1 is also a KKM-map.

Applying Theorem 4.3.1 to f_1 we get $\bigcap_{x \in A} f_1(x) \neq \Phi$, that is, there exists a point $y_* \in A$ such that, $< S(y_*) - x, T(y_*) > \leq 0$, for all $x \in A$.

By assumption 5°) we have that $y_* \in D$ and moreover, $y_* \in D(x_i)$, for every $1 \leq i \leq m$.

Hence $\{D(x)\}_{x \in K}$ has the finite intersection property and the proof is finished. □

If we consider $h(x,y) = \langle S(y) - x, T(y) \rangle$ then from Theorem 6.2.6 we obtain:

Corollary 1

Let $S:K \to K$ and $T:K \to E^*$ be two mappings.
If the following assumptions are satisfied:

1°) the mapping $y \to \langle S(y) - x, T(y) \rangle$ is lower semicontinuous for every $x \in K$,

2°) $\langle S(x), T(x) \rangle \leq \langle x, T(x) \rangle$, for all $x \in K$,

3°) there exists a non-empty compact convex subset $D \subset K$ such that for every $y \in K \backslash D$, there exists a point $x \in D$ with $\langle S(y) - x, T(y) \rangle > 0$

then the problem I.C.P. (T,S,K) has a solution $x_* \in D$. □

If in Corollary 1 of Theorem 6.2.6, $S:K \to K$ is the identity mapping, that is, $S(x) = x$, for every $x \in K$ then we obtain the correct form of Allen's theorem [Al2], that is we have the following result.

Corollary 2

Let $E(\tau)$ be a topological vector space, let $K \subset E$ be a convex cone and let $f:K \to E^*$ be a mapping such that,

1°) for every $y \in K$ the mapping $y \to \langle y - x, f(y) \rangle$ is lower semicontinuous on K,

2°) there exists a non-empty compact convex set $D \subset K$ such that for every $y \in K \backslash D$ there exist $x \in D$ with $\langle y - x, f(y) \rangle > 0$

then the problem C.P.(f,K) has a solution. □

From Carollary 2 we obtain immediately the following classical result.

Theorem 6.2.7

Let $E(\tau)$ be a locally convex space, let $K \subset E$ be a closed convex cone and $f:K \to E^*$ a mapping such that,

1°) the function $(x,y) \to \langle x, f(y) \rangle$ is continuous on $K \times K$,

2°) there exists a non-empty compact convex set $D \subset K$ such that, for every $x \in K \backslash D$, there exists $z \in D$ such that $\langle x - z, f(x) \rangle > 0$

then the problem C.P.(f, K) has a solution.

We suppose now that the Hilbert space $(H, \langle , \rangle)$ is the Euclidean space $(R^n, \langle , \rangle)$.

Let $K \subset R^n$ be a closed convex cone.

Definition 6.2.2

Given $S:K \to K$ we say that $T:K \to R^n$ is strongly K-compositive with respect to S if there exists a scalar $m > 0$ such that for all $x \in K$ we have $\langle S(x), T(x) - T(0) \rangle \geq m\|S(x)\|^2$.

We recall that the level set of order $\lambda \in R$ of the function $x \to \|S(x)\|$ is the set $\{x \in K \mid \|S(x)\| \leq \lambda\}$.

Theorem 6.2.8

Let $K \subset R^n$ be a closed convex cone and $S:K \to K$; $T:K \to R^n$ two mappings.

If the following assumptions are satisfied:

1°) the mapping $y \to \langle S(y) - x, T(y) \rangle$ is lower semicontinuous for every $x \in K$,

2°) $\langle S(x), T(x) \rangle \leq \langle x, T(x) \rangle$, for every $x \in K$,

3°) T is strongly K-copositive with respect to S,

4°) the level sets of order λ, for every $\lambda > 0$ of the function $x \to \|S(x)\|$ are compact subsets of K,

then the problem I.C.P. (T, S, K) has a solution x_*. Moreover, if $T(0) \neq 0$ then

$$x_* \in \text{conv}\,\{x \in K \mid \|S(x)\| \leq \frac{\|T(0)\|}{m}\}$$

Proof

The theorem is a consequence of Corollary 1 of Theorem 6.2.6 if we show that assumption 3°) of this corollary is satisfied.

First, we remark that if $T(0) = 0$ then $x_* = 0$ is a solution of the problem I.C.P. (T, S, K).

So, we suppose $T(0) \neq 0$ and we denote $\rho = \frac{\|T(0)\|}{m} > 0$

From assumption 4°) the set $D = \text{conv}\,\{x \in K \mid \|S(x)\| \leq \rho\}$ is convex compact and since T is strongly K-copositive we have for every $y \in K\setminus D$,

$$\langle S(y), T(y) \rangle \geqslant \langle S(y), T(0) \rangle + m\|S(y)\|^2 > \langle S(y), T(0) \rangle + \|S(y)\|\,\|T(0)\| \geq 0$$

which implies that assumption 3° of Corollary 1 is satisfied for every $y \in K\setminus D$ with $x = 0$.

Hence from Corollary 1 the problem I.C.P. (T, S, K) has a solution $x_* \in D$. □

6.3: The Implicit complementarity Problem and coincidence equations on convex cones.

It is well known that the study of coincidence equations is an important and interesting problem in topology, in fixed point theory and in nonlinear analysis.

In this section we will establish some natural relations between the Implicit Complementarity Problem and the coincidence equations on convex cones.

Consider given two ordered vector spaces (E, K_1), (F, K_2), a bilinear form on F denoted by $\langle , \rangle$ and suppose that F is a vector lattice with respect to the ordering defined by K_2.

Given the mappings $f, \mathcal{G}: E \to F$ we consider the following problems:

$$(P_1): \quad \left\| \begin{array}{l} \text{find } x_o \in K_1 \text{ such that} \\ \mathcal{F}(x_o) = \mathcal{G}(x_o), \text{ where } \mathcal{F}(x) = \sup(o, f(x)); \forall x \in E. \end{array} \right.$$

$$(P_2): \quad \left\| \begin{array}{l} \text{find } x_o \in K_1 \text{ such that} \\ \mathcal{G}(x_o) \in K_2, \ \mathcal{G}(x_o) - f(x_o) \in K_2 \text{ and} \\ < \mathcal{G}(x_o), \mathcal{G}(x_o) - f(x_o) > = 0 \end{array} \right.$$

$$(P_3) \quad \left\| \begin{array}{l} \text{find } x_o \in K_1 \text{ such that} \\ \mathcal{G}(x_o) \in K_2, \mathcal{G}(x_o) - f(x_o) \in K_2 \text{ and} \\ \inf(\mathcal{G}(x_o), \mathcal{G}(x_o) - f(x_o)) = 0 \end{array} \right.$$

Problem (P_1) is a coincidence equation, problem (P_2) is an implicit complementarity problem and problem (P_3) is an order complementarity problem.

Exactly, if $\mathcal{G}(x) = g(x)$ and $f(x) = g(x) - h(x)$ where $g, h: E \to E$, then problem (P_2) is the implicit complementarity problem

$$\underline{\text{I.C.P}(g, h)}: \quad \left\| \begin{array}{l} \text{find } x_o \in K_1 \text{ such that} \\ g(x_o) \in K_2, \ h(x_o) \in K_2 \text{ and} \\ < g(x_o), h(x_o) > = 0 \end{array} \right.$$

<u>Proposition 6.3.1</u>

<u>Problem (P_1) and (P_3) are equivalent. Moreover, if $<, >$ is a K_2-local bilinear form on F then every solution of problem (P_1) is a solution of problem (P_2).</u>

<u>Proof</u>

<u>$(P_1) \Rightarrow (P_3)$</u>. If x_o is a solution of problem (P_1) then $\mathcal{G}(x_o) = \mathcal{F}(x_o) \in K_2$, which implies, $\mathcal{G}(x_o) - f(x_o) \in K_2$ and $0 = \mathcal{G}(x_o) - \mathcal{G}(x_o) = \mathcal{G}(x_o) - \sup(0, f(x_o)) =$ $= \inf(\mathcal{G}(x_o), \mathcal{G}(x_o) - f(x_o))$.

$(P_3) \Rightarrow (P_1)$. If $x_o \in K_1$ is a solution of problem (P_3) then, $0 = \inf(\mathcal{G}(x_o),$ $\mathcal{G}(x_o) - f(x_o)) = \mathcal{G}(x_o) - \mathcal{F}(x_o)$.

If $<, >$ is a K_2-local bilinear form on F and x_o is a solution of problem (P_1) then $x_o \in K_1$ and since (P_1) and (P_3) are equivalent we have $\mathcal{G}(x_o) \in K_2$, $\mathcal{G}(x_o) - f(x_o) \in K_2$ and $\mathcal{G}(x_o), (x_o) - f(x_o) > = 0$.

The following result is a general method which can be used to transform and explicit complementarity problem in a coincidence equation on a convex cone.

Proposition 6.3.2

Let (E, K) be a vector lattice and let $<,>$ be a K-local bilinear form on E. If $\varphi: E \to E$ is an arbitrary mapping and x_* a solution of coincidence equation

(1): $\left\| \begin{array}{l} \text{find } x_o \in E \text{ such that} \\ f([\varphi(x)]^+) = [\varphi(x)]^- , \end{array} \right.$

then $[\varphi(x_*)]^+$ is a solution of the complementarity problem

(2): $\left\| \begin{array}{l} \text{find } x_o \in K \text{ such that} \\ f(x_o) \in K \text{ and } < x_o, f(x_o)> = 0 \end{array} \right.$

Proof

Indeed, if x_* is a solution of equation (1) then we have $[\varphi(x_*)]^+ \in K$, $f([\varphi(x_*)]^+) = [\varphi(x_*)]^- \in K$ and $< [\varphi(x_*)]^+, f([\varphi(x_*)]^+ > = < [\varphi(x_*)]^+, [\varphi(x_*)]^- > = = 0$. □

If $(E, K, <,>)$ is an ordered Hilbert space, where K is a closed convex cone and $<,>$ is the inner product defened on E, then using the projection on K we can associate a coincidence equation to an implicit complementarity problem.

We denote by P_K the projection on K.

Proposition 6.3.3

If $f, g: E \to E$ are two mappings and $x_* \in K$ is a solution of coincidence equation

(3): $g(x) = P_K(g(x) - \alpha f(x))$,

where $\alpha \in R_+ \setminus \{0\}$, then x_* is a solution of implicit complementarity problem

(4): $\left\| \begin{array}{l} \text{find } x_* \in K \text{ such that} \\ g(x_*) \in K,\ f(x_*) \in K^* \text{ and} \\ < g(x_*), f(x_*) > = 0. \end{array} \right.$

Proof

Using Proposition 6.1.1 we have $\alpha f(x_*) = g(x_*) - [g(x_*) - \alpha f(x_*)] \in K^*$, that is $f(x_*) \in K^*$ and $<g(x_*), g(x_*) - [g(x_*) - \alpha f(x_*)]> = 0$, which implies $<g(x_*), f(x_*)> = 0$

The proposition is proved since $g(x_*) \in K$. □

In chapter 5 we introduced the concept of (sm)-compact operator (Definition 5.2.2) and the concept of completely regular (resp. regular) cones (Definition 5.2.1).

In this section we will use again these concepts to study some coincidence equations.

Let $E(\tau)$ be a locally convex space ordered by a closed convex cone $K \subset E$. Consider on E the ordering defined by K.

Let $\mathcal{G}, \Lambda : E \to E$ be two mappings and $A \subset E$ a nonempty subset.

Definition 6.3.1

We say that $f : A \to E$ is $(\mathcal{G},\Lambda)$-monotone increasing if:

1°) $(\mathcal{G} + \Lambda)^{-1}$ is defined and it is monotone increasing,

2°) $f + \Lambda$ is monotone increasing on A.

Using Theorem 5.28 we obtain the following coincidence theorem.

Theorem 6.3.1

Let $E(\tau)$ be a metrizable locally convex space ordered by a normal closed convex cone and let $A \subset E$ be a closed subset.

Suppose that $f : A \to E$ satisfies the following assumptions:

i°) f is $(\mathcal{G}, \Lambda)$-monotone increasing,

ii°) $(\mathcal{G} + \Lambda)^{-1}(f + \Lambda)(A) \subseteq A$,

iii°) $(\mathcal{G} + \Lambda)^{-1}(f + \Lambda)$ is (sm)-compact.

If there exists $x_o \in A$ such that $f(x_o) \le \mathcal{G}(x_o)$ then there exists $x_o \in A$ such that $f(x_*) = \mathcal{G}(x_*)$.

Proof

Indeed, if we put $T = (\mathcal{G} + \Lambda)^{-1}(f + \Lambda)$ we can use Theorem 5.28 and the theorem follows. □

Consider now problem (P_1) with $E = F$, $K_1 = K_2 = K \subset E$ and denote $\mathcal{F}(x) = \sup(0, f(x))$, for every $x \in E$.

Theorem 6.3.2

Suppose that $(E(\tau), K)$ is a locally convex lattice and K is a regular closed convex cone.

If $\mathcal{F}$ is $(\mathcal{G},\Lambda)$-monotone increasing, $(\mathcal{G} + \Lambda)^{-1}(\mathcal{F} + \Lambda)(K) \subseteq K$ and the mappings Λ, $\mathcal{F}(\mathcal{G} + \Lambda)^{-1}$ are continuous, then the following statements are equivalent:

1°) $\mathcal{D} = \{x \in K \mid f(x) \le \mathcal{G}(x) \text{ and } \mathcal{G}(x) \in K\}$ is nonempty,

2°) $\mathcal{N} = \{x \in L \mid \mathcal{F}(x) = \mathcal{G}(x)\}$ is nonempty.

3°) <u>the sequence $\{x_n\}_{n\in N}$ defined by,</u>

$$x_0 = 0$$

$$\mathcal{G}(x_{n+1}) + \Lambda(x_{n+1}) = \mathcal{F}(x_n) + \Lambda(x_n);\ \forall n \in N$$

<u>is convergent and if $x_* = \lim_{n\to\infty} x_n$ then $x_* \in \mathcal{N} \subset \mathcal{D}$ and x_* is the least element of $\mathcal{D}$.</u>

<u>Proof</u>

<u>2°) $\Longrightarrow$ 1°).</u> Indeed, if $\mathcal{N} \neq \Phi$ then there exists $x_0 \in \mathcal{N}$ and we have $\mathcal{G}(x_0) = \sup(0, f(x_0))$ which implies $\mathcal{G}(x_0) \in K$, $\mathcal{G}(x_0) \geq f(x_0)$ and hence $\mathcal{D} \neq \Phi$.

<u>3°) $\Longrightarrow$ 2°).</u> From continuity and the definition of $\{x_n\}_{n\in N}$ we have

$$x_* = (\mathcal{G} + \Lambda)^{-1} [\ \mathcal{F}(x_*) + \Lambda(x_*)\ ]$$

and consequently $\mathcal{N} \neq \Phi$.

<u>1°) $\Longrightarrow$ 3°).</u> Suppose that $\mathcal{D} \neq \Phi$ and consider $y \in \mathcal{D}$. In this case we have $f(y) \leq \mathcal{G}(y)$, $\mathcal{G}(y) \in K$ and hence $\mathcal{F}(y) \leq \mathcal{G}(y)$.

Since $\mathcal{F}$ is $(\mathcal{G}, \Lambda)$-monotone increasing we obtain $\mathcal{F}(y) + \Lambda(y) \leq \mathcal{G}(y) + \Lambda(y)$, which gives $(\mathcal{G} + \Lambda)^{-1} [\ \mathcal{F}(y) + \Lambda(y)\ ] \leq y$, and since $(\mathcal{G} + \Lambda)^{-1} [\ \mathcal{F} + \Lambda\ ](0) \in K$ we obtain, $0 \leq x_1 \leq x_2 \leq \ldots \leq y$.

Since K is a regular cone, there exists $x_* = \lim_{n\to\infty} x_n$ and from continuity we have $x_* = (\mathcal{G} + \Lambda)^{-1}(\mathcal{F} + \Lambda)(x_*)$, or equivalently $\mathcal{F}(x_*) = \mathcal{G}(x_*)$ which means $\mathcal{N} \neq \Phi$.

Obviously, $\mathcal{N} \subset \mathcal{D}$ and we observe that $x_* \leq y$ for every $y \in \mathcal{D}$, which implies that x_* is the least element of $\mathcal{D}$. □

<u>Remark 6.3.1.</u>

The last conclusion of <u>Theorem 6.3.2</u> is very important in the study of the following <u>Indifferent Optimization Problem</u>:

(<u>I.O.P.</u>) $\quad \left\| \begin{array}{l} \min_{x\in\mathcal{D}} \varphi(x), \text{ where} \\ \mathcal{D} = \{x \in K \mid f(x) \leq \mathcal{G}(x) \text{ and } \mathcal{G}(x) \in K\} \text{ and } \varphi: K \to R \text{ is a} \\ \text{monotone increasing mapping.} \end{array} \right.$

Obviously, if x_* is the least element of $\mathcal{D}$ then x_* is a solution of problem (<u>I.O.P.</u>). The problem (<u>I.O.P.</u>) is important in Economics and Mechanics.

The following result extends <u>Theorem 6.3.2</u> to discontinuous case.

<u>Theorem 6.3.3.</u>

<u>Let $E(\tau)$ be a metrizable locally convex space ordered by a normal closed conves cone $K \subset E$.</u>

Suppose that E is a vector lattice with respect to the ordering defined by K. If the following assumptions are satisfied:

i) F is (G, Λ)-monotone increasing,

ii) $(G + \Lambda)^{-1}(F + \Lambda)$ is (sm)-compact,

iii) $(G + \Lambda)^{-1}(F + \Lambda)(K) \subseteq K$,

then the following statements are equivalent:

1°) $\mathcal{D} = \{x \in K \mid f(x) \leq G(x) \text{ and } G(x) \in K\}$ is nonempty,

2°) $N = \{x \in K \mid F(x) = G(x)\}$ is nonempty.

Proof

1° ⟹ 2°). Suppose that $\mathcal{D} \neq \Phi$ and hence there exists $y \in \mathcal{D}$ such that $f(y) \leq G(y)$ and $G(y) \in K$, which implies, $F(y) \leq G(y)$.

Consider now the mapping $g(x) = (G + \Lambda)^{-1}(F + \Lambda)(x)$ and the set $A = \{x \mid 0 \leq x \leq y\}$, which is a closed set since K is a closed convex cone.

Since F is a (G, Λ)-monotone increasing operator we have, $g(y) \leq y$, $g(A) \subseteq A$ and since g is a (sm)-compact operator we can use Theorem 6.3.1 for F, G and A.

Thus, we obtain an element $x_* \in A$ such that $F(x_*) = G(x_*) \in K$ and hence $N \neq \Phi$

2) ⟹ 1). If $N \neq \Phi$, then there exists $x_* \in N$ such that $F(x_*) = G(x_*)$ and obviously the definition of mapping F implies that $\mathcal{D} \neq \Phi$. □

The following result is a localization theorem of solution of problem (I.O.P.).

Theorem 6.3.4.

Suppose satisfied the all assumptions of Theorem 6.3.3. If $\mathcal{D}$ is nonempty then N is nonempty and an element x_* is the least element of $\mathcal{D}$ if and only if x_* is the least element of N. [In this case x_* is also a solution of the problem I.O.P.].

Proof

If $\mathcal{D}$ is nonempty, then it follows from Theorem 6.3.3 that N is nonempty and the definition of F implies that $N \subset \mathcal{D}$. Let x_* be the least element of $\mathcal{D}$. Then we have $F(x_*) \leqslant G(x_*)$ and we can apply Theorem 6.3.1 to F, G and $A = [0, x_*]$.

Thus, we obtain an element $\hat{x} \in A$ such that $F(\hat{x}) = G(\hat{x})$.

We have $\hat{x} \in \mathcal{D}$, $\hat{x} \leq x_*$ and because x_* is the least element of $\mathcal{D}$ we obtain $x_* = \hat{x} \in N \subset \mathcal{D}$ which implies that x_* is the least element of N.

Suppose now that x_* is the least element of N. If we assume that x_* is not the least element of $\mathcal{D}$, then there exists $x_o \in \mathcal{D}$ such that, a) $x_o < x_*$, or b) x_* and x_o are not comparable.

Applying Theorem 6.3.1. to F, G and $A = [0, x_o]$ we obtain that there exists $\hat{x} \in N$ such that $F(\hat{x}) = G(\hat{x})$ and $\hat{x} \leq x_o$.

Since x_* is the least element of N, we have $x_* \leq \hat{x}$ and hence $x_* \leq \hat{x} \leq x_o$ which contradicts a) and b). □

The following result is an existence theorem for the least element of .

Theorem 6.3.5

Let $E(\tau)$ be a metrizable locally convex space ordered by a normal closed convex cone $K \subset E$.

Suppose that E is a lattice with respect to the ordering defined by K.

If the following assumptions are satisfied:

1°) F is (G, Λ)-monotone increasing,

2°) $(G + \Lambda)^{-1}(F + \Lambda)(K) \subseteq K$,

3°) $(G + \Lambda)^{-1}(F + \Lambda)$ is (sm)-compact on intervals,

4°) there exists $x_o \in \mathcal{D} = \{x \in K \mid f(x) \leq G(x) \text{ and } G(x) \in K\}$ such that the set $\{x \in K \mid x \leq x_o \text{ and } F(x) = G(x)\}$ is non-empty and compact,

then $\mathcal{D}$ has a least element.

Proof

Denote $D_* = \{x \in K \mid x \leq x_o$ and $(x) = (x)\}$ and for each $x_* \in D_*$ consider the closed set $D_{x_*} = [0, x_*\} \cap D_*$. Obviously, D_{x_*} is nonempty.

Let $I = \{x_*^1, x_*^2, \ldots, x_*^n\}$ be an arbitrary finite subset of D_*. We prove now that

(3): $\bigcap_{s=1}^{n} D_{x_*}^s \neq \Phi$, where $D_{x_*}^s = [0, x_*^s\} \cap D_*$.

Indeed, for each $s = 1, 2, \ldots, n$ consider $\bar{x}_*^s \in D_{x_*}^s$ and denote,

$$x^{oo} = \inf \{\bar{x}_*^s \mid s = 1, 2, \ldots, n\}.$$

Since $x^{oo} \leq x_*^s$, $\forall s = 1, 2, \ldots, n$; we observe that

$$(G + \Lambda)^{-1} (F + \Lambda) (x^{oo}) \leq (G + \Lambda)^{-1} (F + \Lambda) (\bar{x}_*^s) = \bar{x}_*^s ;$$

for each $s = 1, 2, \ldots, n$, which implies $(G + \Lambda)^{-1} (F + \Lambda) (x^{oo}) \leq x^{oo}$.

Applying now Theorem 6.3.1 to F, G and $A = [0, x^{oo}]$, we obtain an element $z_* \in [0, x^{oo}]$ such that $F(z_*) = G(z_*)$.

Hence $z_* \leq x^{oo} \leq \bar{x}_*^s \leq x_o$; for each $s = 1, 2, \ldots, n$, z_* D_* and $z_* \in \bigcap_{s=1}^{n} D_*^s$ which proves formula (3).

Then, since D_* is a compact set we have,

(4): $\bigcap_{x_* \in D_*} D_{x_*} \neq \Phi$

and we observe that, if $x_{**} \in \bigcap_{x_* \in D_*} D_{x_*}$ then $x_{**} \in D_*$ and $x_{**} \leq x_*$ for each $x_* \in D_*$,

so that x_{**} is the least element of D_*.

But to finish, it is necessary to prove that x_{**} is also the least element of $\mathcal{D}$.

To prove this fact we suppose the contrary, that is, suppose that x_{**} is not the least element of $\mathcal{D}$, hence there exists $x^o \in \mathcal{D}$ such that:

a) or, $x^o < x_{**}$,

b) or x^o and x_{**} are not comparable.

Applying Theorem 6.3.1 to $\mathcal{F}$, $\mathcal{G}$ and $A = [0, \bar{x} = \inf (x^o, x_o)]$ we obtain an element $\tilde{x} \in A$ such that $\mathcal{F}(x) = \mathcal{G}(x)$, $0 \leq \tilde{x} \leq x_o$ and $\tilde{x} \in D_*$. Hence, we have $x_{**} \leq \tilde{x} \leq x_o$, which contradicts a) and b). Consequently, x_{oo} is the least element of $\mathcal{D}$.

Remarks 6.3.2.

1°) Assumptions 3°) and 4°) in Theorem 6.3.5 are verified if $\mathcal{D}$ is nonempty and $(\mathcal{G} + \Lambda)^{-1} (\mathcal{F} + \Lambda)$ is a compact operator (that is, a continuous mapping which maps bounded subsets into relatively compact subsets).

Indeed, if $(\mathcal{G} + \Lambda)^{-1} (\mathcal{F} + \Lambda)$ is a compact operator, then clearly it is a (sm)-compact operator on intervals.

Consider $x_o \in \mathcal{D} \neq \Phi$ and denote, $D_* = \{x \mid 0 \leq x \leq x_o \text{ and } \mathcal{F}(x) = \mathcal{G}(x)\}$.

Since $D_* = \{x \in [0, x_o] \mid (\mathcal{G} + \Lambda)^{-1} (\mathcal{F} + \Lambda)(x) = x\}$ we have that D_* is closed.

Also, D_* is a nonempty set since we can apply Theorem 6.3.1 to $\mathcal{F}$, $\mathcal{G}$ and $A = [0, x_o]$.

Because $D_* \subset \overline{(\mathcal{G} + \Lambda)^{-1} (\mathcal{F} + \Lambda)([0, x_o])}$ we obtain that D_* is a compact set.

2°) Theorem 6.3.5 is true if we have that $\{x \in K \mid x \leq x_o \text{ and } \mathcal{F}(x) = \mathcal{G}(x)\}$ is a nonempty and weakly compact set. We consider now the general case.

Suppose that $(E(\tau_1), K_1)$ and $(F(\tau_2), K_2)$ are two ordered locally convex spaces and consider the coincidence equation:

(5): $\mathcal{F}(x) = \mathcal{G}(x)$, where $\mathcal{F}, \mathcal{G}: K_1 \to K_2$

Definition 6.3.2

We say that $\mathcal{G}$ is Λ-monotone increasing if and only if there exists a continuous mapping $\Lambda : E \to F$ such that, for all $x, y \in K_1$ the inequality $x \leq y$ implies $\mathcal{G}(x) + \Lambda(x) \leq \mathcal{G}(y) + \Lambda(y)$. [Here " $\leq$" is the ordering defined on E by K_1 and on F by K_2].

Definition 6.3.3

We say that $\mathcal{F}$ and $\mathcal{F}$ are Λ-compatible on a subset $D \subseteq K_1$ if and only if, ($\mathcal{G}(x) \leq \mathcal{F}(x)$ and $x \in D$) implies that there exists $x_* \in D$ such that $x_* \leq x$ and $\mathcal{G}(x) + \Lambda(x) = \mathcal{F}(x_*) + \Lambda(x_*)$.

We have now the following general coincidence theorems.

Theorem 6.3.6

Suppose that $K_1 \subset E$ is a completely regular cone and $D \subset K_1$ is a bounded closed set. If the following assumptions are satisfied:

1°) $\mathcal{G}$ is a Λ-monotone increasing nonlinear mapping (possible noncontinuous),
2°) $\mathcal{F}$ is a continuous mapping (possible nonlinear),
3°) $\mathcal{F}$ and $\mathcal{G}$ are Λ-compatible on D,
4°) there exists $x_o \in D$ such that $\mathcal{G}(x_o) \leq \mathcal{F}(x_o)$,

then there exists $x_* \in D$ such that $\mathcal{F}(x_*) = \mathcal{G}(x_*)$.

Proof

From 3°) and 4°) it follows that there exists $x_1 \in D$ such that, $x_1 \leq x_o$ and $\mathcal{F}(x_1) + \Lambda(x_1) = \mathcal{G}(x_o) + \Lambda(x_o)$.

Then, using 1°) we obtain $\mathcal{G}(x_1) + \Lambda(x_1) \leq \mathcal{G}(x_o) + \Lambda(x_o) = \mathcal{F}(x_1) + \Lambda(x_1)$, that is $\mathcal{G}(x_1) \leq \mathcal{F}(x_1)$.

Hence, we can construct a net $\{x_\alpha\}$, where α is an ordinal number using the following operations.

a) If the ordinal number ω has a precedent, then we take $x_\omega \in D$ the element satisfying, $x_\omega \leq x_{\omega-1}$ and $\mathcal{F}(x_\omega) + \Lambda(x_\omega) = \mathcal{G}(x_{\omega-1}) + \Lambda(x_{\omega-1})$.

We have in this case,

$\mathcal{G}(x_\omega) + \Lambda(x_\omega) \leq \mathcal{G}(x_{\omega-1}) + \Lambda(x_{\omega-1}) = \mathcal{F}(x_\omega) + \Lambda(x_\omega)$, and hence $\mathcal{G}(x_\omega) \leq \mathcal{F}(x_\omega)$.

b) If the ordinal number ω has not a precedent, then in this case consider the decreasing net $\{x_\alpha\}_{\alpha<\omega}$.

Since $\{x_\alpha\}_{\alpha<\omega} \subset D$, we have that $\{x_\alpha\}_{\alpha<\omega}$ is bounded and since K_1 is completely regular, there exists $x_\omega = \lim_{\alpha<\omega} x_\alpha$.

Then, we have $x_\omega \in D$, $x_\omega \leq x_\alpha$; $\forall\alpha < \omega$, which implies $G(x_\omega) + \Lambda(x_\omega) \leq$ $\leq G(x_\alpha) + \Lambda(x_\alpha) \leq F(x_\alpha) + \Lambda(x_\alpha)$; $\forall\alpha < \omega$ and since F and Λ are continuous we obtain,

$$G(x_\omega) + \Lambda(x_\omega) \leq F(x_\omega) + \Lambda(x_\omega), \text{ or}$$

$$G(x_\omega) \leq F(x_\omega).$$

Then there exists an ordinal number ω^* such that $G(x_{\omega^*}) = F(x_{\omega^*})$, where $x_{\omega^*} \in D$.

The following result is more particular as the precedent result, but it is constructive.

Definition 6.3.4

We say that $f:E \to F$ is of positive type, if and only if $f(x) \leq f(y)$ implies $x \leq y$.

Theorem 6.3.7

Consider two continuous mappings $F, G:K_1 \to F$ and suppose the following assumptions satisfied:

1°) $K_1 \subset E$ is a closed regular cone,

2°) there exists a continuous mapping $\Lambda:E \to F$ such that $F + \Lambda$ is a mapping of positive type and $G + \Lambda$ is monotone increasing on K_1,

3°) $(G + \Lambda)(K_1) \subseteq (F + \Lambda)(K_1)$,

4°) there exists $x_o \in K_1$ such that $G(x_o) \leq F(x_o)$,

then there exists $x_* \in K_1$ such that $F(x_*) = G(x_*)$.

Proof

Consider the sequence $\{x_n\}_{n\in N}$ defined by the following process.

From 3°) there exists $x_1 \in K$ such that, $(G + \Lambda)(x_o) = (F + \Lambda)(x_1)$, and generally, for every $n \in N$ there exists $x_n \in K_1$ such that,

$$(G + \Lambda)(x_{n-1}) = (F + \Lambda)(x_n).$$

Hence, the sequence $\{x_n\}_{n\in N}$ is well defined.

Since, $(F + \Lambda)(x_1) = (G + \Lambda)(x_o) \leq (F + \Lambda)(x_o)$, and $F + \Lambda$ is a mapping of positive type, we have $0 \leq x_1 \leq x_o$.

If we suppose that

$$0 \leq x_{n-1} \leq x_{n-2} \leq \dots \leq x_1 \leq x_0$$

then we have,

$$(F + \Lambda)(x_n) - (F + \Lambda)(x_{n-1}) = (G + \Lambda)(x_{n-1}) - (G + \Lambda)(x_{n-2}) \leq 0,$$

which implies (using 2°), $0 \leq x_n \leq x_{n-1}$.

Then, for all $n \in N$, $0 \leq x_n \leq x_0$ and $\{x_n\}_{n\in N}$ is a decreasing sequence.

From assumption 1°) it follows that $\{x_n\}_{n\in N}$ is convergent and if we denote, $x_* = \lim_{n\to\infty} x_n$ we obtain by continuity that $F(x_*) = G(x_*)$. □

The next result is a variant for coincidence equations of Knaster - Kantorovich Birkhoff - Tartar theorem.

Theorem 6.3.8

Let $F: K \to E$ be a (G, Λ)-monotone increasing mapping where K is the positive cone of a complete vector lattice E.

If there exists $u_0 \in E$ such that,

1°) $G(0) \leq F(0)$,

2°) $G(u_0) \geq F(u_0)$,

then the set of all points u_* such that $F(u_*) = G(u_*)$ and $0 \leq u_* \leq u_0$ is nonempty and possesses a minimum and a maximum element.

Proof

From assumptions 1°) and 2°) we have $0 \leq (G + \Lambda)^{-1} (F + \Lambda) (0)$ and $(G + \Lambda)^{-1} (F + \Lambda) (u_0) \leq u_0$.

Consider the set,

$$\sum_1 = \{u \quad E \mid 0 \leq u \leq u_0 \text{ and } G(u) \leq F(u)\}.$$

From 1°) we have that $\sum_1 \neq \Phi$ and since E is a complete vector lattice, there exists $u_* = \sup \sum_1$.

For every $u \in \sum_1$ we have $u \leq u_*$ and hence,

$$u \leq (G + \Lambda)^{-1} (F + \Lambda) (u) \leq (G + \Lambda)^{-1} (F + \Lambda) (u_*),$$

which implies,

$$u_* \leq (G + \Lambda)^{-1} (F + \Lambda) (u_*) \text{ , or } G(u_*) \leq F(u_*).$$

We observe now that $0 \leq u_* \leq u_0$ and since F is (G, Λ)-monotone increasing we have, $0 \leq (G + \Lambda)^{-1} (F + \Lambda) (u_*) \leq u_0$.

Then, if we denote $z = (G + \Lambda)^{-1} (F + \Lambda) (u_*)$, we can prove that $z \in \sum_1$.

Indeed, since F is (G, Λ)-monotone increasing we have,

$$(G + \Lambda) (z) = (F + \Lambda) (u_*) \leq (F + \Lambda) (z).$$

But, since $(G + \Lambda)^{-1} (F + \Lambda) (u_*) \in \sum_1$ we have, $(G + \Lambda)^{-1} (F + \Lambda) (u_*) \leq u_*$, or $F(u_*) \leq G(u_*)$.

Consequently, we obtain $G(u_*) = F(u_*)$.

Moreover, since any point u satisfying $F(u) = G(u)$ and $0 \leq u \leq u_0$ belongs to $\sum_1$ it follows from the definition of u_* that u_* is indeed the maximum of all such coincidence points of F and G.

The existence of the minimum coincidence points of F and G on $[0, u_0]$ is shown along the same lines but using the set $\sum_2 = \{u \in E \mid 0 \leq u \leq u_0, G(u) \geq F(u)\}$.

The next result is a generalization of Banach's theorem and it can be used to study some coincidence equations associated to a complementarity problem.

Let $(E, \| \ \|)$ be a Banach space and consider $D \subset E$ as closed subset.

We recall that $G:D \to E$ is a <u>proper mapping</u> if the inverse image of any compact subset of $G(D)$ under G is compact.

<u>Remark 6.3.3</u>

A very important class of proper mappings for practical problems is the class of the mappings of the form $G = L + C$, where L is a linear Fredholm operator, C is a completely continuous nonlinear operator and E is a Hilbert space.

<u>Theorem 6.3.9</u>

<u>Let $(E, \| \ \|)$ be a Banach space and consider $D \subset E$ a closed nonempty subset.</u>

<u>Assume that the continuous mappings $F, G:D \to E$ satisfy the following assumptions:</u>

1°) <u>G is a proper mapping,</u>

2°) <u>$F(D) \subseteq G(D)$,</u>

3°) <u>there exists a constant $0 < \rho < 1$ such that $\| F(x) - F(y) \| \leq \rho \| G(x) - G(y) \|$, for all $x, y \in D$,</u>

<u>then there exists $x_* \in D$ such that $F(x_*) = G(x_*)$.</u>

<u>Moreover, if F or G is one to one then the coincidence point x_* is unique.</u>

<u>Proof</u>

From 2°) we have that there exists a sequence $\{x_n\}_{n \in N}$ defined by $G(x_{n+1}) = F(x_n)$

$\forall n \in N$, where x_o is an arbitrary point of D.

By recurence we have,

$$\| G(x_n) - G(x_{n+1}) \| = \| F(x_{n-1}) - F(x_n) \| \leq \rho^n \| G(x_o) - G(x_1) \|,$$

for all $n \in N$, which implies,

$$\| G(x_n) - G(x_{n+p}) \| \leq \frac{\rho^n}{1-\rho} \| G(x_o) - (x_1) \| ; \forall m, p \in N,$$

and hence, $\{ G(x_n)\}_{n\in N}$ is a Cauchy sequence in E.

Since E is complete $\{ G(x_n)\}_{n\in N}$ is convergent and because is a proper mapping the sequence $\{x_n\}_{n\in N}$ has a convergent subsequence $\{x_{n_k}\}_{k\in N}$.

If $x_* = \lim_{k\to\infty} x_{n_k}$ then from continuity and the definition of $\{x_n\}_{n\in N}$ we have, $F(x_*) = G(x_*)$.

If x_* and x_{**} are two elements of D such that $F(x_*) = G(x_*)$ and $F(x_{**}) = G(x_{**})$ then, from the following relations:

$$\| F(x_*) - F(x_{**}) \| \leq \rho \| G(x_*) - G(x_{**}) \| = \rho \| F(x_*) - F(x_{**})\| ,$$

$$\| G(x_*) - G(x_{**}) \| = \| F(x_*) - F(x_{**}) \| \leq \rho \| G(x_*) - G(x_{**}) \|$$

and the fact that F or G is one to one we obtain that $x_* = x_{**}$ and the theorem is proved.

We will finish this chapter with some considerations about the multivalued implicit complementarity problem.

Let $\langle E, E^*\rangle$ be a dual system of locally convex spaces. Consider given the following mappings:

$m : E \to E$ a point-to-point mapping,

$f : E \to E^*$ a point-to-set mapping.

$T : E \to E$ a cone mapping, that is, for every $x \in E$, $T(x)$ is a closed pointed convex cone.

Consider the multivalued mapping

$$K(x) = m(x) + T(x) = \{u \in E \mid u = m(x) + y,\ y \in T(x)\}$$

The multivalued implicit complementarity problem associated to T, m and f is:

$$\text{M.I.C.P. } (T,m,f): \left\| \begin{array}{l} \text{find } x \in m(x) + T(x) \text{ and} \\ y \in E^* \text{ such that} \\ y \in f(x) \cap [T(x)]^* \text{ and} \\ \langle x - m(x), y \rangle = 0 \end{array} \right.$$

We consider also the multivalued quasivariational inequality associated to K and f, that is, the problem:

M.Q.V.I. (K,f): find $x \in K(x)$ and $y \in f(x)$ such that $< u - x, y > \geq 0$, for all $u \in K(x)$

Proposition 6.3.4

The solution sets of the M.I.C.P. (T,m,f) and the M.Q.V.I.(K,f) are equal.

Proof

Suppose that (x,y) solves M.I.C.P. (T,m,f) and let $u \in K(x) = m(x) + T(x)$.

If $u = m(x) + z$ with $z \in T(x)$, then since $< x - m(x), y > = 0$ and $y \in [T(x)]^*$ it follows that $< u - x, y > = < m(x) + z - x, y > = < z, y > \geq 0$, that is (x, y) solves M.Q.V.I (K,f).

Conversely, let (x, y) solve M.Q.V.I. (K,f).

Then $x - m(x) \in T(x)$ which implies $2(x - m(x)) \in T(x)$ because T(x) is a cone.

Hence $u = m(x) + 2(x - m(x)) = 2x - m(x) \in K(x)$.

Consequently, we have

(6): $\quad 0 \leq < u - x, y > = < x - m(x), y >$.

Moreover, since 0 T(x) we have that $m(x) \in K(x)$ and hence,

(7): $\quad 0 \leq < m(x) - x, y >$.

But from (6) and (7) we deduce $< x - m(x), y > = 0$.

Finally, let $z \in T(x)$ be an arbitrary element. We have $m(x) + z \in K(x)$ which implies $0 \leq < m(x) + z - x, y > = < z, y >$, that is $y \in [T(x)]^*$ and so (x, y) solves the problem M.I.C.P. (T,m,f). □

Remark 6.3.4

The last proposition is important since it establishes a relation between the problem M.I.C.P. and the theory of quasivariational inequalities.

We remark also that in [A242] we find some existence theorems for the problem M.I.C.P. (T,m,f) in R^n but we note that this problem is not studied in an infinite dimensional space.

Comments

Theorem 6.1.2 was proved by G. ISAC [C26] and Theorem 6.1.3 by M.A. Noor [C77].

The all results presented in sections 6.2 and 6.3 were proved by G. ISAC exception, Proposition 6.3.4 which was obtained by Pang [A242].

CHAPTER 7
ISOTONE PROJECTION CONES AND COMPLEMENTARITY

In this chapter we introduce the concept of isotone projection cone in a Hilbert space and we apply this notion to the study of the Complementarity Problem.

7.1. Isotone projection cones.

Let $(H, \langle\, ,\rangle)$ be a Hilbert space and let $K \subset H$ be a closed pointed convex cone.

We denote by "$\leq$" the ordering defined by K, that is, "$x \leq y$" $\Longleftrightarrow$ "$y - x \in K$".

We recall that the dual of K is $K^* = \{y \in H \mid \langle x, y\rangle \geq 0;\ \forall x \in K\}$ and its polar is $K^o = \{y \in H \mid \langle x, y \rangle \leq 0\ ;\ \forall x \in K\}$.

It is known that $K^* = - K^o$ and K is closed if and only if $K = (K^o)^o$.

When the closed convex cone $K \subset H$ is well defined we say that $(H, \langle\, ,\rangle, K)$ is an ordered Hilbert space.

The properties of H as ordered vector space are strongly dependent of the following situations:

1°) $K \subset K^*$; 2°) $K = K^*$; 3°) $K \subset K^*$.

Definition 7.1.1.

A convex cone $K \subset H$ is said to be:

i) sub-adjoint, if $K \subset K^*$,

ii) self-adjoint, if $K = K^*$,

iii) super-adjoint, if $K \supset K^*$,

An ordered Hilbert space $(H, \langle\, ,\rangle, K)$ is called a vector lattice if the upper and lower bounds exist for any two elements $x, y \in H$. Let them be denoted by $x \vee y$ and $x \wedge y$ (as in chapter 5).

In this case we can define the absolute value for every $x \in H$ by

$|x| = x \vee (-x)$.

We say that the norm $\|\cdot\|$ of H is absolute if for every $x \in H$ we have $\| |x| \| = \| x \|$.

Definition 7.1.2

We say that an ordered Hilbert space $(H, \langle\, ,\rangle, K)$ is a Hilbert lattice if:

1°) H is a vector lattice,

2°) the norm $\|\cdot\|$ of H is absolute,

3°) $(\forall x, y \in H)(0 \leq x \leq y$ implies $\| x \| \leq \| y \|)$.

Remark 7.1.1.

We can show that $(H, <,>, K)$ is a Hilbert lattice, if and only if,

i) H is a vector lattice,

ii) $(\forall x, y \in H)(\, |x| \le |y|$ implies $\| x \| \le \| y \|.)$

Examples.

1°) If $H = R^n$, $< x, y > = \sum_{i=1}^{n} x_i y_i$, $x = (x_i)$, $y = (y_i)$ and $K = R^n_+$ then $(H, <,>, K)$ is a Hilbert lattice.

2°) If $H = L^2(\Omega, \mu)$, $K = \{f \mid f \geqslant 0 \text{ a.e.}\}$ where (Ω, μ) is a measure space with a positive Borel measure, is a Hilbert lattice.

3°) Let H be a Hilbert space and $\{e_i\}_{i=1}^{n}$ an orthonormal basis in it. Denote, $K = \{x \in H \mid < e_i, x > \le 0 \, ; \forall i \in N\}$.

We remark that K is a closed pointed convex cone.

For $x = \sum_{i=1}^{\infty} \alpha_i e_i$ and $y = \sum_{i=1}^{\infty} \beta_i e_i$ we can define the lattice operations by,

$$x \wedge y = \sum_{i=1}^{\infty} (\max\{\alpha_i, \beta_i\}) e_i$$

$$x \vee y = \sum_{i=1}^{\infty} (\min\{\alpha_i, \beta_i\}) e_i$$

We have, $|x| = \sum_{i=1}^{\infty} - |\alpha_i| e_i$, which implies $\| \, |x| \, \|^2 = \sum_{i=1}^{\infty} \alpha_i^2 = \| x \|^2$, that is, the norm $\| \cdot \|$ of H is absolute.

Further, the relations $o \le x \le y$ are equivalent to $\alpha_i \le 0$ and $\beta_i - \alpha_i \le 0$, for every $i \in N$, that is to $\beta_i \le \alpha_i \le 0$, for every $i \in N$, which implies $\sum_{i=1}^{\infty} \alpha_i^2 \le \sum_{i=1} \beta_i^2$ and hence $\| x \| \le \| y \|$.

This shows that $(H, <,>, K)$ is a Hilbert lattice.

Remark 7.1.2

1°) If $(H, <,>, K)$ is a Hilbert lattice then K is a normal cone.

2°) If $(H, <,>, K)$ is a Hilbert lattice then there exists a neighborhood basis of zero for the norm topology consisting of solid sets. We recall that a subset A of the vector lattice H is solid if $y \in A$ whenever $x \in A$ and $|y| \le |x|$.

3°) If $(H, <,>, K)$ is a Hilbert lattice, then the following mappings are continuous:

$$(x, y) \to x \wedge y \,;\; (x, y) \to x \vee y \,;\; x \to x^+ ;\; x \to x^- \text{ and } x \to |x| .$$

4°) If $(H, <,>, K)$ is a Hilbert lattice then $K = K^*$.

Let $K \subset H$ be a pointed closed convex cone. We denote by P_K the projection onto K

Definition 7.1.3

We say that K is isotone projection, if and only if, for every $x, y \in H$, $y - x \in K$ implies that $P_K(y) - P_K(x) \in K$.

We say that two closed convex cones $K, Q \subset H$ are mutually polar if $K = Q^o$ (which implies $K^o = Q$).

The following result has many interesting applications in the theory of ordered Hilbert spaces and also in mechanics.

Theorem 7.1.1 [Moreau][C87]

If K and Q are two mutually polar convex cones in H and $x, y, z \in H$ then the following statements are equivalent:

i) $z = x + y$, $x \in K$, $y \in Q$ and $\langle x, y \rangle = 0$,

ii) $x = P_K(z)$ and $y = P_Q(z)$.

Proposition 7.1.1

The closed convex cone $K \subset H$ is isotone projection, if and only if, $P_{K^o}(y) - P_{K^o}(x) \leq y - x$, for every $x, y \in H$ such that $x \leq y$ (where "$\leq$" is the ordering defined by K).

Proof

From Moreau's decomposition theorem [Theorem 7.1.1] we have for every $x, y \in H$ such that $x \leq y$, $x = P_K(x) + P_{K^o}(x)$; $y = P_K(y) + P_{K^o}(y)$, which imply $P_K(x) = x - P_{K^o}(x)$ and $P_K(y) = y - P_{K^o}(y)$. Hence, $P_K(x) \leq P_K(y)$ if and only if $P_{K^o}(y) - P_{K^o}(x) \leq y - x$. □

Corollary

If $(H, \langle , \rangle, K)$ is an ordered Hilbert space which is a vector lattice and for every $x, y \in H$ we have $\| P_{K^o}(y) - P_{K^o}(x) \| \leq \| y - x \|$, then K is isotone projection

Theorem 7.1.2

If K is isotone projection then it is sub-adjoint.

Proof

We must show that $K \subset K^*$.

Indeed, from Moreau's decomposition theorem we have $P_K^{-1}(0) = - K^*$ $(= K^o)$.

If we suppose that $x \in K$, then $- x \leq 0$ and since P_K is isotone it follows that $P_K(- x) \leq P_K(o) = 0$, that is, $P_K(- x) = 0$.

Accordingly, $- x \in P_K^{-1}(o) = - K^*$, that is $x \in K^*$. □

Remark 7.1.3

If H is the euclidean space $(R^n, < , >)$ and $n = 2$ then we can show that $K \subset R^2$ is isotone projection if and only if $K \subset K^*$. This fact is not true if $n > 2$.

Example [C84]

Consider $H = R^3$ with euclidean structure and let K the closed convex cone generated by the vectors (1, 0, 0), (0, 1, 0) and (1, 1, 1).

We can prove that $K = \{(x_1, x_2, x_3) \in R^3 \mid o \leq x_3 \leq \min(x_1, x_2)\}$ and we observe that K is minihedral and $K \subset K^*$.

Consider $u = (3, 3, 3) \in K$ and $v = (2, 1, 2)$.

We observe that $u - v = (1, 2, 1) \in K$ and we can show that $P_K(u) - P_K(v) \notin K$. □

Theorem 7.1.3

If $K \subset H$ is isotone projection then it is normal and regular.

Proof

Suppose $o \leq x \leq y$. Then $y - x \in K$ and since K is sub-adjoint we have $y - x \in K^*$ which implies, $\langle y - x, y \rangle \geq 0$ and $\langle y - x, x \rangle \geq 0$ and finally, $\| x \|^2 \leq \langle x,y \rangle \leq \| y \|^2$, which has as consequence the fact that K is normal. □

Let $K \subset H$ be a closed convex cone.

Definition 7.1.4

We say that a subset F of K is a face of K if:

1°) F is a convex cone,

2°) $x \in K$, $y \in F$ and $y - x \in K$ imply $x \in F$.

Definition 7.1.5

The cone $K \subset H$ is called correct if for each face $F \subset K$ we have $P_{\overline{sp}(F)}(K) \subseteq K$, where $\overline{sp}(F)$ is the closed linear span of F.

In our paper [C82] we proved the following very nice geometrical property of isotone projection cone.

We give this result without proof since the prodof is very long [C82].

Theorem 7.1.4

If K is a generating isotone projection cone in H then it is lattícial and correct.

Proof

The proof is given in [C82] □

Theorem 7.1.5

Let $(H, <,>)$ be a Hilbert space and let $K \subset H$ be a polyhedral cone defined by a_i; $i = 1, 2, \ldots, n$, that is,

$K = \{x \in H \mid < a_i, x > \leq 0 ; \forall i = 1, 2, \ldots, n\}$.

If $< a_i, a_j > = 0$ for all $i \neq j$ then K is isotone projection.

Proof

Since for each i we suppose that $a_i \neq 0$ we can consider that $\| a_i \| = 1$; $i = 1, 2, \ldots, n$.

Obviously, K is a closed convex cone.

If $x \in H$ and $x_p = P_K(x)$ then we have,

(1): $$\| x - x_p \|^2 = \min_z \{\| x - z \|^2 \mid < a_i, z > \leq 0 ; 1, 2, \ldots, n\}.$$

Since $h(x) = \| x \|^2$ is convex and differentiable, the conditions Kuhn - Tucker for problem (1) are satisfied and hence there exist $\lambda_1, \lambda_2, \ldots, \lambda_n \geq 0$ such that,

(2): $$\left\| \begin{array}{l} x - x_p = \sum_{i=1}^{n} \lambda_i a_i \\ \lambda_i < a_i, x_p > = 0 ; i = 1, 2, \ldots, n \end{array} \right.$$

If $y \in H$ and $y_p = P_K(y)$ then we have also that there exist $\mu_1, \mu_2, \ldots, \mu_n \geq 0$ such that,

$$(3): \quad \left\| \begin{array}{l} y - y_p = \sum_{i=1}^{n} \mu_i a_i \\ \mu_i < a_i , y_p > = 0 \; ; \; i = 1, 2, \ldots, n. \end{array} \right.$$

We suppose now that $y - x \in K$ and we will show that, $y_p - x_p \in K$ that is,

$$(4): \quad < a_i , y_p - x_p > \leq 0 \; ; \; i = 1, 2, \ldots, n.$$

Indeed, for an arbitrary $i \in \{1, 2, \ldots, n\}$ we have the following situations.

(a): < If $a_i, x_p > = 0$ then $< a_i , y_p - x_p > = < a_i , y_p > \leq 0$.

(b): If $< a_i , x_p > < 0$ then we have $\lambda_i = 0$, (from (2)), which implies, $< a_i , x - x_p > = 0$, (since $< a_i , a_j > = 0 \; ; \; \forall i \neq j$).

Moreover, $< a_i , y - y_p > = \mu_i$ (from (3)) and hence we have,

$<a_i, y_p - x_p > = < a_i, y_p - y > + < a_i, y - x > + < a_i, x - x_p > =$

$= - \mu_i + < a_i, y - x > + 0 \leq 0$, and finally we deduce $<a_i, y_p - x_p > \leq 0$, for all $i = 1, 2, \ldots, n$. □

Theorem 7.1.6

If $(H, <, >, K)$ is a Hilbert lattice, then K is isotone projection and moreover, $P_K(x) = x^+$, for all $x \in H$.

Proof

First we show that if $(H, <, >)$ is a Hilbert lattice then:

i) $(\forall x, y \in K) \; (< x, y > \geq 0)$,

ii) if $x, y \in K$ and $x \wedge y = 0$ then $< x, y > = 0$.

Indeed, for every $x, y \in K$ we have, $|x - y| = (x - y)^+ + (x - y)^- =$ $= 0 \vee (x - y) + 0 \vee (y - x) \leq x + y$, which implies (since H is a Hilbert lattice),

$$(5): \quad \| x - y \| \leq \| x + y \|.$$

On the other hand we have, $\| x + y \|^2 - \| x - y \|^2 = 4 < x,y >$, and using formula (5) we obtain $< x, y > \geq 0$, that is, statement (i).

Suppose now that $x, y \in K$ and $x \wedge y = 0$.

Since in a vector lattice we have,

$$(6): \quad x \perp y \iff \big| |x| - |y| \big| = |x| + |y|$$

we obtain $|x - y| = x + y$ and since H is a Hilbert lattice we deduce,

$\| x + y \| = \| x - y \|$, which implies, $< x,y > = 1/4[\| x + y \|^2 - \| x - y \|^2] = 0$, that is, statement (ii) is true.

Let $x \in H$ be an arbitrary element.

We have $x = x^+ - x^-$, $x^+ \wedge x^- = 0$ and hence (using iii)), $< x^+, x^- > = 0$ or

$<x^+, -x^-> = 0$.

Since $x^- \in K$ we get for every $k \in K$, $<-x^-, k> = -<x^-, k> \leq 0$ (using i)), which gives $-x^- \in K^o$.

Now, using Moreau's decomposition Theorem for x, K and K^o we obtain that $x^+ = P_K(x)$ (since $x = x^+ + (-x^-)$ and $<x^+, -x^-> = 0$).

Because $y - x \in K$ implies $y^+ - x^+ \in K$ we deduce that P_K is isotone with respect to the order defined by K, that is K is isotone projection.

Remark 7.1.4

The fact that for a Hilbert lattice $(H, <,>, K)$ we have $P_K(x) = x^+$ for every $x \in H$, is important in optimization and numerical analysis.

Theorem 7.1.7

Let $(H, <,>, K)$ be an ordered Hilbert space.
If, for every subspace $L \subset H$ such that $\dim L \leq 4$ the convex cone $K_L = K \cap L$ is isotone projection in L, then K is isotone projection in H.

Proof

Let $x, y \in H$ such that $y - x \in K$.
Then there exists $k_o \in K$ such that $y = x + k_o$
We denote, $z_1 = P_K(x)$, $z_2 = P_K(y)$ and let L be the subspace generated in H be $\{x, k_o, z_1, z_2\}$.

We observe that $y \in L$. If we put $K_L = K \cap L$ we can prove that $P_{K_L}(x) = P_K(x)$ and $P_{K_L}(y) = P_K(y)$.

But since we have $P_{K_L}(y) - P_{K_L}(x) \in K_L$ we obtain $P_K(y) - P_K(x) \in K$ (since $K_L \subset K$). □

The next resuslt gives a complete characterization of isotone projection cones in the euclidean space $(R^n, <,>)$.

Theorem 7.1.8

Let K be a closed generating cone in R^n. Then the following assertions are equivalent:

(i): K is isotone projection,

(ii): K is correct and latticial,

(iii): K is polyhedral and correct,

(iv): there exists a set of vectors $\{u_i \mid i \in I\}$ with the property that $\langle u_i, u_j \rangle \leq 0$, for all $i, j \in I$, $i \neq j$ and such that $K = (\{u_i \mid i \in I\})^o$,

(v): K is latticial and $P_K(x) \leq x^+$ for every $x \in R^n$, where $x^+ = \sup(o, x)$.

Proof

The proof of this result is very long and the reader finds this proof in our paper [C83]. □

Remark 7.1.5

We do not know if in an infinite dimensional Hilbert space a closed convex cone is isotone projection if and only if it is correct and latticial.

7.2. Isotone projection cones and the Complementarity Problem

In this section we will study the Complementarity Problem in a Hilbert space ordered by an isotone projection cone.

Let $(H, \langle , \rangle)$ be a Hilbert space and let $K \subset H$ be a closed convex cone.

Given a mapping $h: K \to H$ we consider the following complementarity problem associated to h and K,

C.P. (h, K): find $x_o \in K$ such that $h(x_o) \in K^*$ and $\langle x_o, h(x_o) \rangle = 0$.

Proposition 7.2.1

Let $(H, \langle , \rangle)$ be a Hilbert space $K \subset H$ a closed convex cone and $h: K \to H$ a mapping.

The following statements are equivalent:

1°) the problem C.P.(h, K) has a solution,

2°) the mapping $\Psi(x) = P_K(x - h(x))$, defined for every $x \in K$, has a fixed point in K

3°) the mapping $\Phi(x) = P_K(x) - h(P_K(x))$, defined for every $x \in H$, has a fixed point in H.

Proof

1°) <==> 2°). This equivalence is a direct consequence of Proposition 6.1.1.

2°) ==> 3°). We suppose that $x_* \in K$ is a solution of the problem C.P.(h, K) and we denote, $x_o = x_* - h(x_*)$.

By Theorem 7.1.1 (since x_* is a solution of the problem C.P.(h, K)) we deduce that $P_K(x_o) = x_*$ and finally, $\Phi(x_o) = P_K(x_o) - h(P_K(x_o)) = x_* - h(x_*) = x_o$, that is x_o is a fixed point for Φ.

3°) => 1°). If x_o is a fixed point of Φ then $x_o = P_K(x_o) - h(P_K(x_o))$.

We denote, $x_* = P_K(x_o)$ and we have $x_* \in K$, $x_o = x_* - h(x_*)$ and finally $x_* - x_o = h(x_*)$.

By Proposition 6.1.1 we have that $h(x_*) \in K^*$ and $< x_*, h(x_*) > = 0$, that is, x_* is a solution of the problem C.P.(h, K). □

Corollary

Supposing the same assumptions as in Proposition 7.2.1 and $h(x) = x - f(x)$, where $f:K \to H$, then the following statements are equivalent:

1°) the problem C.P.(h, K) has a solution,

2°) the mapping $\Psi(x) = P_K(f(x))$, defined for all $x \in K$, has a fixed point in K,

3°) the mapping $\Phi(x) = f(P_K(x))$, defined for all $x \in H$, has a fixed point in H.

Remark 7.2.1

Proposition 7.2.1 implies that we can solve the problem C.P.(h, K) if we are able to find a fixed point for the mapping Φ or for the mapping Ψ, but this problem is not so simple because many known fixed point theorems are not applicable in this case.

In this sense we present now some results.

First, we will use our fixed point theorem, Theorem 5.2.8.

Theorem 7.2.1

Let $(H, <, >)$ be a Hilbert space ordered by a closed convex cone $K \subset H$.

If $f:K \to H$ is monotone increasing with respect to the ordering defined by K (not necessary continuous), K is isotone projection and there exists $x_o \in K$ such that $P_K(f(x_o)) \leq x_o$, then the problem C.P.(h, K) has a solution (here $h(x) = x - f(x)$, $\forall\, x \in K$).

Proof

Since K is isotone projection it is normal and regular (Theorem 7.1.3).

The operator $T(x) = P_K(f(x))$ is monotone increasing and (sm)-compact with respect to $A = [0,x] = \{x \in K \mid 0 \leq x \leq x_o\}$.

The proof is finished if we apply Theorem 5.2.8 and the Corollay of Proposition 7.2.1. □

If $(H, \langle , \rangle)$ is a Hilbert space, $K \subset H$ a closed convex cone and $f:K \to H$ a mapping we denote,

$$F = \{x \in K \mid f(x) \le_K x\}$$

$$F_* = \{x \in K \mid f(x) \le_{K^*} x\}$$

$$N = \{x \in K \mid P_K(f(x)) = x\}.$$

We recall that F_* is the feasible set associated to C.P.(h, K), where $h(x) = x - f(x)$.

Theorem 7.2.2

Let $(H, \langle , \rangle)$ be a Hilbert space ordered by an isotone projection cone $K \subset H$ and let $f:K \to H$ be a continuous and monotone increasing mapping.

If we consider the statements:

1°) F is nonempty,

2°) F_* is nonempty,

3°) N is nonempty [which is equivalent to the fact that C.P.(h, K) has a solution, where $h(x) = x - f(x)$],

4°) the sequence $\{x_n\}_{n \in N}$ defined by,

$$(\theta : \left\| \begin{array}{l} x_0 = 0 \\ x_{n+1} = P_K(f(x_n)) \ ; \ \forall n \in N, \end{array} \right.$$

is convergent and if $x_* = \lim_{n \to \infty} x_n$ then $x_* \in N \subset F_*$ and x_* is the least element of F,

then we have 1°) => 4°) => 3°) => 1°).

If in addition we have that $F_* \subset F$ then we have 1°) <=>2°) <=>3°) <=>4°).

Proof

1° => 4°). If we suppose that F is nonempty let $y \in F$ be an arbitrary element.

In this case we have, $f(y) \le y$ which implies that $P_K(f(y)) \le y$ and from the definition of the sequence $\{x_n\}_{n \in N}$ we deduce,

$$0 \le x_1 \le x_2 \le \dots \le x_n \le \dots \le y.$$

Since K is regular, there exists $x_* = \lim_{n \to \infty} x_n$ and from continuity and using (θ) we obtain, $x_* = P_K(f(x_*))$, that is $x_* \in N$ and from Corollay of Proposition 7.2.1 we have that the problem C.P.(h, K) has a solution which is exactly x_* and so, we have $N \subset F_*$.

Finally, we observe that $x_* \leq y$ for every $y \in F$ and hence x_* is the least element of F.

4° => 3°). Indeed, if the sequence $\{x_n\}_{n \in N}$ defined by (θ) is convergent and $x_* = \lim_{n\to\infty} x_n$, then from continuity and the construction of $\{x_n\}_{n\in N}$ we have, $x_* = P_K(f(x_*))$, that is, N is nonempty.

3°) => 2°). If N is nonempty then Corollary of Proposition 7.2.1 implies that the problem C.P.(h, K) has a solution and hence F_* is nonempty.

Obviously, if $F_* \subseteq F$ then we have that 2°) => 1°) and the proof is finished. □

Proposition 7.2.2

Let $(H, <,>)$ be a Hilbert space ordered by a closed isotone projection cone $K \subset H$ and let $f:K \to H$ be a continuous monotone increasing map.

If there exist $x_o, y_o \in H$ such that $x_o \leq y_o$, $x_o \leq f(P_K(x_o))$ and $f(P_K(y_o)) \leq y_o$ then the problem C.P.(h, K) has a solution.

Proof

Indeed, the sequence $\{x_n\}_{n\in N}$ defined by $x_{n+1} = f(P_K(x_n))$, for all $n \in N$ is monotone increasing and for all $n \in N$ we have $x_o \leq x_n \leq y_o$. Using the fact that K is isotone projection we have that K is regular, which implies that $\{x_n\}_{n\in N}$ is convergent.

If $x_* = \lim_{n\to\infty} x_n$, by continuity we obtain $f(P_K(x_*)) = x_*$ and we apply the corollary of Proposition 7.2.1. □

Let $(H, <,>)$ be a Hilbert space.

If A is a subset of H we denote by $\alpha(A)$the measure of noncompactness of A, that is,

$\alpha(A) = \inf \{r > 0 \mid$ A can be covered by a finite family of subsets of H whose diameters $< r\}$.

Let D be a subset of H and $f:D \to H$ a continuous mapping.

We say that f is condensing if for every non-compact bounded set $A \subset D$ we have $\alpha(f(A)) < \alpha(A)$.

The following classical fixed point theorems will be cited in this section.

Theorem 7.2.3 [Browder] [C79]

Let $(E, \|\ \|)$ be an uniformly convex Banach space and let $D \subset E$ be a bounded closed convex subset.

If $T:D \to D$ is a non-expansive mapping, then T has a fixed point.

Theorem 7.2.4 [Sadovski} [C89]

Let $(E, \| \ \|)$ be a Banach space and let $D \subset E$ be a closed bounded convex subset. If $T:D \to D$ is a condensing mapping, then T has a fixed point.

In the next results are defined some interesting iterative methods for the general complementarity problem.

Theorem 7.2.5.

Let $(H, \langle , \rangle)$ be a Hilbert space ordered by an isotone projection cone K. Let $f:K \to H$ be a mapping. We suppose that f has a decomposition of the form:
$f(x) = f_1(x) + f_2(x) + d, \ \forall x \in K,$
where f_1 is monotone decreasing, f_2 is monotone increasing and $d \in H$.

Given $x_0, y_0 \in H$ with $x_0 \leq y_0$ consider the sequences $\{x_n\}_{n\in N}$, $\{y_n\}_{n\in N}$ defined by

$$x_{n+1} = P_K(x_n) - f_1(P_K(x_n)) - f_2(P_K(y_n)) - d$$
$$y_{n+1} = P_K(y_n) - f_1(P_K(y_n)) - f_2(P_K(x_n)) - d$$
$$n = 0, 1, 2, \ldots$$

If the following assumptions are satisfied:

i°). $x_0 \leq x_1$ and $y_1 \leq y_0$,

ii°). denoting $\Phi(x) = P_K(x) - f_1(P_K(x)) - f_2(P_K(x)) - d$ one of the following assumptions are satisfied:

a) Φ is nonexpansive,

b) Φ is condensing,

c) Φ is continuous and $\dim H < +\infty$

then there exists $\hat{x}$ such that:

1°). $x_n \leq \hat{x} \leq y_n$, for every $n \in N$,

2°). $x_* = P_K(\hat{x})$ is a solution of the problem G.C.P.(f, K),

3°). $\| x_* - P_K(x_n) \| \leq \| y_n - x_n \|$, for every $n \in N$,

4°). if $\lim_{n\to\infty} \| y_n - x_n \| = 0$ then $\lim_{n\to\infty} P_K(x_n) = x_*$, or $P_K(u) = x_*$, where $\lim_{n\to\infty} x_n = u$.

Proof

By Proposition 7.2.1 we obtain the existence of the element $\hat{x}$, if we show that Φ has a fixed point satisfying 1°).

First, we prove the following relation:

(5): $(\forall n \in N)(x_n \leq x_{n+1} \leq y_{n+1} \leq y_n)$.

Indeed, by assumption i°) we have for n = 0,

$x_o \le x_1 \le y_1 \le y_o$.

Supposing (5) true for n we obtain (using the fact that P_K is isotone),

$x_{n+1} \le x_{n+2}$; $y_{n+2} \le y_{n+1}$ and

$x_{n+2} = P_K(x_{n+1}) - f_1(P_K(x_{n+1}) - f_2(P_K(y_{n+1})) - d \le P_K(y_{n+1}) - f_1(P_K(y_{n+1})) -$

$- f_2(P_K(x_{n+1}) - d = y_{n+2}$.

Hence, we have

$x_o \le x_1 \le \dots \le x_n \dots \le y_n \le \dots \le y_1 \le y_0$.

We show now that,

(6): $\Phi([x_n, y_n]) \subseteq [x_n, y_n]$, for every $n \in N$.

Indeed, let $x \in [x_n, y_n]$ be an arbitrary element.

Since f_2 and P_K are increasing and f_1 is decreasing we have,

$x_{n+1} = P_K(x_n) - f_1(P_K(x_n)) - f_2(P_K(y_n)) - d \le$

$\le P_K(x) - f_1(P_K(x)) - f_2(P_K(x)) - d \le$

$\le P_K(y_n) - f_1(P_K(y_n)) - f_2(P_K(x_n)) - d = y_{n+1}$,

that is $\Phi([x_n, y_n]) \subseteq [x_{n+1}, y_{n+1}] \subseteq [x_n, y_n]$, for every $n \in N$.

Since K is regular there exist $u = \lim_{n\to\infty} x_n$, $v = \lim_{n\to\infty} y_n$ and we have $u \le v$.

Moreover, we can show that

(7): $\Phi([u, v]) \subseteq [u, v]$.

Indeed, for every $n \in N$ and $x \in [u, v]$ we have,

$x_{n+1} = P_K(x_n) - f_1(P_K(x_n)) - f_2(P_K(y_n)) - d \le$

$\le \Phi(x) \quad P_K(y_n) - f_1(P_K(y_n)) - f_2(P_K(x_n)) - d \le$

$\le y_{n+1}$ and computing the limit we deduce, $u \le \Phi(x) \le v$.

The set $[u, v] = \{x \in H \mid u \le x \le v\}$ is convex. Since K is normal and closed we have that [u, v] is closed and bounded.

From assumption ii°) and using Browder's Theorem or Sadovski's Theorem or Brouwer's Theorem we obtain that Φ has a fixed point in [u, v] and conclusion 1°) is proved.

Conclusion 2°) is a consequence of Proposition 7.2.1.

To prove conclusion 3°) we use the fact that P_K is isotone and we have (from 1°) and 2°)),

$P_K(x_n) \le x_* \le P_K(y_n)$,

which implies,

(8): $0 \le x_* - P_K(x_n) \le P_K(y_n) - P_K(x_n)$.

Since K is normal with the property that the normality constant $\beta = 1$, we deduce (from (8)),

$$\| x_* - P_K(x_n) \| \le \| P_K(y_n) - P_K(x_n) \| \le \| y_n - x_n \|,$$

for every $n \in N$.

Now, from 3°) and the continuity of P_K we have 4°) and the proof is finished. □

We have a similar method for the Implicit Complementarity Problem.

If $f, h: K \to H$ are given we consider the following implicit complementarity problem:

I.C.P. (f, I-h, K): $\left\| \begin{array}{l} \text{find } x_o \in K \text{ such that} \\ f(x_o) \in K,\ x_o - h(x_o) \in K^* \\ \text{and } \langle f(x_o), x_o - h(x_o) \rangle = 0. \end{array} \right.$

Theorem 7.2.6.

Let $(H, \langle , \rangle)$ be a Hilbert space ordered by an isotone projection cone $K \subset H$. Suppose given two continuous monotone decreasing mappings $f, h: K \to H$. Given $x_o, y_o \in K$ with $x_o \le y_o$, consider the sequences $\{x_n\}_{n\in N}$, $\{y_n\}_{n\in N}$ defined defined by:

$$\begin{cases} x_{n+1} = P_K[x_n - h(x_n) - f(x_n)] + h(y_n), \\ y_{n+1} = P_K[y_n - h(y_n) - f(y_n)] + h(x_n) \\ n = 0, 1, 2, \ldots \end{cases}$$

If the following assumptions are satisfied:

i°) $x_o \le x_1$ and $y_1 \le y_o$

ii°) If $\dim H = +\infty$ then the mapping $\Phi(x) = h(x) + P_K[x - h(x) - f(x)]$ is nonexpansive or condensing,

then the problem I.C.P.(f, I-h, K) has a solution $x_* \in K$ such that $(\forall n \in N)$ $(x_n \le x_* \le y_n)$.

Moreover, if $\lim_{n\to\infty} \| y_n - x_n \| = 0$ then $\lim_{n\to\infty} x_n = x_*$.

Proof.

The proof is similar to the proof of Theorem 7.2.5 but using the mapping $\Phi(x) = h(x) + P_K[x - h(x) - f(x)]$.

Since $x_o, y_o \in K$ then the fixed point x_* of Φ satisfying $x_n \le x_* \le y_n$, for every $n \in N$ is exactly a solution of the problem I.C.P.(f, I - h, K). □

Definition 7.2.1.

Given $\alpha \in R$ such that $0 < \alpha < 1$ and $T_1, T_2 : H \to H$ two mappings we say:

1°) T_1 is (α)-concave if for every $x \in H$ and every λ such that $0 < \lambda < 1$ we have

$\lambda^{\alpha} T_1(x) \le T_1(\lambda x)$

2°) T_2 is $(-\alpha)$-convex if for every $x \in H$ and every λ such that $0 < \lambda < 1$ we have

$T_2(\lambda x) \le \lambda^{\alpha} T_2(x)$.

Remark 7.2.2

a) Notions similar to (α)-concave and $(-\alpha)$-convex operators were studied by Krasnoselski [C86] and Potter [C88].

b) We can show that T_1 is (α)-concave (or T_2 is $(-\alpha)$-convex) if and only if

$T_1(\mu x) \le \mu^{\alpha} T_1(x)$ (or $T_2(\mu x) \ge \mu^{-\alpha} T_2(x)$), for every $x \in H$ and $\mu > 1$.

Theorem 7.2.7.

Let $(H, \langle , \rangle)$ be a Hilbert space, $K \subset H$ an isotone projection cone and $f : K \to H$. Suppose that the mapping $\Phi(x) = P_K(x) - f(P_K(x))$ associated to the complementarity problem C.P.(f, K) has a decomposition of the form $\Phi(x) = T_1(x) + T_2(x)$, where T_1 is increasing and (α)-concave and T_2 is decreasing and $(-\alpha)$-convex (T_1, T_2 are not necessary continuous).

Given $u_0 \in K$ and $\mu_o > 1$ such that $\mu_o^{\alpha-1} u_0 \le T_1(u_0) + T_2(u_0) \le \mu_o^{1-\alpha} u_0$, we consider the sequences $\{x_n\}_{n\in N}$ defined by:

$x_0^{-1} = \mu_o^{-1} . u_0$

$y_0 = \mu_o . u_0$

$x_n = T_1(x_{n-1}) + T_2(x_{n-1})$; for every $n \in N$,

$y_n = T_1(y_{n-1}) + T_2(x_{n-1})$; for every $n \in N$.

Then the following statements are true:

1°) the sequences $\{x_n\}_{n\in N}$, $\{y_n\}_{n\in N}$ are convergent,

2°) $\lim_{n\to\infty} x_n = \lim_{n\to\infty} y_n$.

3°) the element $x_* = \lim_{n\to\infty} x_n = \lim_{n\to\infty} y_n$ is a solution of the problem C.P.(f,K),

4°) $\| x_* - x_n \| \le \mu_o (1 - \frac{1}{\mu_o^{2\alpha^n}}) \| u_0 \|$, for every $n \in N$.

Proof

We will prove this theorem by several steps.

a) For every $n \in N$, we have
$[x_n, y_n] \subseteq [x_{n-1}, y_{n-1}]$.

Indeed, $x_o \le y_o$ and from assumptions we have,

$$x_1 = T_1(x_o) + T_2(y_o) = T_1(\mu_o^{-1}u_o) + T_2(\mu_o u_o),$$

$$T_1(\mu_o^{-1}u_o) \ge \mu_o^{-\alpha}T_1(u_o),$$

$$T_2(\mu_o u_o) \ge \mu_o^{-\alpha}T_2(u_o),$$

which imply,

$$x_1 \ge \mu_o^{-\alpha}[T_1(u_o) + T_2(u_o)] \ge \mu_o^{-\alpha}[\mu_o^{\alpha-1}.u_o] = \mu_o^{-1}u_o = x_o.$$

$y_1 = T_1(y_o) + T_2(x_o)$ and from assumptions,

$$T_1(\mu_o u_o) \le \mu_o^{\alpha}\, T_1(u_o),$$

$$T_2(\mu_o^{-1}u_o) \le \mu_o^{\alpha}\, T_2(u_o),$$

which imply,

$$y_1 = T_1(y_o) + T_2(x_o) \le \mu_o^{\alpha}\,[T_1(u_o) + T_2(u_o)] \le \mu_o^{\alpha}\,[\mu_0^{1-\alpha}\, u_o] = \mu_o u_o = y_o.$$

Now, we prove that $x_1 \le y_1$. Indeed we have,

$$x_1 = T_1(x_o) + T_2(y_o) = T_1(\mu_o^{-1}\, u_o) + T_2(\mu_o u_o)$$

$$y_1 = T_1(y_o) + T_2(x_o) = T_1(\mu_o u_o) + T_2(\mu_o^{-1}\, u_o).$$

But, since $\mu_o u_o^{-1} \le \mu_o u_o$, T_1 is increasing and T_2 decreasing we have,

$$T_1(\mu_o^{-1} u_o) \le T_1(\mu_o u_o),$$

$$T_2\mu_o u_o) \le T_2(\mu_o^{-1}u_o),$$

whence $x_1 \le y_1$. Finally we have,

(5): $x_o \le x_1 \le y_1 \le y_o$

From (5) and by induction since T_1 is increasing and T_2 decreasisng we obtain

(6): $x_{n-1} \le x_n \le y_n \le y_{n-1}$; for all $n \in N$.

b) The sequence $\{x_n\}_{n\in N}$ are convergent.

Because K is isotone projection it is regular and hence $\{x_n\}_{n\in N}$, $\{y_n\}_{n\in N}$ are convergent.

c) If $x = \lim_{n\to\infty} x_n$ and $y_* = \lim_{n\to\infty} y_n$ then $x_* = y_*$.

Indeed, we have

(7): $x_n \le x_* \le y_* \le y_n$; for all $n \in N$ which implies,

$$x_{n+1} = T_1(x_n) + T_2(y_n) \le T_1(x_*) + T_2(y_*) \le T_1(y_n) + T_2(x_n) = y_{n+1},$$

and computing the limit we get

(8): $x_n \le x_* \le T_1(x_*) + T_2(y_*) \le y_* \le y_n$; for all $n \in N$.

Now, we prove that for every $n \in N$ we have,

(9): $\mu_o^{-2\alpha^n} y_n \le x_n$.

We prove this relation by induction.

Indeed, for $n = 0$ we have,

$$\mu_o^{-2}.y_o = \mu_o^{-2}(\mu_o u_o) = \mu_o^{-1} u_o = x_o$$

If (9) is true for n we have

$$\mu_o^{-2\alpha^n}.y_{n+1} = \mu_o^{-2\alpha^{n+1}}.[T_1(y_n) + T_2(x_n)] = \mu^{-2\alpha^{n+1}}.T_1(y_n) + \mu_o^{-2\alpha^{n+1}}.T_2(x_n) \le$$

$$\le T_1(\mu_o^{-2\alpha^n}.y_n) + \mu_o^{-2\alpha^{n+1}}.T_2(\mu_o^{-2\alpha^n}.y_n) \le T_1(x_n) + \mu_o^{-4\alpha^{n+1}}.T_2(y_n) \le$$

$$\le T_1(x_n) + T_2(y_n) = x_{n+1}$$

Hence (9) is true for every $n \in N$.

From (8) and (9) we get

(10): $0 \le y_* - x_* \le y_n - x_n \le (1-\mu_o^{-2\alpha^n})\, y_n \le (1-\mu_o^{-2\alpha^n})\, y_o$,

and because the ordered Hilbert space $(H, <,>, K)$ is Archimedean we obtain $y_* = x_*$.

So, we have that x_* is a fixed point for the mapping Φ and from <u>Proposition 7.2.1</u> (3°) we have that x_* is a solution of the problem <u>C.P.(f,K)</u> because $x_* \in K$.

To show conclusion 4°) we observe that from (8) we have $0 \le x_* - x_n \le y_n - x_n$, for every $n \in N$ and using the normality of K and (10) we obtain $\| x_* - x_n \| \le$ $\le \mu_o(1 - \frac{1}{\mu^{2\alpha^n}}) \| u_o \|$. □

<u>Remark 7.2.3</u>

For the problem <u>I.C.P. (f, I-h, K)</u> we have a similar theorem to <u>Theorem 7.2.7</u> but in this case we use the mapping $\Phi(x) = h(x) + P_K[x - h(x) - f(x)]$.

7.3. <u>Mann's interations and the Complementarity Problem</u>

The importance of Mann's interations in the Fixed Point Theory is well known since for some nonexpansive mappings it is possible to have that Picard's interations are not convergent, while Mann's iterations are convergent.

In this section we will use the following result.

Theorem 7.3.1. [Ishikawa][C85]

Let D be a closed subset of a Banach space E and let T be a nonexpansive mapping from D into a compact subset of E.

If there exist $x_1 \in D$ and $\{t_n\}_{n=1}^{\infty}$ such that:

1°) $0 \le t_n \le \alpha < 1$, for all $n \in N$,

2°) $\sum_{n=1}^{\infty} t_n = +\infty$,

3°) $x_{n+1} = (1 - t_n) x_n + t_n T(x_n) \in D$, for all $n \in N$

then $\{x_n\}_{n\in N}$ is convergent to a fixed point $x_* \in D$ of T. □

We consider now the problem C.P.(f, K), where K is a closed convex cone in a Hilbert space, $(H, \langle\ ,\ \rangle)$ and $f(x) = x - h(x)$, with $h:K \to H$.

We recall that h is compact if every bounded set of K is applied by h in a relatively compact set.

We have the following iterative method.

Theorem 7.3.2

If K is an isotone projection cone in a Hilbert space $(H, \langle\ ,\ \rangle)$ and the following assumptions are satisfied:

1°) h is nonexpansive and compact,

2°) there exists $b \in H$ such that $P_K(b) \ne 0$ and $h(x) \le b$ for every x satisfying $0 \le x \le P_K(b)$,

then the sequence $\{x_n\}_{n\in N}$ defined by:

$$x_{n+1} = (1 - t_n) x_n + t_n P_K(h(x_n)), \text{ for all } n \in N;$$

where $0 \le t_n \le \alpha < 1$, $\sum_{n=1}^{\infty} t_n = +\infty$ and x_1 is an arbitrary element such that $0 \le x_1 \le P_K(b)$,

is convergent to a solution of the problem C.P. (I-h, K).

Proof

From the Corollary of Theorem 7.2.1. we have that the problem C.P.(I-h, K) has a solution if the mapping $T(x) = P_K(h(x))$ has a fixed point in K.

If we consider $D = \{x \in K \mid 0 \le x \le P_K(b)\}$, the properties of P_K and the assumption that K is isotone projection, then we observe that all assumptions of Ishikawa's Theorem are satisfied for T and D, whence $\{x_n\}_{n\in N}$ is convergent to a solution of the problem C.P.(I-h, K). □

Remark 7.3.1

The iterative method defined in Theorem 7.3.2. can be applied to the complementarity problem C.P.(aI-h, K), where $a > 0$ and h is a mapping satisfying the Lipschitz condition with the constant $L > 0$.

In this case $x_{n+1} = (1-t_n)x_n + t_n P_K[\frac{1}{a} h(x_n)]$ (with t_n defined in Theorem 7.3.2) and the sequence $\{x_n\}_{n\in N}$ is convergent to a solution of the problem C.P.(aI-h, K) if $L \leqslant a$.

7.4. Projective metrics and the Complementarity Problem

In this last section we will present a new direction of application of isotone projection cones in the domain of iterative methods for complementarity problems.

Let $(E, \| \ \|)$ be a Banach space and let $K \subset E$ be a closed pointed convex cone.

In this case the partial ordering "$\leq$" defined by K is Archimedean, that is $nx \leq y$ for $n = 1, 2, 3, \ldots$ then $x \leq 0$.

Definition 7.4.1.

We say that a function $d_p : K \times K \to R \cup \{+\infty\}$ is a projective metric on K if:

1°). $d_p(x,y) = d_p(y,x)$; for all $x, y \in K$,

2°). $d_p(x,y) \leq d_p(x,z) + d_p(z,y)$; for all $x, y, z \in K$,

3°). $d_p(x,y) \geq 0$, for all $x, y \in K$,

4°). $d_p(x,y) = 0$, if and only if $x = \lambda y$ for some $\lambda \in R_+\setminus\{0\}$.

Remark 7.4.1

We can show that for every projective metric d_p we have $d_p(\lambda x, \mu y) = d_p(x,y)$, for all $\lambda, \mu \in R_+\setminus\{0\}$.

We say that $x, y \in K\setminus\{0\}$ are "comparable" if there exist real numbers $\alpha > 0$, $\beta > 0$ such that $\alpha x \leq y \leq \beta x$.

If $x, y \in K\setminus\{0\}$ are comparable we define numbers $m(x,y)$ and $M(x,y)$ by:

$m(x,y) = \sup \{\lambda \geq 0 \mid x \geq \lambda y\}$

$M(x,y) = \inf \{\lambda \geq 0 \mid x \geq \lambda y\}$.

Now we can define Hilbert's projective metric d_H on K which is the most important projective metric.

This metric is defined by:

i). $d_H(0,0) = 0$,

ii). $d_H(x,0) = d_H(0,y) = +\infty$,

iii). $d_H(x,y) = \ln \frac{M(x,y)}{m(x,y)}$, if x and y are comparable,

iv). $d_H(x,y) = +\infty$, if x and y are not comparable.

Initially this concept was defined by Hilbert in 1903 in a particulary case in [C80].

It is interesting to remark that if K is a cone with nonempty interior and if $\overset{o}{K}$ is its interior then $(\overset{o}{K}, d_H)$ is a pseudo-metric space and $(\overset{o}{K} \cap U, d_H)$ is a metric space, where U denotes the unit sphere of E.

Since this distance has many and interesting applications in nonlinear analysis, probably it has also interesting applications in the complementarity theory.

We consider here another distance on K more flexible and which is a generalization of a variant of Hilbert's projective metric defined by Thompson [C90]

Let $E(\tau)$ be a locally convex space and let $K \subset E$ be a closed pointed convex cone.

We suppose that K is <u>normal</u>, that is, the topology τ is defined by a family of seminorms $\{p_\alpha\}_{\alpha\in A}$ such that $0 \leq x \leq y$ implies $p_\alpha(x) \leq p_\alpha(y)$, for all $\alpha \in A$.

<u>Lemma 7.4.1.</u>

<u>If $K \subset E$ is a normal cone and $x \leq \lambda y$, $y \leq \lambda x$, $p_\alpha(x) \leq m$ and $P_\alpha(y) \leq m$ then $p_\alpha(x-y) \leq 3m(\lambda-1)$.</u>

<u>Proof</u>

Since $x - y \leq (\lambda-1)y$ and $y-x \leq (\lambda-1)x$ we have that there exist $u, v \in K$ such that,

$x - y + u = (\lambda-1)y$ and

$y - x + v = (\lambda-1)x$.

Applying seminorm p_α we deduce

$$p_\alpha(u) \leq p_\alpha(u+v) = p_\alpha(x - y + u + y - x + v) =$$

$$= p_\alpha[(\lambda-1)y + (\lambda-1)x] \leq (\lambda-1)[p_\alpha(y) + p_\alpha(x)] \leq 2m(\lambda-1)$$

which implies,

$$p_\alpha(x-y) = p_\alpha(x - y + u - u) \leq p_\alpha(x - y + u) + p_\alpha(u) \leq$$

$$\leq (\lambda - 1)[m + 2m] = 3m(\lambda - 1). \quad \square$$

We say that two elements x and y of $K\setminus\{0\}$ are "equivalent" if they are comparable, this defines an equivalence relation on $K\setminus\{0\}$ and divides $K\setminus\{0\}$ into disjoint subsets which we call <u>"components of K"</u>.

If $x \in K\setminus\{0\}$ we set

$C_x = \{y \in K\setminus\{0\} \mid y$ is comparable with $x\}$.

We remark that $C_x \cup \{0\}$ satisfies all conditions to be a cone, except that $C_x \cup \{0\}$ need not be closed.

Let $x_* \in K\setminus\{0\}$ be an arbitrary element.

If $x, y \in C_{x_*}$ we define:

$\alpha(x,y) = \alpha = \inf \{\lambda \mid x \leq \lambda y\}$

$\beta(x,y) = \beta = \inf \{\mu \mid y \leq \mu x\}$.

<u>Remarks 7.4.2.</u>

1°). Since K is closed we have $x \leq \alpha y$ and $y \leq \beta x$,

2°). $\alpha \neq 0$ and $\beta \neq 0$.

Indeed, if $\alpha = 0$ or $\beta = 0$ we have $x = y = 0$ which is impossible since $x, y \in C_{x_*}$

We consider the function $d_o : C_{x_*} \times C_{x_*} \to R$ defined by $d_o(x,y) = \ln\{\max(\alpha(x,y), \beta(x,y))\}$.

<u>Lemma 7.4.2.</u>

<u>The function d_o is a distance in C_{x_*}</u>.

<u>Proof</u>

If $x = y$ then by definition we have $d_o(x,y) = 0$.

Conversely, if $d_o(x,y) = 0$ then $\max\{\alpha(x,y), \beta(x,y)\} = 1$, which implies $\alpha = 1$ or $\beta = 1$ and finally $x \leq y$ and $y \leq x$ that is $x = y$ (since K is pointed).

If $x \neq y$ then $\alpha > 1$ or $\beta > 1$ that is $d_o(x,y) > 0$.

Consider now $x,y,z \in C_{x_*}$ arbitrary elements. We have

$x \leq \alpha_1 y$; $y \leq \beta_1 x$

$x \leq \alpha_2 z$; $z \leq \beta_2 x$

$z \leq \alpha_3 y$; $y \leq \beta_3 z$.

Since $x \leq \alpha_2\alpha_3 y$ we deduce $\alpha_1 \leq \alpha_2\alpha_3$ (from the definition of α and β) and if we suppose $\alpha_1 \geq \beta_1$ then we get $d_o(x,y) = \ln \alpha_1 \leq \ln(\alpha_2\alpha_3) = \ln \alpha_2 + \ln \alpha_3$ $\leq \ln\{\max(\alpha_2, \beta_2)\} + \ln\{\max(\alpha_3, \beta_3)\} = d_o(x,z) + d_o(z,y)$. Similarly, if $\beta_1 \geq \alpha_1$. □

<u>Theorem 7.4.1</u>

<u>If K is closed normal and sequentially complete then for every $x \in K\setminus\{0\}$ C_x is</u>

<u>a metric complete space with respect to the distance d_o</u>.

<u>Proof</u>

Let $\{x_n\}_{n\in N}$ be a Cauchy sequence of elements of C_x with respect to the distance d_o. We put

$\alpha_{pq} = \inf \{\lambda \mid x_p \le \lambda x_q\}$; $p, q \in N$.

<u>The sequence $\{x_n\}_{n\in N}$ is a bounded sequence with respect to the topology τ.</u>

Indeed, since $\{x_n\}_{n\in N}$ is a Cauchy sequence we have that there exists $n_o \in N$ such that $d_o(x_p,x_q) < 1$, for all $p, q \ge n_o$ which implies

$\max(\alpha_{pq},\alpha_{qp}) < \exp(1)$; for all $p,q \ge n_o$.

In particular we have

$\alpha_{pn_o} < \exp(1)$; for all $p \ge n_o$

which implies $x_p \le \exp(1)\, x_{n_o} \le 3x_{n_o}$

and finally $p_\alpha(x_p) \le 3p_\alpha(x_{n_o})$

If we put $m = \max \{p_\alpha(x_1), \ldots, p_\alpha(x_{n_o}), 3p_\alpha(x_{n_o})\}$ we obtain $p_\alpha(x_n) \le m$ and since p_α is an arbitrary seminorm of the family $\{p_\alpha\}_{\alpha\in A}$ we deduce that $\{x_n\}_{n\in N}$ is bounded.

<u>The sequence $\{x_n\}_{n\in N}$ is a τ-Cauchy sequence.</u>

Indeed, let p_α be an arbitrary seminorm of the family $\{p_\alpha\}_{\alpha\in A}$

If $\varepsilon > 0$ then there exists $\delta_\varepsilon > 0$ such that $\exp(\delta_\varepsilon) < 1 + \frac{\varepsilon}{M}$, where $M = 3m$.

Since $\{x_n\}_{n\in N}$ is a Cauchy sequence with re4spect to d_o we have that there exists $n_\varepsilon \in N$ such that,

$d_o(x_p,x_q) < \delta_\varepsilon$; for all $p,q \ge n_\varepsilon$ which implies, $\max(\alpha_{pq}, \alpha_{qp}) < 1 + \frac{\varepsilon}{M}$

and hence,

$x_p \le (1 + \frac{\varepsilon}{M})\, x_q$ and $x_q \le (1 + \frac{\varepsilon}{M})\, x_p$.

From <u>Lemma 7.4.1</u> we have $p_\alpha(x_p - x_q) \le 3m(1 + \frac{\varepsilon}{M} - 1) = \varepsilon$; for all $p,q \ge n_\varepsilon$

So, we have that $\{x_n\}_{n\in N}$ is a τ-Cauchy sequence.

Because K is sequentially complete we have that there exists $x_* = (\tau) - \lim_{n\to\infty} x_n$.

<u>The proof is finished if we show that x_* is also the limit of $\{x_n\}_{n\in N}$ but with with respect to the distance d_o.</u>

If $\varepsilon > 0$ is given then since $\{x_n\}_{n\in N}$ is a Cauchy sequence with respect to d_o we have $d_o(x_p,x_q) < \varepsilon$ for p,q big enough.

So, we have $x_p \leq \exp(\varepsilon)x_q$ and $x_q \leq \exp(\varepsilon)\, x_p$ and since K is closed and $x_* = (\tau) - \lim_{n\to\infty} x_n$ we deduce that $x_p \leq \exp(\varepsilon)\, x_*$ and $x_* \leq \exp(\varepsilon)\, x_p$, for every p big enough.

Finally we obtain that $x_* \in C_x$ and $d_o(x_p,x_*) \leq \varepsilon$ for every p big enough and the theorem is proved. □

We consider now a Hilbert space (H, <, >) ordered by an isotone projection cone $K \subset H$.

Given a mapping $h:K \to H$ we consider the problem <u>C.P.(I-h, K)</u>.

We have the following result.

<u>Theorem 7.4.2</u>

<u>Let (H,< , >) be a Hilbert space ordered by an isotone projection cont $K \subset H$</u>.
<u>If the mapping $h:K \to H$ satisfies the following assumptions</u>:

1°). <u>h is isotone with respect to the ordering defined by K,</u>

2°). <u>there exist p_1, p_2 $\in]\,0,\,1\,[$ such that $h(\lambda x) \leq \lambda^p h(x)$, for every $x \in K$ and $\lambda \in R_+$, where $p = p_1$ if $\lambda < 1$ amd $p = p_2$ if $\lambda > 1$,</u>

3°). <u>there exist $x_o \in K\setminus\{0\}$ and $\lambda,\ \mu > 0$ such that $\mu x_o \leq P_K(h(x_o)) \leq \lambda x_o$, then the problem C.P.(I-h, K) has a solution $x_* \in C_{x_o}$ which is unique in this component and the sequence $\{x_n\}_{n\in N}$ defined by $x_{n+1} = P_K(h(x_n))$, for every $n \in N$ and x_1-arbitrary in C_{x_o} is convergent to x_*.</u>

<u>Moreover, $\{x_n\}_{n\in N}$ is convergent to x_* with respect to the norm $\|\cdot\|$ of H and we have,</u>

$$d_o(x_n,x_*) \leq \frac{p^n}{1-p}\, d_o(x_o,x_1),\ \text{where } p = \max(p_1,\ p_2)$$

<u>Proof</u>

We consider the mapping $T(x) = P_K(h(x))$

Since K is isotone projection by assumption 1°) we obtain that T is isotone with respect to the ordering defined by K.

Using assumptions 2°) and 3°) we can show that T is a p-contraction with respect to d_o on the component C_{x_o}. Also, we remark that $T(C_{x_o}) \subseteq C_{x_o}$.

Since (C_{x_o}, d_o) is a complete metric space, the theorem is a consequence of Banach's contraction Theorem (because a fixed point of T is a solution of the problem C.P.(I-h, K).

The proof is finished since we can show that every Cauchy sequence with respect to d_o is a Cauchy sequence with respect to the norm $\| \ \|$ of H.

Remark 7.4.3

It is interesting to apply other fixed point theorems to the mapping T and the complete metric space (C_{x_o}, d_o) to obtain new existence theorems for the problem C.P.(I-h, K).

Comments

The concept of isotone projection cone was defined by G. Isac and studied by G. Isac and A.B. Német [C81], [C82], [C83], [C31], [C84].

The all results on the Complementarity Problem presented in this chapter were obtained by G. Isac and A.B. Német [C81], [C82], [C83], [C31], [C84].

CHAPTER 8

TOPICS ON COMPLEMENTARITY PROBLEMS

We present in this chapter some subjects nonconsidered in precedent sections and which can be probably developed in future researches.

8.1 The basic theorem of complementarity

In 1971 Eaves proved a very nice result about the General Complementarity Problem in R^n [A81].

This result is known as the basic theorem of complementarity problem. Eaves' theorem was generalized in 1976 to point-to-set mappings by Saigal [A266].

We present in this section Saigal's generalization.

Let E be the n-dimensional Euclidean space (that is, $E = R^n$ and $\|x\|^2 = \langle x, x \rangle$, where $\langle x, y \rangle = \sum_{i=1}^{n} x_i y_i$).

Let $K \subset R^n$ be a pointed closed convex cone.

Since every pointed convex cone K in R^n is "well based" we have that the dual K^* is closed convex cone with a nonempty interior.

Let d be an element of $\mathrm{Int}(K^*)$.

For every $r \in R_+ \setminus \{0\}$ we denote $D_r^d = \{x \in K \mid \langle d, x \rangle \leq r\}$. Since K is a closed pointed convex cone we have that D_r^d is compact for every $r \in R_+ \setminus \{0\}$.

Let $C \subset R^n$ be a nonempty subset and suppose given a point-to-set mapping $f: C \to R^n$.

Following Eaves' definition we say that $x_o \in C$ is a stationary point of the pair (f, C) if and only if there is a $y_o \in f(x_o)$ such that $\langle u - x_o, y_o \rangle \geq 0$, for all $u \in C$.

Supposing given a point-to-set mapping $f: R^n \to R^n$ we consider the following general multivalued complementarity problem:

$$\text{G.M.C.P}\ (f, K): \quad \left\| \begin{array}{l} \text{find } x_o \in K \text{ and } y_o \in R^n \\ \text{such that } y_o \in f(x_o) \cap K^* \\ \text{and } \langle x_o, y_o \rangle = 0. \end{array} \right.$$

Proposition 8.1.1.

Every stationary point of (f, K) is a solution of G.M.C.P. (f, K) and conversely.

Proof.

Let (x_o, y_o) be a solution of the problem G.M.C.P. (f, K). We have $x_o \in K$, $y_o \in R^n$, $y_o \in f(x_o) \cap K^*$ and $< x_o, y_o > = 0$.

Since $y_o \in K^*$ we obtain that $< u - x_o, y_o > \geq 0$, for all $u \in K$.

Conversely, let x_o be a stationary point of (f, K). Then there exists $y_o \in f(x_o)$ such that $< x_o, y_o > \leq < u, y_o >$, for all $u \in K$.

Since $0 \in K$, $< x_o, y_o > \leq 0$ and $x_o \in K$ implies $\lambda x_o \in K$, for all $\lambda \geq 0$, that is, we have $< x_o, y_o > \leq < \lambda x_o, y_o >$, for all $\lambda \geq 0$. Hence, we deduce $(\lambda-1) < x_o, y_o > \geq 0$ and considering a $\lambda > 1$ we obtain $< x_o, y_o > \geq 0$.

So, finally we have $< x_o, y_o > = 0$.

Because $< x_o, y_o > = 0$ we have $< u, y_o > \geq 0$, for all $u \in K$, that is $y_o \in K^* \cap f(x_o)$ and the proof is finished.□

Proposition 8.1.2.

If x_o is a stationary point of (f, D_r^d) for some $r > 0$ and $< d, x_o > < r$ then x_o is a solution of G.M.C.P. (f, K).

Proof

Let x_o be a stationary point of (f, D_r^d), that is, we suppose that there is a $y_o \in f(x_o)$ such that $< u - x_o, y_o > \geq 0$, for all $u \in D_r^d$.

To show that (x_o, y_o) is a solution of G.M.C.P.(f,K) it is sufficient to show that x_o is a stationary point for (f, K) (with the same y_o) and to use Proposition 8.1.1.

So, we must show that $< u - x_o, y_o > \geq 0$, for all $u \in K$.

Consider $u \in K$ such that $< d, u > > r$. Since $x_o \in D_r^d$ and $< d, x_o > < r$ there is $\lambda \in]\, 0, 1\, [$ such that $\lambda x_o + (1 - \lambda) u \in D_r^d$, which implies

$< \lambda x_o + (1 - \lambda) u - x_o , y_o > \geq 0$ and finally $< u - x_o , y_o > \geq 0$, for all $u \in K$ and the proposition is proved. □

From Proposition 8.1.2. we deduce that it is important to study the set of stationary points for (f, D^d_r), $r \geq 0$.

In this sense we need to extend Theorem 3.1. [Hartman-Stampacchia] to upper semicontinuous point-to set mappings [defined in Chap. 4].

Theorem 8.1.1

Les $C \subset R^n$ be a nonempty compact convex set.

If $f:C \to R^n$ is an upper semicontinuous point-to-set mapping with $f(x)$ nonempty contractible and compact for each $x \in C$, then there is an $x \in C$ and a $y \in f(x)$ such that $< u - x, y > \geq 0$, for all $u \in C$.

Proof

First, for a set M we denote by $P^*(M)$ the class of all nonempty subset of M.

Let $E \subset R^n$ be an arbitrary compact convex set containing $f(C) = \bigcup_{x \in C} f(x)$.

Define a mapping $T:C \times E \to P^*(C) \times P^*(E)$ by $T(x, y) = (\Pi_C(y), f(x))$, where $\Pi_C(y) = \text{sol.} \min_{u \in C} < u, y >$ (the set of solutions of $\min_{u \in C} < u, y >$).

We remark that T is upper semicontinuous with $T(x, y)$ convex compact for every $(x, y) \in C \times E$. Hence, we can apply Theorem 4.2.11 [Eilenberg-Montgomery] and we obtain that T has a fixed point. Les (x_o, y_o) be this fixed point for T. We have $y_o \in f(x_o)$ and $x_o \in \Pi_C (y_o)$, that is, we have $< u - x_o, y_o > \geq 0$, for all $u \in C$ and the theorem is proved. □

The following result is necessary to prove the principal result of this section.

Theorem 8.1.2 [Mas-Colell]

Let C be a compact convex subset of R , $r \geqslant 0$ and $F:C \times [0, r] \to P^*(C)$ be upper semicontinuous and $F(x, t)$ be contractible and compact for each $(x, t) \in C \times [0, r]$

Then there is a connected set $D \subset C \times [0, r]$ which intersects both $C \times \{0\}$ and $C \times \{r\}$ such that $x \in F(x, t)$ for each $(x, t) \in D$.

Proof

The proof is in [C48].

Remark 8.1.1.

Theorem 8.1.2. is a generalization for point-to-set mappings of a Browder's Theorem [C7]:

Theorem 8.1.3 [Basic Theorem]

Let $f:K \to R^n$ be an upper semicontinuous mapping with $f(x)$ contractible and compact for each $x \in K$ and let $d \in \operatorname{Int}(K^*)$ be an arbitrary element.

Then there is a closed connected set S in K such that:

i°). each $x \in S$ is a stationary point of (f, D^d_r) for $r = \langle d, x \rangle$,

ii°). for each $r \geq 0$ there is an $x \in S$ which is a stationary point of (f, D^d_r)

Furthermore S can be chosen so that it is maximal or minimal.

Proof

Let S_r be the set of all stationary points of (f, D^d_r) which is nonempty by Theorem 8.1.1. and let S be the maximal connected component of $\bigcup_{r\geq 0} S_r$ containing 0.

Obviously, i°) is satisfied. We show now that ii°) is also true for S.

Indeed, let $r \geq 0$ given. Define $E_r \subset R^n$ to be any compact convex set containing $\bigcup_{r\geq 0} f(D^d_r)$ and $F:D^d_r \times E_r \times [0, r] \to P^*(D^d_r) \times P^*(E_r)$ as the mapping

$F(x, y, t) = (\Pi_t(y), f(x))$, where $\Pi_t(y) = \Pi_{D^d_t}(y) = \text{sol} \min_{u\in D^d_t} \langle u, y \rangle$.

By Mass-Colell's Theorem [Theorem 8.1.2.] there is a connected component $D \subset D^d_r \times E_r \times [0, r]$ intersecting both $D^d_r \times E_r \times \{0\}$ and $D^d_r \times E_r \times \{r\}$ such that for each $(x,y,t) \in D$, x is a stationary point of (f, D^d_t). Thus $M = \{x \mid (x,y,t) \in D\} \subset S$ and $M \cap S_r \neq \Phi$ and conclusion ii°) is verified.

To show that S can be chosen minimal (or maximal) we apply Zorn's lema to the collection (ordered by inclusion) of all sets which are closed, connected and satisfy i°) and ii°). □

Definition 8.1.1.

We say that $U \subset K\setminus D$ separates D from ∞ if each unbounded connected subset of K which intersects D also intersects U.

We consider now the case of polyhedral pointed convex cones in R^n.

Thus we suppose $K = \{x \in R^n | Ax \geq 0\}$ and that $Ax = 0$ implies $x = 0$, where $A \in M_{mxn}(R)$.

It is known that in this case $K^* = \{y \in R^n | y = u^t A,\ u \in R^n_+\}$

Proposition 8.1.3.

Let x be a stationary point of (f, D^d_r) for some $r \geq 0$ and $d \in Int(K^*)$.

Then there exist $0 \leq u \in R^n$, $t \geqslant 0$ and $y \in f(x)$ such that:

$$(1): \quad \left\| \begin{array}{l} Ax \geq 0 \\ u^t A = y + td \\ t(r - < d, x >) = 0 \\ u^t Ax = 0 \end{array} \right.$$

Proof

Since x is a stationary point of (f, D^d_r) then there is a $y \in f(x)$ such that x solves the linear programming problem $\min_{x \in D} < y,x >$, where $D = \{x | Ax \geq 0$ and $< d,x > \leq r\}$.

Applying to this linear program the duality theory (of Linear Programming) we obtain relations (1). □

Remark 8.1.2.

If in (1) we have $t = 0$ then we obtain that x is a solution of the problem G.M.C.P. (f,K). □

Proposition 8.1.4.

Let $K \subset R^n$ be a closed, polyhedral pointed convex cone and that U separates D^d_r from ∞ for some $r \geq 0$ and $d \in Int(K^*)$.

If for each $x \in U$ there is a $w \in D^d_r$ for which $< w-x,y > \leq 0$ for all $y \in f(x)$, then the problem G.M.C.P.(f,K) has a solution.

Proof

Let S be the closed connected set defined by Theorem 8.1.3.

If S is bounded, then we have a stationary point x_o of some (f, D_r^d) with $< d, x_o >< r$ and by Proposistion 8.1.2. the problem G.M.C.P. (f,K) has a solution.

Thus, we assume now that S is unbounded. Since U separated D_r^d for some $r \geq 0$, $S \cap U$ is nonempty.

Let $\bar{x} \in S \cap U$. Since $\bar{x} \in S$ we have that (1) holds.

Hence there is $0 \leq u \in R^n$, $t \geq 0$ and $y \in f(\bar{x})$ such that

(2): $\quad 0 = u^t A\bar{x} = < \bar{x},y > + t < d, \bar{x} >.$

Since $\bar{x} \in U$ there exists $w \in D_r^d$ such that $< w-\bar{x}, y > \leq 0$, for all $y \in f(\bar{x})$.

Thus $u^t Aw = < w,y > + t < d,w >$, which implies

$$u^t.Aw - t < d,w > = < w,y > \leq < \bar{x},y >$$

and using (2) we obtain

$$u^t.Aw - t < d,w > \leq < \bar{x},y > = - t < d,\bar{x} > ,$$

that is we have,

$$u^t.Aw \leq t < d,w > - t < d,\bar{x} >.$$

Since $u^t Aw \geq 0$, $0 \leq < d,w > \leq r$ and $< d,\bar{x} > > r$ we deduce

$$0 \leq t < d,w > - t < d,\bar{x} > < 0,$$

which is impossible if $t > 0$.

Hence, we must have $t = 0$ and from Remark 8.1.2. we obtain that the problem G.M.C.P. (f,K) has a solution. □

As immediate corollary of basic theorem [Theorem 8.1.3.] we have the following classical result.

Theorem 8.1.4 [Eaves]

Given the continuous function $f:R_+^n \to R$ then there is a closed connected set $S \subset R_+^n$ such that:

i°) *each $x \in S$ is a stationary point of (f,D_r^d), where $r = < x,d >$,*

ii°) *for each $r \geq 0$ there is an $x \in S$ which is a stationary point of (f,D_r^d).*

Moreover, S can be chosen so that it is maximal or minimal. □

It is interesting to know if Theorem 8.1.3 is true in infinite dimensional spaces for a general cone or for locally compact cones.

8.2. The Multivalued Order complementarity Problem

We consider in this section the Multivalued Order Complementarity Problem defined in Chapter 1.

Let E be a vector lattice with the partial ordering denoted by "$\leq$" and $K = \{x \in E | x \geq 0\}$.

Given a point-to-set mapping $f: K \to E$ such that $f(x)$ is nonempty for every $x \in K$, the Multivalued Order Complementarity Problem associated to f and K is:

$$\text{M.O.C.P. } (f,K): \quad \left\| \begin{array}{l} \text{find } x_o \in K \text{ and } y_o \in E \\ \text{such that } y_o \in f(x_o) \text{ and} \\ x_o \wedge y_o = 0 \end{array} \right.$$

This problem was considered in 1984 by Fujimoto [A101].

Now, we consider this problem for a class of mappings more general as the class considered by Fujimoto but we follow the ideas used in [A101].

First, we need to extend Tarski's Theorem to point-to-set mappings.

In this sense, we suppose that L is a complete lattice, that is, for every subset A of L there exist (in L) sup A and inf A. The order on L is denoted by "$\leq$".

Let $T: L \to L$ be a point-to-set mapping with $T(x)$ nonempty for every $x \in L$. For every $x \in L$ we denote $[T(x)]_s = \{z \in L | z \leq u \text{ for some } u \in T(x)\}$.

Definition 8.2.1.

We say that T is isotone (monotone increasing) if $x, y \in L$ and $x \leq y$ imply that for every $u \in T(x)$ there is a $v \in T(y)$ such that $u \leq v$.

We recall that a subset $A \subset L$ is inductively ordered if every totally ordered subset of A has a majorant in A (an upper bound in A).

We use the following form of Zorn's Lemma.

If X is an inductively ordered set then X has a maximal element.

[see: C. Berge: Espaces topologiques. (Functions multivoques). Dunod (1966)].

The following result is an extension of Tarski's Theorem.

Theorem 8.2.1.

If L is a complete lattice and $T: L \to L$ an isotone point-to-set mapping with $T(x)$ nonempty for every $x \in L$ and such that $[T(x)]_s$ is inductively ordered for every $x \in L$, then there exists a point $z_* \in L$ such that $z_* \in T(z_*)$.

Proof

Consider the set $D = \{x | x \in L, x \leq u \text{ for some } u \in T(x)\}$. D is nonempty since $\inf L \in D$.

We show now that D is inductively ordered. Indeed, let $B \subset D$ be a totally ordered set.

For any $x \in B$ there exists $y \in T(x)$ such that $x \leq y$. Since T is isotone and $x \leq \sup B$ there exists $z \in T(\sup B)$ such that $y \leq z$, which implies that $B \subset [T(\sup B)]_s$.

Because $[T(\sup B)]_s$ is inductively ordered then by assumptions we obtain that B has an upper bound x_* in $[T(\sup B)]_s$. Certainly $\sup B \leq x_*$.

On the other hand, from the definition of $[T(\sup B)]_s$, there exists an element $x_o \in T(\sup B)$ such that $x_* \leqslant x_o$, which implies that $\sup B \in D$ and that D is inductively ordered.

Hence, by Zorn's Lemma there exists a maximal element $z_* \in D$.

We show now that $z_* \in T(z_*)$.

Indeed, since $z_* \in D$ we can find $u_* \in T(z_*)$ such that $z_* \leq u_*$.

From assumption that T is isotone we have that there exists an element $u_o \in T(u_*)$ such that $u_* \leq u_o$, which implies $u_* \in D$, but since z_* is maximal we have $z_* = u_* \in T(z_*)$ and the proof is finished. □

Remark 8.2.1.

The assumption that "T is isotone" alone is insufficient to have the conclusion of Theorem 8.2.1.

Example

Consider $L = [0,1] \subset \mathbb{R}$ and $T(x) = \{y \mid 0 < y < 1,\ y \neq x\}$. T is isotone but without fixed point.

Let E be a complete vector lattice (that is, E is a vector lattice such that for every subset $A \subset E$, $\sup A$ exists whenever A has an upper bound.

Given a point-to-set mapping $T_o : K \to E$ with $T_o(x)$ nonempty for every $x \in K$ we consider the problem M.O.C.P. $(I-T_o, K)$, that is, the problem

$$\text{M.O.C.P.}(I-T_o,K):\quad \left\| \begin{array}{l} \text{find } x \in K \text{ and } y \in E \\ \text{such that } y \in T_o(x) \\ \text{and } x \wedge (x-y) = 0 \end{array} \right.$$

Definition 8.2.2.

We say that a point-to-set mapping $f:K \to E$ is Λ-isotone if there exists a point-to-point mappint $\Lambda:E \to E$ such that:

i°) $(I+\Lambda)^{-1}$ exists and it is isotone,

ii°) if $x \leq y$ then for every $u \in f(x) + \Lambda(x)$ there exists $v \in f(y) + \Lambda(y)$ such that $u \leq v$.

Remark 8.2.2.

Every isotone point-to-set mapping is Λ-isotone with $\Lambda(x) = 0$ for every $x \in E$.

Given T_o, we define the point-to-set mapping T_o^+ by:

$T_o^+(x) = \{y^+ \mid y \in T_o(x)\}$, where $y^+ = o \vee y$.

Theorem 8.2.2.

Let E be a complete vector lattice and let $T_o:K \to E$ be a point-so-set mapping with $T_o(x)$ nonempty for every $x \in K$.

If the following assumptions are satisfied:

1°) T_o^+ is Λ-isotone,

2°) $[(I+\Lambda)^{-1}(T_o^+(x) + \Lambda(x))]_s$ is inductively ordered for every $x \in K$,

3°) there exists an element $x_o \in K$ such that for every $y \in T_o(x_o)$ we have $y \leq x_o$ and for every x such that $0 \leq x \leq x_o$ we have

$(I+\Lambda)^{-1}(T_o^+(x) + \Lambda(x)) \geq 0$

then the problem M.O.C.P.$(I-T_o, K)$ has a solution.

Proof

We consider the point-to-set mapping $T:K \to E$ defined by

$T(x) = (I+\Lambda)^{-1}(T_o^+(x) + \Lambda(x))$; for all $x \in K$, and we remark the following:

a) T is isotone,

b) $[T(x)]_s$ is inductively ordered for every $x \in K$,

c) there exists $x_o \in K$ such that for every $y \in T(x_o)$ we have $y \leq x_o$,

d) for every x such that $0 \leq x \leq x_o$ we have $T(x) \geq 0$.

We consider the set $D = \{x \in K | x \le x_o\}$, where x_o is the element given in assumption 3°). D is a complete lattice, T is isotone with respect to K and it maps D into itself.

Hence, we can apply <u>Theorem 8.2.1.</u> and we obtain an element x_* D such that $x_* \in T(x_*)$, which implies that there exists $z_* \in T(x_*)$, such that

$x_* = z_* \in (I+\Lambda)^{-1} (T_o^+ (x_*) + \Lambda(x_*))$.

So, there exists $y_* \in T_o(x_*)$ such that $x_* = (I+\Lambda)^{-1} (y_*^+ + \Lambda(x_*))$, that is,

$x_* = y_*^+$ and finally we have, $x_* \Lambda(x_*-y_*) = x_* + 0\ \Lambda(-\ y_*) = x_* - y_* \vee 0 = 0$,

that is, x_* is a solution of the problem M.O.C.P. $(I-T_o, K)$.

<u>Corollay</u> [Fujimoto]

<u>Let E be a complete vector lattice and let $T_o: K \to E$ be a point-to-set mapping with $T_o(x)$ nonempty for every $x \in K$.</u>

<u>If the following assumptions are satisfied:</u>

1°) <u>T_o is isotone,</u>

2°) <u>$[T_o(x)]_s$ is inductively ordered for every $x \in K$,</u>

3°) <u>there exists an element $x_o \in K$ such that for every $y \in T_o(x_o)$ we have $y \le x_o$,</u>

<u>then the problem M.O.C.P. $(I-T_o, K)$ has a solution.</u>□

8.3. <u>Some classes of matrices and the Linear Complementarity Problem.</u>

For the Linear Complementarity Problem many authors have studied methods of finding a solution.

In this sense we remark the methods proposed by Cottle and Dantzig [A53], Lemke [A173-A178], Eaves [A82], Garcia [A103], Karamardian [A158], Saigal [A262-A264], Murty [A226-A227] etc. For these methods we recommend the excellent book [<u>K.G. Murty</u>: Linear complementarity, linear and nonlinear programming. <u>Heldermann Verlag, Berlin</u> <u>(1988)</u>].

About the solvability of the Linear Complementarity Problem several classes of matrices were defined, but probably the first important class is the class Q defined in relation with Lemke's method.

We consider the space R^n with the euclidean structure, $A \in M_{nxn}(R)$ and the closed convex cone R^n_+.

The following problem is considered in this section:

$$\text{L.C.P.}(A,q,R^n_k): \quad \begin{cases} \text{find } x_o \in R^n_+ \text{ such that} \\ Ax_o + q \in R^n_+ \text{ and} \\ \langle x_o, Ax_o + q\rangle = 0, \end{cases}$$

where q is an arbitrary element in R^n.

For the problem L.C.P.(A,q,R_+) the feasible set is $F(A,q,R^n_+) =$

$= \{x \in R^n_+ \mid Ax + q \geq 0\}$.

We say that the problem L.C.P.(A,q,R^n_+) is feasible if $F(A,q,R^n_+)$ is nonempty.

Let $A = (a_{ij}) \in M_{nxn}(R)$ be an arbitrary matrix. We recall the definition of principal minor of A (defined also in chapter 1.)

If $J \subset \{1, 2, ..., n\}$ then the <u>principal submatrix</u> of A determined by the subset J is the matrix $A_{JJ} = \{a_{ij}\}_{i,j \in J}$.

The <u>principal minor</u> of A determined by J is the determinant of A_{JJ}.

We denote by A_{i*} the ith row of A and by A_{*j} the jth column.

For all $I \subset \{1, 2, ..., n\}$ we denote by A_{I*} the submatrix of A consisting of all rows A_{i*} for $i \in I$ and by A_{*I} the submatrix of A consisting of all columns A_{*j} for $j \in I$.

A characteristic of Lemke's method is the fact that it shows one of the following two situations:

i) for every $q \in R^n$ the problem L.C.P.(A,q,R^n_+) has a solution,

ii) the problem L.C.P.(A,q,R^n_+) has no solution.

Since this result the following class of matrices is important.

<u>Definition 8.3.1.</u>

<u>We say that a matrix $A \in M_{nxn}(R)$ is a Q – matrix if the problem L.C.P.(A,q,R^n_+) has a solution for all $q \in R^n$.</u>

<u>Definition 8.3.2.</u>

<u>We say that a matrix $A \in M_{nxn}(R)$ is a Q_0 – matrix if the problem L.C.P.(A,q,R^n_+) has a solution whenever it is feasible.</u>

Obviously, we have $Q \subset Q_o$.

Definition 8.3.3

A matrix $A \in \mathcal{M}_{nxn}(R)$, whether symmetric or not is said to be a P-matrix if and only if all its principal minors are strictly positive.

We remark that P-matrices were initially used in economics and studied by Gale and Nikaido. [D. Gale and H. Nikaido: The Jacobian matrix and global univalence of mappings. Math. Ann. 159 (1965), 81-93].

For the linear complementarity problems the class of P-matrices is important since the following classical result: "The problem L.C.P.(A,q,R^n_+) has a unique solution for every $q \in R^n$ if and only if A is a P-matrix."

Definition 8.3.4.

We say that $A = (a_{ij}) \in \mathcal{M}_{nxn}(R)$ is a Z-matrix if $a_{ij} \leq 0$ for all $i \neq j$.

Definition 8.3.5

A matrix which is a P- and Z-matrix simultaneously is called an M-matrix (or a Minkowski matrix.)

In many problems, important in economics or in some problems studied in mechanics it is important to know if the problem L.C.P.(A,q,R^n_+) has a solution which is also the least element of $\mathcal{F}(A,q,R^n_+)$.

In these problems the class of M-matrix are very important [64].

We give without proof the following classical result.

Theorem 8.3.1. [Cottle-Veinott] [A64].

The following statements about $A \in \mathcal{M}_{nxn}(R)$ are equivalent:

i) A is a Minkowsk matrix,

ii) for every $q \in R^n$, $\mathcal{F}(A,q,R^n_+)$ has a least element which is the unique solution of the problem L.C.P.(A,q,R^n_+). □

Definition 8.3.6 [A125]

We say that a matrix $A \in \mathcal{M}_{nxn}(R)$ is adequate if:

i°) det $A_{II} \geq 0$ for all $I \subset \{1, 2, \ldots, n\}$

ii°) if det $A_{II} = 0$ for some $I \subset \{1, 2, \ldots, n\}$ then the columns of A_{*I} are linearly dependent,

iii°) if det $A_{II} = 0$ for some $I \subset \{1, 2, \ldots, n\}$ then the rows of A_{I*} are linearly dependent.

Proposition 8.3.1.

A nonsingular matrix $A \in \mathcal{M}_{nxn}(R)$ is adequate if and only if it is a P-matrix.

Proof

If A is nonsingular and adequate then det $A > 0$. Let $I \subset \{1, 2, \ldots, n\}$ be a proper subset and suppose det $A_{II} = 0$.

Then the columns of A_{*I} are linearly dependent by ii°) of Definition 8.3.6.

Hence, the equation $Ax = 0$ must have a nontrivial solution and this contradicts the nonsingularity of A. The converse is trivial. □

The following result is an improvement of a result obtained by Ingleton [A125].

Theorem 8.3.2 [Cottle]

Let $A \in \mathcal{M}_{nxn}(R)$ be an adequate matrix.

If $c = Av$ for some vector v, then there exists a unique y such that for some x we have, $y = Ax + c$, $x \geq 0$, $y \geq 0$ and $\langle x,y \rangle = 0$.

Moreover, when A is nonsingular x is unique.

Proof

The proof is long and this is the paper goal of the [A44]. □

Corollary

If $A \in \mathcal{M}_{nxn}(R)$ is a P-matrix then for every $q \in R^n_+$ the problem L.C.P.(A,q,R^n_+) has a solution and this solution is unique. □

Remark 8.3.1

If we denote by $\mathcal{P}$ the class of P-matrices then from the Corollary of Theorem 8.3.2 we have the $\mathcal{P} \subset \mathcal{Q}$.

The converse of the precedent corollary is also true. This is a consequence of one of the next results which is also a constructive method to decide if a matrix is a P-matrix.

In this sense we note that for the mathematical programming and for practical problems a very important goal is to characterize every class of matrices constructively, that is, to determine in a finite number of (algebraic) steps if a given matrix $A \in \mathcal{M}_{nxn}(R)$ is an element of a determined class.

Definition 8.3.7 [158]

A matrix $A \in \mathcal{M}_{nxn}(\mathbb{R})$ is regular if the system:

$$\left\| \begin{array}{l} A_{i*}x + t = 0 \quad \text{for} \quad i \in I_+(s), \\ A_{i*}x + t \geq 0 \quad \text{for} \quad i \in I_o(x) \\ 0 \neq x \geq 0, \ t \geq 0 \end{array} \right.$$

is inconsistent (where $I_+(x) = \{i | x_i > 0\}$ and $I_o(x) = \{i | x_i = 0\}$)

The next result establishes a relation between regular matrices and Q-matrices.

Theorem 8.3.3 [Karamardian]

If $A \in \mathcal{M}_{nxn}(R)$ is a regular matrix then the problem L.C.P. (A,q,R^n_+) has a solution for every $q \in R^n$, that is A is a Q-matrix.

Proof

This result is a consequence of several results proved in [A158]. □

Theorem 8.3.4

Let $A \in \mathcal{M}_{nxn}(R)$ be an arbitrary matrix. If the problem L.C.P.(A,q,R^n_+) has a unique solution for each $q \in \{A_{*1}, \dots, A_{*n}, e\}$, where $e = \begin{bmatrix} 1 \\ \vdots \\ 1 \end{bmatrix}$ then A is a Q-matrix.

Proof

We prove this theorem by two steps.

a) If the problem L.C.P.(A,q,R^n_+) has a unique solution whenever $q \in \{A_{*1}, \dots, A_{*n}\}$ then $x_* = 0$ is the unique solution corresponding to $q = 0$.

Indeed, we suppose that the problem L.C.P.$(A,0,R^n_+)$ has another solution $z^* \neq 0$. Then without loss of generality we can assume that z^* has the form

$$z^* = (z^*_1, z^*_2, \ldots, z^*_k, 0, 0, \ldots, 0) \text{ with } z^*_i > 0 \ ; \ 1 \leq i \leq k \quad \text{and}$$

$$y^* = (0, 0, \ldots, 0, y^*_{k+1}, \ldots, y^*_n) \text{ with } y^*_i \geq 0, \ k+1 \leq i \leq n.$$

Moreover, we can assume that $z^*_1 = 1$.

Now, considering the problem L.C.P.(A,q,R^n_+) corresponding to $q = A_{*1}$, we can verify that $\bar{z}^1 = (0, z^*_2, \ldots, z^*_k, 0, \ldots, 0)$ and $\bar{z}^2 = (1, 2z^*_2, \ldots, 2z^*_k, 0, \ldots, 0)$ are two different solutions of the problem L.C.P.(A,A_{*1}, R^n_+), which is impossible.

b). <u>If the problem L.C.P.(A,q,R^n_+) has a unique solution whenever $q \in \{A_{*1}, \ldots, A_{*n}, e\}$ then A is a regular matrix.</u>

Indeed, for every $x \in R^n_+$ let $I_+(x)$ and $I_o(x)$ as defined in <u>Definition 8.3.7.</u>

From a) and the fact that the problem L.C.P.$(A, e R^n_+)$ has a unique solution we have that the system

$$\left\| \begin{array}{l} A_{*i}x + t = 0, \text{ for } i \in I_+(x) \\ A_{*i}x + t \geq 0, \text{ for } i \in I_o(x) \\ 0 \neq x \geq 0, \ t \in \{0,1\} \end{array} \right.$$

is inconsistent, which implies that A is a regular matrix.

Now by <u>Theorem 8.3.3</u> we obatain that A is a Q-matrix. □

An interesting method to study when a matrix is a P-matrix was developed by Habetler and Kostreva in [A116].

We denote by K_i $(i = 1, 2, \ldots, 2^n)$ the 2^n orthants of R^n and given a matrix $A \in \mathcal{M}_{nxn}(R)$ and a vector $q \in R^n$ we consider for every $i = 1, 2, \ldots, 2^n$ the problem, L.C.P.(A,q,K_i).

We remark that for every K_i we have $K^*_i = K_i$, also we can transform the 2^n problems L.C.P.(A,q,K_i) into 2^n linear complementarity problems on R^n_+ ; corresponding to each orthant K_i there exists a unique diagonal matrix D_i with plus or minus ones down the diagonal and such that $x \in K_i$ if and only if $D_i x \geq 0$.

Then 2^n problems L.C.P.(A,q,K_i) can be written as the 2^n usual complementarity problems L.C.P.$(D_i A D_i, D_i q, R^n_+)$.

Given $q \in R^n$ and $A \in M_{nxn}(R)$, suppose $x \in R^n$ satisfies the complementarity condition $(Ax + q)_i x_i = 0$; $i = 1, 2, \ldots, n$.

The vector $z = x + (Ax + q)$ is called the complementary point of (A,q) associated with x.

The problem L.C.P.(A,q,K_i) is then equivalent to showing the existence of a complementary point in K_i.

Let $\sum$ denote the set of all complementary points for a given q and A.

Definition 8.3.8

1°) We say that $\sum$ is pigeonholed if the interior of each orthant in R^n contains one and only one complementary point.

2°) We say that $A \in M_{nxn}(R)$ is a nondegenerate matrix if all the principal minors of A are nonzero.

3°) Given z, a complementary point of (A,q) we say that z is a nondegenerate complementary point if it has no zero components.

4°) A vector $q \in R^n$ is called nondegenerate with respect to A if for each $x \in R^n$ at most n of the 2n variables $(y(x), x)$ are equal to zero, where $y(x) = Ax + q$. (This is equivalent to saying that the vector q does not lie in any subspace generated by n - 1 or less column vectors of $\{I_{*1}, I_{*2}, \ldots, I_{*n}, -M_{*1}, \ldots, -M_{*n}\}$).

If q is nondegenerate with respect to A, the complementary points of (A,q) are nondegenerate.

However, the converse need not be true.

Also, pigeonholing of $\sum$ does not imply that q is nondegenerate with respect to A.

Proposition 8.3.2

If $A \in M_{nxn}(R)$ is a P-Matrix and $q \in R^n$ then:

i). the problem L.C.P.(A,q,K_i) has a unique solution for each $i = 1, 2, \ldots, 2n$,

ii). if q is such that (A,q) has no degenerate complementary points, then the complementary points are pigeonholed.

Proof

Since A is a P-matrix we have that the problem L.C.P.(A,q,R^n_+) has a unique solution for each $q \in R^n$.

Our proposition is a consequence of the fact that, if A is a P-matrix then for each $i = 1, 2, \ldots, 2^n$ the matrix $D_i A D_i$ is also a P-matrix and of the fact that the problem L.C.P.($D_i A D_i$, $D_i q$, R_+) <u>is equivalent to the problem L.C.P.(A, q, K_i)</u>. □

The next result shows that pigeonholing for any one q such that (A,q) has only nondegenerate complementary points characterizes A as a P-matrix.

Given $A \in \mathcal{M}_{nxn}(R)$ we define: $\mathcal{A}_A = \{q \in R^n \mid (A,q)$ has only nondegenerate complementary points$\}$,

$\mathcal{H}_A = \{q \in R^n \mid \sum$ corresponding to (A,q) is pigeonholen$\}$.

From <u>Definition 8.3.8</u> we have that $\mathcal{H}_A \subset \mathcal{A}_A$

A consequence of <u>Proposition 8.3.2</u> and of the principal results proved in [A116] is the following result.

<u>Theorem 8.3.5</u> [Habetler-Kostreva].

<u>$A \in \mathcal{M}_{nxn}(R)$ is a P-matrix if and only if $\mathcal{H}_A = \mathcal{A}_A$</u>.

In conclusion, if we examine all the complementary points of (A,q) for a given q and there are no degenerate ones, then A is a P-matrix if and only if the complementiary points are pigionholed.

We will finish the section with an interesting open problem on Q-matrices.

We denote by e_i; $i = 1, 2, \ldots, n$ the standard basis of R^n.

In 1979 Kelly and Watson proved the following result [A159].

<u>Proposition 8.3.3</u>

<u>Let $A \in \mathcal{M}_{2x2}(R)$ be a nondegenerate matrix. Then A is a Q-matrix if and only if L.C.P.(A, $-e_1$, R^2_+) and L.C.P.(A, $-e_2$, R^2_+) have solutions.</u>

<u>Proof</u> [A159]

We prove the sufficiency only, since the necessity is trivial.

So, we suppose that A is not a Q-matrix but we assume that the problem L.C.P.(A, $-e_1$, R^2_+) has a solution.

If we denote by $\mathcal{C}(-A_{*1}, v)$, where $v \in \{e_2, -A_{*2}\}$ the convex cone generated by $\{-A_{*1}, v\}$, that is $\mathcal{C}(-A_{*1}, v) = \{\alpha_1(-A_{*1}) + \alpha_2 v \mid \alpha_1, \alpha_2 \geq 0\}$, then we have $-e_1 \in \mathcal{C}(-A_{*1}, v)$.

Suppose $v = e_2$, in which case $-A_{*1}$ must lie in the third quadrant.

Since A is not a Q-matrix, $-A_{*2}$ must lie in the first or second quadrant with det $A < 0$. But then the problem L.C.P.$(A, -e_2, R^2_+)$ has no solution. If $v = -A_{*2}$ then $-e_1 \in \mathcal{C}(-A_{*1}, -A_{*2})$ (the convex cone generated by $\{-A_{*1}, -A_{*2}\}$) and since A is not a Q-matrix we have that $-A_{*1}$ must lie in the third quadrant, which was the previous case.

Since Proposition 8.3.3 we have the following conjecture.

Conjecture [Kelly-Watson] [A159]

Let $A \in M_{nxn}(\mathbb{R})$ be a nondegenerate matrix.

Then A is a Q-matrix if and only if the problem L.C.P.$(A, -e_i, R^n_+)$ has a solution for $i = 1, 2, \ldots, n$.

It seems that this conjecture is not solved. However, several authors have studied this problem and several interesting relations with the spherical geometry are established. [A159], [A61], [A278], [A171].

8.4. Some results about the cardinality of solution set

We present in this section some results giving interesting informations about the cardinality of solution set for some complementarity problems.

Let R^n denote n-dimensional Euclidean space with the inner product

$$< x,y > = \sum_{i=1}^{n} x_i y_i.$$

Given $A \in M_{nxn}(R)$ and $q \in R^n_+$ we consider the problem L.C.P.(A,q,R^n_+) and we denote by $n(A,q,R^n_+)$ the cardinal of solution set of this problem.

We suppose that A is positive semi-definite.

Since a complementarity problem is equivalent to a variational inequality (see chap. 3) we have, when A is positive semi-definite, that our problem L.C.P(A,q,R^n) is equivalent to a variational inequality associated to a monotone operator and as in [B7] is showed, the solution set is in this case a convex set.

So, if A is positive semi-definite then $n(A,q,R^n_+)$ is either zero, one or infinity. In this sense, we can decide if $n(A,q,R^n_+)$ is zero, one or infinity using the following procedure proposed by Kaneko in [A146].

This procedure is based on the complementarity pivot algorithm or on some extensions of this algorithm as presented in the book [C52].

The validity of this procedure is proved in [C37].

I. We solve L.C.P.(A,q,R^n_+) by a <u>pivoting method</u>. If there is no solution, stop; $n(A,q,R^n_+) = 0$. Otherwise, let $I \subset \{1, 2, \ldots, n\}$ be the index set such that $i \in I$ if and only if $\bar{w}_i = \bar{x}_i = 0$, where $\bar{x}$ is the solution of L.C.P.(A,q,R^n_+) computed above and $\bar{w} = q + A\bar{x}$.

If $I = \Phi$ (the empty set), then stop; $n(A,q,R^n_+) = 1$. Otherwise proceed to the next step.

II. Let $\bar{A}$ be the principapl pivotal transform of A at the final tableau in the algorithm performed in step I. □

Solve the problem L.C.P.(B,d,R^{s+1}_+) by a standard algorithm, with $d = \begin{bmatrix} 0 \\ 1 \end{bmatrix}$ and $B = \begin{bmatrix} \bar{A}_{II} & -e \\ e^T & 0 \end{bmatrix}$, where e is the vector of ones and s is at most n, the dimension of A.

$\bar{A}_{II}$ denotes the submatrix of $\bar{A}$ with rows in I and columns also in I.

Suppose that a solution (x^*, λ^*) is computed, where $x^* \in R^{|I|}$ and $\lambda^* \in R$.

Then we have: i°) if $\lambda^* = 0$ then $n(A,q,R^n_+) = \infty$; ii°) if $\lambda^* > 0$ then $n(A,q,R^n_+) = 1$. □

Remark 8.4.1

Solving the problem L.C.P.(L,d,R^{s+1}_+) is a reasonable fact since L is positive semi-definite (and hence a standard method can be used) and the dimension s+1 of L, in many cases is considerable smaller than n since it is the number of degenerate components in the computed solution of L.C.P.(A,q,R^n_+).

Suppose now that $A \in \mathcal{M}_{nxn}(R)$ is an arbitrary matrix and $q \in R^n$.

Proposition 8.4.1 [Tamir]A283]

<u>Let $A \in \mathcal{M}_{nxn}(R)$ be an arbitrary matrix such that $n(A,q,R^n_+) \geq 1$ for all $q \leq 0$. Then, for any $q \in R^n$ the inequality $n(A,q,R^n_+) < +\infty$ implies $n(A,q,R^n_+) \leq 2^n - 1$.</u>

Proof

For any $I \subseteq \{1, 2, \ldots, n\}$ consider the complementary cone gerated by the n columns $-A_{*i}$; $i \in I$ and e_i; $i \in \bar{I}$ (the complement of I with respect to $\{1, 2, \ldots, n\}$), where e_i is the ith unit vector.

The correspondence between subsets of $\{1, 2, \ldots, n\}$ and the complementary cones is not necessarily one to one.

Consider the problem:

(1): find a solution $\bar{x}$ of the problem L.C.P.(A,q,R^n_+) such that $\bar{x}_i > 0$ for $i \in I$ and $\bar{x}_i = 0$ for $i \in \bar{I}$

Obviously, if there are two different solutions for problem (1) and for some $I \subseteq \{1, 2, \ldots, n\}$ then $n(A,q,R^n_+)$ is not finite since every convex combination of these solutions is also a solution.

Hence, if $n(A,q,R^n_+)$ is finite each set $I \subseteq \{1, 2, \ldots, n\}$ contributes at most one solution to L.C.P.(A,q,R^n_+) and $n(A,q,R_+) \leq 2^n$.

If $q \ngeq 0$ then the cone defined by $I = \Phi$ does not contribute a solution and $n(A,q,R^n_+) \leq 2^n - 1$.

We remark that $n(A,0,R^n_+)$ is finite only if $n(A,0,R^n_+) = 1$.

Consider now a nonzero, nonnegative q.

By assumption L.C.P.$(A, -q, R^n_+)$ has a solution.

Defining a cone to be nondegenerate if it has a nonempty interior, we observe that the union of all the degenerate complementary cones and the proper faces of the nondegenerate cones is closed and nowhere dense in R^n.

Thus, there exists a sequence $\{q^k\}$ converging to $-q$, where $q^k \leqslant 0$ for every $k \in N$ and q^k belongs to a nondegenerate cone.

Since there is a finite number of complemenitary cones, we can assume without loss a generality (if necessary we can choose a subsequence) that for every $k \in N$, q^k belongs to the same nondegenerate cone. We have that $-q$ is in that cone.

It is then clear that q is not contained in this nondegenerate cone, since otherwise we would have that the n generators are linearly dependent, which is a contradiction to the nondegeneracy assumption. Hence we have $n(A,q,R^n_+) \leq 2^n - 1$. □

Corollary

If $A \in \mathcal{M}_{n\times n}(R)$ is a nondegenerate Q-matrix then $n(A,q,R^n_+) \leq 2^n - 1$.

The cardinality of solution set can be studied also by the topological degree.

This idea is a very nice and deep subject to develop in complementarity theory. Some results in this sense were obtained in 1981 by Kojima and Saigal [A167].

Let $(H,\langle\ ,\ \rangle)$ be a Hilbert space and let $K \subset H$ be a closed convex cone.

Given a mapping $f:K \to H$ and $q \in H$ we consider the following complementarity problem:

$$C.P.(f-q,K): \left\| \begin{array}{l} \text{find } x_o \in K \text{ such that} \\ f(x_o) - q \in K^* \text{ and} \\ \langle x_o, f(x_o) - q \rangle = 0. \end{array} \right.$$

From Proposition 7.2.1 we have that the problem C.P.(f−q, K) has a solution if and only if the mapping $\Phi(x) = P_K(x) - f(P_K(x))$ has a fixed point, that is, if and only if, the equation

(2): $\quad \Phi(x) = x$

has a solution in H. (If x_o is a solution of (2) then $x_* = P_K(x_o)$ is a solution of the problem C.P.(f−q, K)).

But, equation (2) is equivalent to the following equation:

(3): $\quad x - P_K(x) + f(P_K(x)) = q$

If we denote, $F(x) = x - P_K(x) + f(P_K(x))$ we obtain finally that the problem C.P.(f−q, K) is equivalent to solving equation:

(4): $\quad F(x) = q$

In nonlinear analysis it is well known that equation (4) can be studied by the topological degree.

In this sense Kojima and Saigal obtained in [A167] several interesting results when $H = {}^nR$ with Euclidean structure and $K = R^n_+$. In this case $P_K(x) = {}^+x$ for every $x \in R^n$.

We give as information, without proof one of the principal results proved in [A167].

For this, we introduce the following conditions:

(5): $f(0) = 0$ and f is continuous differentiable.

(6): for any sequence $\{x^k\}_{k=1}^{\infty}$ in R^n_+ such that $\|x^k\| \to \infty$ exists a subsequence $\dot{J}$ of N such that either there is a j with $f_j(x^k) \to -\infty$ for k in $\dot{J}$; or there is a j with $x^k_j > 0$ for all $k \in \dot{J}$ and $f_j(x^k) \to +\infty$ for k in J.

Theorem 8.4.1 [Kojima-Saigal] [A167]

Let f satisfy (5) and (6). Also, for each x let Df(x) have all principal minors negative. Then if:

i° $Df(0) \nless 0$ then the problem C.P.($f-q$, R^n_+) has a unique solution for each $q \nleq 0$ and $q = 0$, three solutions for $q < 0$ and either one or two solutions for $0 \neq q \leq 0$, $q_i = 0$ for at least one i.

ii° $Df(0) < 0$ then the problem C.P.($f-q$, R^n_+) has no solution for $q \nless 0$, one solution for $q \leq 0$, $q_i = 0$ for at least one i and two solutions for $q < 0$.

Proof

See [A167] □

Another method to study the solution set of a complementarity problem is to use the trnversality theory [R. Abraham and J. Robbin: Transversal mappings and flows. Benjamin, New York, 1967].

We note that it is very interesting to develop this method.

In this sense we give now a result obtained by Saigal and Simon [A267].

This result is that, for almost all continuously differentiable maps $f:R^n \to R^n$ the solution set of the problem C.P.(f, R^n_+) is discrete.

We suppose that f is a C^1 map from R^n into R^n since every C^1 map from R^n_+ into R^n can be extended as a C^1 map on all of R^n [R. Abraham and J. Robbin, Appendix A].

We denote by $C^1(R^n, R^n)$ the vector space of C^1 functions from R^n into R^n with the topology of uniform convergence of both f and its derivative Df on compact sets.

Definition 8.4.1

A subset D of a topological space is residual if D contains the intersection of a countable number of open dense sets.

Remark 8.4.2

The intersection of a countable number of residual sets is residual.

Since $C^1(R^n, R^n)$ is a complete metric space, by the Baire Category Theorem any residual subset of it is dense.

Definition 8.4.2

We say that a property is generic if it holds for a residual subset of a topological space.

We introduce now some notations.

$Df(x)$ denotes the nxn matrix $\left(\frac{\partial f_i(x)}{\partial x_j}\right)$

If A is a non-empty subset of {1, 2, ..., n}, we denote:

$$R^A = \{(x_1, x_2, \ldots, x_n) \in R^n | x_i = 0 \text{ for } i \notin A\}$$

$$R_+^A = R^A \cap R_+^n .$$

$R^A \oplus R^B$ is the direct sum of R^A and R^B where B is also a subset of {1, 2, ..., n}

$|A|$ is the cardinality of A which is also equal to the dimension of R^A.

Let A, B, C, D $\subset$ {1, 2, ..., n}, $A \cup B = C \cup D = \{1, 2, \ldots, n\}$ and $A \cap B = C \cap D = \Phi$.

So, $R^n = R^A \oplus R^B = R^C \oplus R^D$ and hence each $x \in R$ has a unique decomposition of the form $x = x_A + x_B$, where $x_A \in R^A$ and $x_B \in R^B$ or $x = x_C + x_D$, where $x_C \in R^C$ and and $x_D \in R^D$.

For simplicity we abbreviate $x_A + x_B$ to (x_A, x_B) and $x_C + x_D$ to $[x_C, x_D]$.

If $f \in C^1(R^n, R^n)$ is given we write $f: R^A \oplus R^B \to R^C \oplus R^D$ as

$$f(x_A, x_B) = [f_C(x_A, x_B), f_D(x_A, x_B)],$$

where $f_C(x_A, x_B)$ is the component of $f(x_A, x_B)$ in R^C.

Let $\tilde{f} : R^A \to R^C$ denote the map $x_A \to f_C(x_A, 0)$ and let $D\tilde{f}_C(x_A)$ be its derivative. We consider $Df_C(x_A)$ as an nxn matrix of partial derivatives with $|D|$ rows of zeroes and $|B|$ columns of zeroes.

<u>Proposition 8.4.2</u>

<u>$x \in R_+^n$ is a solution of the problem C.P.(f, R_+^n) if and only if there are non-empty subsets A and B of {1, 2, ..., n} such that $A \cap B = \Phi$ and $x \in R_+^A \cap f^{-1}(R_+^B)$.</u>

<u>Proof</u>

The proof is immediate. □

<u>Definition 8.4.3</u>

<u>Let $f: \mathcal{M} \to \mathcal{N}$ be a C^1 map from manifold $\mathcal{M}$ to manifold $\mathcal{N}$. Let U be a submanifold of $\mathcal{M}$ and V a submanifold of $\mathcal{N}$.</u>

Then $f(U)$ is transversal to V if and only if for $x \in U$ and $y = f(x)$, either $y \notin V$ or $y \in V$ and the tangent space $T_y N$ of y in N is the sum (not necessary direct sum) of the tangent space $T_y V$ of y along V and the image of the tangent space $T_x U$ of x along U under the derivative of f, that is $T_{f(x)} N = T_{f(x)} V + Df(x)(T_x U)$.

Remark 8.4.3

Given a C^1 map $f: R^A \oplus F^B \to R^C \oplus R^D$ and $U \subset R^A$, $f(U)$ is transversal to R^D if, whenever $x_A \in U$ and $f_C(x_B, 0_B) = 0$ the rank of $Df_C(x_A)$ is $|C|$.

The following result is necessary for the next theorem but its proof is long. [A267].

Proposition 8.4.3

Let A, B, C, D be subsets of $\{1, 2, \ldots, n\}$ with $A \cap D = \Phi$ and $R^A \oplus R^B = R^n = R^C \oplus R^D$.

Then $F = \{f \in C^1(R^n, R^n) \mid f(R^A)$ is transversal to $R^D\}$ is a residual subset of $C^1(R^n, R^n)$.

If $A \cup D \neq \{1, 2, \ldots, n\}$ then $f \in F$ implies $f(R^A) \cap R^D = \Phi$.

Proof

See [A267]. □

Theorem 8.4.2. [Saigal-Simon]

For some residual set of g in $C^1(R^n, R^n)$ the solution set for the problem C.P.(g, R^n_+) is discrete.

Proof. [A267]

Let A, D be subsets of $\{1, 2, \ldots, n\}$ with $A \cap D = \phi$. Let

$F^{A,D} = \{f \in C^1(R^n, R^n) \mid f(R^A)$ is transversal to $R^D\}$

Let G be the intersection of the $F^{A,D}$ for all such pairs A,D. Since each $F^{A,D}$ is a residual set, G is a residual subset of $C^1(R^n,R^n)$. If $f \in F^{A,D}$ and $A \cup D \neq \{1, 2, \ldots, n\}$ then $R^A \cap f(R^D)$ is empty.

If $A \cup D = \{1, 2, \ldots, n\}$ and $x \in R^A \cap f^{-1}(R^D)$, then $A \equiv C$, $\tilde{f}_C: R^A \to R^A$, $\tilde{f}_C(x) = 0$ and $D\tilde{f}_C(x)$ is non-singular.

By the Inverse Function Theorem, there is an open set $W \subset R^A$ around x on which $\tilde{f}_C$ is a diffeomorphism (one-to-one and onto f(W)), and so x is the only point in W with $\tilde{f}_C = 0$, that is x is the only point in $R^A \cap f^{-1}(R^D)$. Therefore $R^A \cap f^{-1}(R^D)$ is discrete. (It clearly may be an infinite set).

By Proposition 8.4.2 we have that $R^A \cap f^{-1}(R^D)$ is the solution set for the problem C.P.(f, R^n_+).□

Since $x + f(x) > 0$ for a solution x of the problem C.P.(f, R^n_+) is equivalent to $A \cup D = \{1, 2, \ldots, n\}$ and $A \cap D = \phi$ for each pair A and D for which $x \in R^A \cap f^{-1}(R^D)$, and since $D\tilde{f}_C(x)$ is the principal minor of Df(x) corresponding to the indices for which $x_i > 0$, we obtain from Theorem 8.4.2 the following resuslt.

Corollary

Let $f \in C^1(R^n_+, R^n)$. Suppose each solution $x \in R^n_+$ to the problem C.P.(f, R^n_+) has the following properties:

i) $x + f(x) > 0$,

ii) if $B \subset \{1, 2, \ldots, n\}$ is the set of indices i such that $x_i > 0$, the principal minor of the matrix $D\tilde{f}(x)$ corresponding to B is non-zero.

Then the solution set of the problem C.P.(f, R^n_+) is discrete and the set of all such maps f is a residual subset of $C^1(R^n_+, R^n)$. □

8.5. Alternative theorems and complementarity problems.

Some alternative theorems for variational inequalities can be used to obtain new existence theorems for the complementarity problem.

This idea was introduced recently by J.M. Borwein [C4] and probably is susceptible to be developed.

Let $\langle E, E^* \rangle$ be a dual system of Banach spaces where E* is the dual norm of E.

Let $K \subset E$ be a closed convex cone and K* the dual of K.

We suppose given a mapping $f: K \to E^*$.

Definition 8.5.1

We say that f is of monotone-type on a subset C of K if either:

i°) $\langle f(x),x \rangle$ is weakly lower semi-continuous on C or

ii°) f is monotone and hemicontinuous on C.

Remark 8.5.1

Both classes of maps are pseudomonotone in the sense of Brezis [C5], [C6].

Given an arbitrary element $q \in E^*$, we consider the following complementarity problem:

$$C.P.(f,q,K): \quad \left\| \begin{array}{l} \text{find } x_o \in K \text{ such that} \\ f(x_o) + q \in K^* \text{ and} \\ \langle x_o, f(x_o) + q \rangle = 0. \end{array} \right.$$

If two continuous sublinear functionals $g,h:E \to R$ are given, we define $K(g) = \{y \in K | g(y) \le 1\}$ and we introduce the following hypothesis:

$$H(g,h,q): \quad \left\| \begin{array}{l} \text{the variational inequality} \\ \langle f(x) + (1-h(x))q, y-x \rangle \ge 0, \ \forall y \in K(g) \\ \text{is solvable for some } x \in K(g). \end{array} \right. \qquad (H)$$

Theorem 8.5.1

For f,g,h given as above such that $f(0) = 0$, each of the following implies (H):

1°) f is of monotone-type on K(g) and K(g) is norm compact,

2°) f is of monotone-type on K(g), K(g) is weakly compact and h is weakly continuous.

Proof

a. If $\langle f(x),x \rangle$ is weakly lower semicontinuous we may apply Fan's minimax inequality theorem [C15] to the function $\Phi(x,y) = \langle f(x) + (1-h(x)) q, y-x \rangle$ and we obtain a point $x \in K(g)$ with $\Phi(x,y) \ge 0$ for all $y \in K(g)$.

b. Suppose now that f is monotone and hemicontinuous on K(g).

Let $D(y) = \{x \in K(g) | \langle f(y) + (1-h(x))q, y-x \rangle \ge 0\}$ for $y \in K(g)$. Let M be a finite subset of K(g) and since conv (M) is compact, the argument of (a) implies the existence of a solution x in K(g) to $\langle f(x) + (1-h(x))q, y-x \rangle \ge 0$.

Since f is monotone on K(g) we can show that x lies in D(y) for each $y \in M$. Because each D(y) is weakly compact there must be some point x_o in each D(y).

Let $x(t) = x_o + t(y-x_o)$ for $t \in (0,1)$.

Then $x(t) \in K(g)$ and we have $t \langle f(x(t)) + (1-h(x_o))q, y-x_o \rangle =$

$= \langle f(x(t)) + (1-h(x_o))q, x(t) - x_o \rangle \ge 0.$

Dividing be t and computing the limit when t tends to 0 we obtain that x_o solves the desired variational inequality.

If $g:E \to \overline{R}$ is a convex function we denote by $\partial g(x)$ the <u>subgradient</u> of g at x, that is,

$\partial g(x) = \{x^* \in E^* | <x^*,y-x> \leq g(y)-g(x) \text{ for } y \in E\}$.

Theorem 8.5.2 [Alternative]

Suppose that H(g,h,q) holds. Then either one can solve

1° $<f(x) + (1-h(x))q, y-x> \geq 0$ for all $y \in K$, for some $x \in K(g)$,

or one can solve

2°) $<f(x) + (1-h(x))q + tp, y-x> \geq 0$ for all $y \in K$, for some $x \in K$ with $g(x) = 1$, some $t \geq 0$ and $p \in \partial g(x)$.

Proof

We denote by x the solution given by H(g,h,q) and we consider the convex program:

$\Phi(x) = \min \{\Phi(y) | y \in K \text{ and } g(y) \leq 1\}$ where $\Phi(y) = <f(x) + (1-h(x))q, y-x>$;

As is indicated in [C4], since $g(0) = 0$ Slater's condition holds and using a special convex multiplier theorem, we have a real number $t \geq 0$ such that

(1): $\Phi(y) + tg(y) \geq \Phi(x) + tg(x)$ for all $y \in K$ and such that $t(g(x) - 1) = 0$.

Since $\Phi(x) = 0$ we obtain that 1°) is true whenever $g(x) < 1$.

If $g(x) = 1$, we obtain 2°) computing subgradients in (1). We deduce

$0 \in \partial \Phi(x) + t\, \partial g(x) - (K-x)^*$, which is exactly 2°). □

To obtain from Theorem 8.5.2 some existence theorems for the problem C.P.(f,q,K) we generally have g sublinear and h = g or h = 0.

Obviously, linearity of h ensures weak continuity.

We say that $f:K \to E^*$ is <u>coercive</u> on K if $\frac{<f(x),x>}{\|x\|}$ tends to $+\infty$ as $\|x\|$ tends to $+\infty$ in K.

Theorem 8.5.3

1°) If H(g,o,q) is satisfied then either one can solve C.P.(f,q,K) or one can solve C.P.(f,q + tp,K) by some $x \in K$ with $g(x) = 1$ where $t > 0$ and p is in $\partial g(x)$.

2°) If f is of monotone-type on K, E is reflexive and there exists $r_o > 0$ such that if $x \in K$ and $\|x\| = r_o$ one can find $y(x) \in K$ with $\|y(x)\| < r_o$ such that $<f(x) + q, x-y(x)> \geq 0$, then C.P.(f,q,K) has a solution.

3°) If f is of monotone-type and coercive on K then C.P.(f,q,K) is solvable for all $q \in E^*$.

Proof

1°) This part is exactly Theorem 8.5.2 if we use the fact that C.P.(f,q,K) is equivalent to a variational inequality. We consider in this case $h = 0$.

2°) If E is reflexive the $H(r\|.\|,0,q)$ is satisfied for any $r > 0$.

We choose $r = \frac{1}{r_o}$ and suppose that C.P.(f,q,K) has no solution.

Using part 1°) we obtain a solution x of the problem C.P.(f,q + tp,K) with $\|x\| = r_o, t > 0$, $\|p\| = 1$ and $p(x) = r_o$

But, if $\|y\| < r_o$ and $y \in K$ we have $\langle f(x) + q, x-y \rangle \le tp(y-x) \le t(\|y\| - \|x\|) < 0$

which is impossible (since our assumptions).

3°) If f is of monotone-type and coercive on K then we remark that the all assumptions of part 2°) are satisfied with $y(x) = 0$, for r_o sufficiently large.

We consider now an existence theorem for the problem C.P.(f,q,K) obtained by Gowda [C18] as consequence of Theorem 8.5.1. The proof in some sense is also similar to the proof of Theorem 8.5.2.

Suppose given $f:K \to E^*$ and $q \in E^*$ and we denote by $\langle E,E^* \rangle$ the natual duality.

Recall that f is copositive on K if $\langle f(x),x \rangle \ge 0$ for all $x \in K$ and we say that f

is ρ-positive homogeneous on K(where $\rho > 0$) if $f(\lambda x) = \lambda^{\rho} f(x)$, for all $\lambda \in R_+$ and for all $x \in K$.

We denote $S_o = \{x \in K | f(x) \in K^* \text{ and } \langle f(x),x\rangle = 0\}$ and we recall that $\sigma(E^*,E)$ is the weak * topology on E^*.

Theorem 8.5.4

Let $K \subset E$ be locally compact convex cone (not necessary pointed) and $f:K \to E^*$ copositive, positive homogeneous, and of monotone type on K.

If $0 \ne x \in S_o$ implies $\langle q,x \rangle > 0$ then the problem C.P.(f,q,K) has a solution. Moreover, if f is continuous from the topology of E to the weak * topology of E^*, then the solution set is bounded (and hence compact).

Proof

Since K is locally compact but not necessary pointed $(K \cap (-K) \ne \{0\})$ then from Theorem 3.12.8 of the book [C32] we have that $K = L + M$, where M is the finite dimensional subspace $K \cap (-K)$ and L is a convex cone with a compact base B given B $B = \{x \in L | \langle e,x \rangle = 1, \text{ for some } e \in K^*\}$.

So, every $x \in K$ has a representation of the form $x = \lambda b + m$, where $\lambda \ge 0$, $b \in B$ and $m \in M$.

Consider the mapping $g(x) = \lambda + \|m\|$, for every $x \in K$ and we remark that g is continuous. The set $\{x \in K | g(x) \leq 1\}$ is convex compact.

Since the set $\{x \in K | g(x) \leq 1$ and $< q,x > \leq 0\}$ is convex compact we can apply <u>Theorem 8.5.1</u> and we obtain that the variational inequality,

$\hat{H}(g,g,q)$: $< f(x) + (1-g(x))q, y - x > \geq 0$, for all $y \in K$ such that $g(y) \leq 1$ and $< q,y > \leq 0$.

is solvable for some $x \in K$ with $g(x) \leq 1$ and $< q,x > \leq 0$.

Let x be the solution defined by $\hat{H}(g,g,q)$ and consider the function $\Phi(y) = < f(x) + (1-g(x))q, y - x >$, for $y \in K$.

Since zero solves the problem C.P.(f,q,K) when $q \in K^*$, we suppose that $q \notin K^*$.

Then there is a $k \in K$ such that $g(k) < 1$ and $q,k > < 0$.

Thus the convex program $\Phi(x) = \min \{\Phi(y) | g(y) \leq 1, < q,y > \leq 0$ and $y \in K\}$, satisfies the Slater condition.

Hence using <u>Theorem 2 pg 68</u> of [C24] we obtain two numbers $t \geq 0$ and $s \geq 0$ such that,

(2): $\Phi(x) \leq \Phi(y) + t(g(y) + s < q,y >$; $\forall y \in K$ and $t((g) - 1) = 0 = s < q,x >$.

If $g(x) < 1$, then $t = 0$ and (2) implies $< f(x) + (1-g(x) + s)q, y - x > \geq 0$ (since $\Phi(x) = 0$ and $s < q,x > = 0$).

Also, $g(x) < 1$ gives $1 - g(x) + s > 0$ and using the homogeneity of f we obtain that $(1-g(x) + s)^{-\frac{1}{\rho}} x$ solves the problem <u>C.P.(f,q,K)</u>.

Now suppose that $g(x) = 1$.

Writing (2) in the subdiferential form [C24 Theorem 2' pg 69], we get

$$0 \in \partial \Phi(x) + t \, \partial g(x) + sq - (K-x)^*.$$

In this case we obtain,

(3): $< f(x) + tp + sq, y-x > \geq 0$; $\forall y \in K$, for some $p \in \partial g(x)$.

From this, we get $< f(x) + tp + sq, x > = 0$, that is, $< f(x),x > + t < p, x > = 0$.

Since $p \in \partial g(x)$ we have $< p,x > \geq g(x) = 1$; hence, $< f(x),x > = 0$ (since f is copositive) and $t < p,x > = 0$, which leads to $t = 0$.

If $s = 0$, then (3) gives $f(x) \in K^*$ and $< f(x),x > = 0$, that is we have $x \in S_o$. Since $g(x) = 1$ and $< q,x > \leq 0$, we have a contradiction.

Hence $s > 0$ and in this case (3) shows (since $t = 0$ and f is positive homogeneous) that $s^{-\frac{1}{\rho}} x$ solves the problem <u>C.P.(f,q,K)</u>.

Suppose that $f:K \to (E^*, \sigma(E^*,E))$ is continuous. From this it follows that the solution set of C.P.(f,q,K) is closed.

Suppose that the solution set is unbounded. In this case we can choose an unbounded sequence $\{x_n\}$ (consisting of nonzero elements) in the solution set. Using the decomposition $K = L \oplus M$, we can write $x_n = \lambda_n b_n + m_n$ and observe from the non-negativity of λ_n that $g(x_n) = \lambda_n + \|m_n\|$ is an unbounded sequence in R_+.

The sequence $\{g(x_n)^{-1} x_n\}$ is contained in the compact set $\{x \in K | g(x) = 1\}$, hence $\{g(x_n)^{-1} x_n\}$ has a convergent subsequence $\{g(x_{n_k})^{-1} x_{n_k}\}$. We denote

$\bar{x} = \lim_{k \to \infty} \dfrac{x_{n_k}}{g(x_{n_k})}$ and we have $\bar{x} \in K$ and $g(\bar{x}) = 1$.

Then $< f(x_{n_k}) + q, y >\geq 0$, (for all $y \in K$ and for all k) leads to

$< f\left(\dfrac{x_{n_k}}{g(x_{n_k})}\right) + qg(x_{n_k})^{-\rho}, y >\geq 0$, which, upon taking limits (and using the continuity of f) gives

(4): $< f(\bar{x}), y >\geq 0, \ \forall y \in K$

Also,

(5): $< f(x_{n_k}), x_{n_k} > + < q, x_{n_k} > = 0 \ (\forall k)$ (since x_{n_k} are solutions), leads to

$< q, x_{n_k} >\leq 0$ (by copositivity) and hence to $< q, g(x_{n_k})^{-1} x_{n_k} >\leq 0$.

Computing the limit we obtain

(6): $< q, \bar{x} >\leq 0$.

From (5) we have,

$$< f\left(\frac{x_{n_k}}{g(x_{n_k})}\right), \frac{x_{n_k}}{g(x_{n_k})} > + \frac{<q, \frac{x_{n_k}}{g(x_{n_k})}>}{g(x_{n_k})} = 0;$$ $\forall k$ and by the continuity of f

we deduce,

(7): $< f(\bar{x}), \bar{x} > = 0$

We see that $\bar{x} \in S_o$ and $< q, \bar{x} >\leq 0$.

This contradicts our hypothesis, since $g(x) = 1$ implies that $x \neq 0$.

Thus the solution set of the problem C.P.(f,q,K) is bounded and the proof is finished. □

8.6. Again on the Implicit Complementarity Problem.

We present in this section some results about the implicit complementarity problem obatained recently by G. Isac and D. Goeleven.

Let $< E,E^* >$ be a dual system of Banach spaces. We consider on E^* the strong topology. Let $K \subset E$ be a pointed closed convex cone.

Given $r > 0$ we denote $K_r^{\leq} = \{x \in K \mid \|x\| \leq r\}$ and $K_r^{<} = \{x \in K \mid \|x\| < r\}$.

Given a subset $D \subset E$ and the mappings $S:D \to$ and $T:D \to E^*$ we consider the following problems:

S.V.I.(T,S,K,D): find $x_o \in D$ such that $< x - S(x_o), T(x_o) > \geq 0; \forall x \in K$

I.C.P.(T,S,K) find $x_o \in D$ such that $T(x_o) \in K^*$ and $< S(x_o), T(x_o) > = 0$

By a similar proof to the proof of Proposition 6.2.1 we have the following result.

Proposition 8.6.1

If $S:D \to K$ and $T:D \to E^*$ are given then the problem S.V.I.(T,S,K,D) is equivalent to the problem I.C.P.(T,S,K,D).

We use in this section the following classical result.

Theorem 8.6.1

A mapping $f:D \to 2^D$, where $D \subseteq E$, will have a fixed point if the following conditions are satisfied:

(i): *E is locally convex space and the set D is nonempty, compact and convex,*

(ii): *the set $f(x)$ is nonempty and convex for all $x \in D$ and the preimages $f^{-1}(y)$ are relatively open with respect to D for all $y \in D$.*

Proof

A proof of this theorem is in [E. Zeidler: Nonlinear functional analysis. Tome 1 Springer-Verlag, p.453].

We denote by S.V.I(R,S,D) the following variational inequality:

find $x_o \in D$ such that

$< x - S(x_o), T(x_o) > \geq 0; \forall x \in D.$

Theorem 8.6.2

Let $D \subset E$ be a nonempty compact convex set, $S:D \to K$ and $T:D \to E^*$ continuous mappings.

If, for every $x \in D$ we have $< S(x), T(x) > \leq < x, T(x) >$, then the problem S.V.I(T,S,D) has a solution.

Proof

If the problem S.V.I.(T,S,D) does not have a solution then,

(1); $(\forall x \in D)(\exists u \in D)(< u - S(x), T(x) > < 0)$

Let $f: D \to D$ be the point-to-set mapping defined by:

$$f(x) = \{u \in D \mid < u-S(x), T(x) > < 0\}$$

For every $x \in D$, $f(x)$ is nonempty and convex.

Since T and S are continuous, the mapping $v \to < x - S(v), T(v) >$ is continuous and we have that $f^{-1}(y) = \{x \in D \mid y \in f(x)\} = \{x \in D \mid < y - S(x), T(x) > < 0\}$ is relatively open with respect to D.

Hence, by Theorem 8.6.1 there is an element $x_* \in D$ such that $x_* \in f(x_*)$, that is $< x_* - S(x_*), T(x_*) > < 0$, which is impossible since for every $x \in D$ we have $<S(x), T(x) > \le < x, T(x) >$. □

We denote by I.C.P.(T,S,K) the problem: find $x_o \in K$ such that $S(x_o) \in K$, $T(x_o) \in K^*$ and $< S(x_o), T(x_o) > = 0$.

Theorem 8.6.3

Let $K \subset E$ be a pointed locally compact cone and $S: K \to E$, $T: K \to E^*$ continuous mappings. If the following assumptions are satisfied:

1°): there is $r > 0$ such that $S(K_r^{\le}) \subset K$,

2°): there is an element $u_o \in K$ such that $\|S(u_o)\| < r$, $ss(u_o) \in K$ and $< x - S(u_o), T(x) > \ge 0$, for all $x \in K$ satisfying $r \le \|x\| \le \max(r, r_o)$, where $\sup\{\|S(u)\| \mid u \in K_r^{\le}\} \le r_o$,

3°): $< S(x), T(x) > \le < x, T(x) >$; for all $x \in K_r^{\le}$,

then the problem I.C.P.(T,S,K) has a solution $x_* \in K_r^{\le}$ such that $\|S(x_*)\| \le \max(r, r_o)$.

Proof

Since K is locally compact we have that $K_r^{\le}$ is a convex compact set.

Applying Theorem 8.6.2 with $D = K_r^{\le}$ we obtain an element $x_* \in K_r^{\le}$ such that,

(2): $< x - S(x_*), T(x_*) > \le 0$; $\forall x \in K_r^{\le}$.

We remark that $S(x_*) \in K$. Two cases are posssible.

(a) $\|S(x_*)\| < r$. If $x \in K$ is an arbitrary element then there is a sufficiently small $\lambda \in]0,1[$ such that $w = \lambda x + (1 - \lambda) S(x_*) \in K_r$.

If in (2) we put $x = w$, we have $\lambda < x - S(x_*), T(x_*) > \geq 0$, that is, $< x - S(x_*), T(x_*) > \geq 0$, for all $x \in K$, and by Proposition 8.6.1 we obtain that x_* is a solution of the problem I.C.P.(T,S,K).

(b) $\|S(x_*)\| \geq r$. In this case we have $r \leq \|S(x_*)\| \leq \max(r, r_o)$ and by assumption 2°) we obtain

(3): $\quad < S(x_*) - S(u_o), T(x_*) > \geq 0$ and since for every $x \in K_r^{\leq}$ we have,

(4): $\quad < x - S(x_*), T(x_*) > \geq 0$

From (3) and (4) we deduce,

$< x - S(u_o), T(x_*) > = < x - S(x_*) - S(u_o), T(x_*) > =$

$= < x - S(x_*), T(x_*) > + < S(x_*) - S(u_o), T(x_*) > \geq 0$, that is, we have

(5): $\quad < x - S(u_o), T(x_*) > \geq 0; \forall x \in K_r^{\leq}$.

If $x \in K$ is an arbitrary element, then there is a sufficiently small $\lambda \in]0,1[$ such that $v = \lambda x + (1-\lambda) S(u_o) \in K_r^{\leq}$.

Now, if we put $x = v$ in (5) we obtain,

(6): $\quad < x - S(u_o), T(x_*) > \geq 0, \forall x \in K$.

Since $\|S(u_o)\| < r$ we can put $x = S(u_o)$ in (2) and we deduce,

(7): $\quad < S(u_o) - S(x_*), T(x_*) > \geq 0$.

From (6) and (7) we obtain

(8): $\quad < x - S(x_*), T(x_*) > \geq 0; \forall x \in K$ and because $S(x_* \in K$ from (8) and Proposition 8.6.1 we have that x_* is a solution of the problem I.C.P.(T,S,K).

If $X, Y \subset E$ are subsets we say that $\{X_n\}_{n\in N}$ (resp. $\{Y_n\}_{n\in N}$) is a filtration of X (resp. of Y) if for every $n \in N$, X_n(resp. Y_n) is a subset of X (resp. of Y) such that $n \leq m$ implies $X_n \subseteq X_m$ (resp. $y_n \leq Y_m$) and $X = \overline{\bigcup_{n\in N} X_n}$ (resp. $Y = \overline{\bigcup_{n\in N} Y_n}$).

If $f: X \to y$ is a mapping we say that f is subordonate to the filtrations $(\{X_n\}_{n\in N}, \{Y_n\}_{n\in N})$ if there is $n_o \in N$ such that for every $n \geq n_o$ we have $f(X_n) \subseteq Y_n$.

Given $r > 0$ and $K(K_n)_{n\in N}$ a Galerkin cone in E we have that $\{K_{nr}^{\leq}\}_{n\in N}$ is a filtration of $K_r^{\leq}$ where $K_{nr}^{\leq} = \{x \in K_n \mid \|x\| \leq r\}$

Theorem 8.6.4

Let E be a relexive Banach space and $K(K_n)_{n\in N}$ a Galerkin cone in E.

Let $S:K \to E$, $T:K \to E^*$ be completely continuous mappings.

If the following assumptions are satisfied:

1°) there is $r > 0$ such that S is subordonate to the filtrations $\{K^{\leq}_{nr}\}_{n\in N}$ and $\{K_{nr}\}_{n\in N}$,

2°) there is a natural number m and $u_o \in K_m$ such that $\|S(u_o)\| < r$, $S(u_o) \in K_m$ and for every $n \geq \max(n_o, m)$, $< x - S(u_o), T(x) \geq 0$, for all $x \in K_n$, satisfying $r \leq \|x\| \leq \max(r, r_n)$, where $r_n \geq \sup\{\|S(x)\| \mid x \in K_{nr}\}$,

3°) $< S(x), T(x) > \leq < x, T(x) >$; for all $x \in K_r$

then the problem I.C.P.(T,S,K) has a solution x_* such that $\|x_*\| \leq r$.

Proof

We remark that for every $n \geq \max(n_o, m)$ the all assumptions of Theorem 8.6.3 are satisfied for every problem I.C.P.(T,S,K_n) and hence we have a solution x_n^* of this problem. Since, for every x_n^* we have $\|x_n^*\| \leq r$, we have that $\{x_n^*\}$ is a bounded sequence. Because E is reflexive $\{x_n^*\}$ has a weakly convergent subsequence $\{x_{n_k}^*\}$. We denote again this subsequence by $\{x_n^*\}$ and we put $x_* = (w)\text{-}\lim x_n^*$. We have $x_* \in K$, $\|x_*\| \leq r$ and $S(x_*) \in K$.

Let $x \in K$ be an arbitrary element. For every $n \geq \max(n_o, m)$ we have

(9): $\quad < P_n(x) - S(x_n^*), T(x_n^*) \geq 0,$

and since S and T are strongly continuous we compute the limit in (9) and we deduce,

$$< x - S(x_*), T(x_*) > \geq 0; \text{ for all } x \in K.$$

Now, by Proposition 8.6.1 we have that x_* is a solution of the problem I.C.P.(T,S,K).

Theorem 8.6.5

Let $K \subset E$ be a pointed locally compact convex cone and $S:K \to$, $T:K \to E^*$ continuous mappings. If the following assumptions are satisfied:

1°) $(\forall x \in K)\ (< S(x), T(x) > \leq < x, T(x) >)$

2° there is a number $r > 0$ such that for every $x \in K$ with $r \leq \|x\|$ there is $v_x \in K$ such that $\|v_x\| < r$ and $< S(x) - v_x, T(x) > > 0$,

then the problem I.C.P.(T,S,K) has a solution x_* such that $\|x_*\| < r$.

Proof

We denote $D_n = \{x \in K \mid \|x\| \leq n\}$. Since K is locally compact we have that D_n is a convex compact set for every $n \in N$.

We apply Theorem 8.6.2 with $D = D_n$ and we obtain a solution x_n, for every $n \in N$, for the problem S.V.I.(T,S,D_n), that is,

$$(10) \quad \left\| \begin{array}{l} \text{for every } n \in N \text{ there is } x_n^* \in D_n \text{ such that} \\ \langle S(x_n^*) - v, T(x_n^*) \rangle \leq 0; \text{ for all } v \in D_n \end{array} \right.$$

The sequence $\{x_n^*\}_{n \in N}$ is bounded.

Indeed, if no, then $(\forall k > 0)(\exists n \in N)(\|x_n^*\| \geq k)$

If $k \geq r$ then there is $n \in N$ such that $n \geq \|x_n^*\| \geq k \geq r$. For this x_n^*, by assumption 2°) there is an element $v_{x_n^*} \in K$ such that $\|v_{x_n^*}\| < r$ and

$$(11) \quad \left\| \quad \langle S(x_n^*) - v_{x_n^*}, T(x_n^*) \rangle > 0 \right.$$

But, since $\|v_{x_n^*}\| < r < \|x_n^*\| \leq n$ we have from (9) $\langle S(x_n^*) - v_{x_n^*}, T(x_n^*) \rangle \leq 0$ which is a contradiction of (11).

Hence, $\{x_n^*\}$ is bounded and since K is locally compact the sequence $\{x_n^*\}$ has a norm convergent subsequence $\{x_{n_k}\}_{k \in N}$. Let $x_* = \lim_{k \to \infty} x_{n_k}$. We show now that x_* is a solution of the problem I.C.P.(T,S,K).

Indeed, if $v \in K$ is an arbitrary element then there is $m \in N$ such that for every $n \geq m$ we have $v \in D_n$ and for every $n_k \geq m$, $v \in D_{n_k}$ and

$$\langle S(x_{n_k}^*) - v, T(x_{n_k}^*) \rangle \leq 0.$$

Using the continuity of S and T we obtain,

$$\langle S(x_*) - v, T(x_*) \rangle \leq 0; \ \forall v \in K,$$

that is, x_* is a solution of the problem S.V.I.(T,S,K) which by Proposition 8.6.1 is equivalent to the problem I.C.P.(T,S,K).

Obviously, by assumption 2°) we must have $\|x_*\| < r$.

Corollary 1

Let $K \subset E$ be a pointed locally compact cone and $S:K \to K$, $T:K \to E^*$ continuous mappings.

If the following assumptions are satisfied:

1°). $(\forall x \in K)(\langle S(x), T(x) \rangle \leq \langle x, T(x) \rangle)$

2°). there is $r > 0$ such that for every $x \in K$ with $r \leq \|x\|$ we have $\langle S(x), T(x) \rangle > 0$,

then the problem I.C.P.(T,S,K) has a solution x_* such that $\|x_*\| < r$.

Proof

We apply Theorem 8.6.5 with $v_x = 0$ for every $x \in K$ satisfying $r \leq \|x\|$.

Corollary 2.

Let $K \subset E$ be a pointed locally compact cone and $S:K \to K$, $T:K \to E^*$ continuous mappings. If the following assumptions are satisfied:

1°). $(\forall x \in K)(\langle S(x), T(x) \rangle \leq \langle x, T(x) \rangle)$,

2°). there is $r_o > 0$ and $u_o \in K$ such that for every $x \in K$ with $r_o \leq \|x\|$ we have $\langle S(x) - u_o, T(x) \rangle > 0$,

then the problem I.C.P.(T,S,K) has a solution x_* such that $\|x_*\| < 1 + \max(r_o, \|u_o\|)$.

Proof

If we denote, $r = \max(r_o, \|u_o\|) + 1$ then we have $r > r_o$ and $r > \|u_o\|$.

Now, we can apply Theorem 8.6.5 since assumption 2°) of this theorem, is satisfied with $v_x = u_o$, for every $x \in K$ satisfying $r \leq \|x\|$.

Remark

Condition 2°) of Corollary 2 is satisfied if T is semicoercive with respect to S in the following sense:

$$(\exists u_o \in K)\left(\lim_{\|x\| \to +\infty} \frac{\langle S(x) - u_o, T(x) \rangle}{\|x\|} = +\infty \right)$$

Using a similar proof to the proof of Theorem 8.6.4 we obtain from Theorem 8.6.5 the following result.

Theorem 8.6.6

Let E be a reflexive Banach space and $K(K_n)_{n\in N}$ a Galerkin cone in E.

Let $S:K \to K$, $T:K \to E^*$ be completely continuous mappings.

If the following assumptions are satisfied:

1°). S is subordonate to the filtration $(K_n)_{n\in N}$,

2°). $(\forall x \in K)\ (< S(x), T(x) > \leq <x, T(x)>)$

3°). there is a $r > 0$ such that for every $n \geq n_o$ and every $x \in K_n$ with $r \leq \|x\|$ there is a $v_x \in K_n$ such that $\|v_x\| < r$ and $< S(x) - v_x, T(x) > > 0$,

then the problem I.C.P.(T,S,K) has a solution x_* such that $\|x_*\| \leq r$. □

8.7. Some new complementarity problems.

We present in this section some new complementarity problems but little studied till now.

A.) Let $(E, \| \|)$ be a Banach space. For $x, y \in E$ we define

$N'(x,y) = \lim_{t\to o^+} (\|x+ty\| - \|x\|)/t.$

In general we have

$$\lim_{t\to 0^-} (\|x+ty\| - \|x\|)/t = -N'(x,-y) \leq N'(x,y).$$

Definition 8.7.1

We say that E is smooth when the norm is Gateaux differentiable at every $x \in E\setminus\{0\}$ that is, $N'(x,y) = -N'(x,-y)$, for $x, y \in E\ \{0\}$.

If E is smooth then we have that $N'(x,y)$ is linear in y. [See: R.T. Tapia: A characterization of inner product spaces. Proc. Amer. Math. Soc. 41 (1973), 569-574.]

If $T:E \to E$ then for any $x \in E$ the application $T'_x:y \to N'(T(x),y)$ belongs to E^*.

So, we can define $T':E \to E^*$ by $x \to T'_x$. The application T' is monotone if and only if the following condition holds.

(1): $N'(T(x),x-y) - N'(T(y),x-y) \geq 0$, for all $x, y \in E$.

Condition (1) can be used as a definition of monotonicity for operators for E to E.

The mapping N' can be used also to define a "semi-inner product" on E.

Indeed, if we put $< x,y > = \|x\|\ N'(x,y)$, for every $x, y \in E$ we obtain a semi-inner product on E. If E is a Hilbert space $<,>$ is exactly the inner product of E.

Let $K \subset E$ be a pointed closed convex cone and $T:K \to E$ a mapping.

Definition 8.7.2

The complementarity problem associated to T, K and E is

$$\text{C.P.}(T,K,E)\left\|\begin{array}{l} \text{find } x_o \in K \text{ such that} \\ N'(T(x_o),y) \geq 0; \text{ for all } y \in K \text{ and} \\ N'(T(x_o),y) = 0 \end{array}\right.$$

Given $x \in E$, we denote by P_K the (possible empty) set of best approximation to x from K. We have the following result.

Proposition 8.7.1

We have that $x_o \in P_K(x)$ if and only if:

i°). $-N'(x_o - x, x_o) \leq 0 \leq N'(x - x_o, x_o)$,

ii°). $N'(x_o - x, y) \geq 0$, for all $y \in K$.

Proof

See pg. 362 in [I. Singer: Best approximation in normed linear spaces. Springer-Verlag, (1970).]

Proposition 8.7.2

Let $(E, \|\ \|)$ be a smooth Banach space, $K \subset E$ a pointed closed convex cone and $T:K \to E$. Then $x_o \in P_K[x_o - T(x_o)]$ if and only if x_o is a solution of the problem C.P. (T,K,E).

Proof

The proposition is a consequence of Proposition 8.6.1 since when E is smooth condition i°) of Proposition 8.7.1 reduces to $N'(x - x_o, x_o) = 0$. □

The problem C.P.(T,K,E) is interesting and it is worth to be studied.

This problem was defined by Baronti [C1].

Probably a similar definition can be used in a general semi-inner product space.

B.) The complementaritiy problems considered in this section are important for the vector optimization (Pareto optimization). We consider that the research in this direction must be developed.

Let (E,K_1) and (F,K_2) be ordered Banach spaces such that Int $K_2 \neq 0$.

We denote by L(E,F) the set of linear continuous operators from E into F.

We denote by $<,>$ the vector bilinear form, $< T,x > = T(x)$, for every $T \in L(E,F)$ and $x \in E$. We define the following duals of K_1 with respect to K_2:

i° the weak dual defined by,

$$[K_1]^{w*}_{K_2} = \{T \in L(E,F) \mid \langle T,x \rangle \nless 0, \text{ for all } x \in K_1\}$$

ii°) the strong dual defined by,

$$[K_1]^{S*}_{K_2} = \{T \in L(E,F) \mid \langle T,x \rangle \geq 0, \text{ for all } x \in K_1\}$$

The weak dual is a cone, generally not convex, but the strong dual is a convex cone and we have $[K_1]^{S*}_{K_2} \subseteq [K_1]^{w*}_{K_2}$.

Given $f:E \to L(E,F)$ we may consider the following Vector Complementarity Problems:

a) Weak Vector Complementarity Problem,

$$\text{W.V.C.P.}(f,K_1,K_2): \quad \left\| \begin{array}{l} \text{find } x_o \in K_1 \text{ such that} \\ f(x_o) \in [K_1]^{w*}_{K_2} \text{ and} \\ \langle f(x_o), x_o \rangle \ngtr 0. \end{array} \right.$$

b) Positive Vector Complementarity Problem,

$$\text{P.V.C.P.}(f,K_1,K_2): \quad \left\| \begin{array}{l} \text{find } x_o \in K_1 \text{ such that} \\ f(x_o) \in [K_1]^{S*}_{K_2} \text{ and} \\ \langle f(x_o), x_o \rangle \ngtr 0. \end{array} \right.$$

c) Strong Vector Complementarity Problem,

$$\text{S.V.C.P.}(f,K_1,K_2): \quad \left\| \begin{array}{l} \text{find } x_o \in K_1 \text{ such that} \\ f(x_o) \in [K_1]^{S*}_{K_2} \text{ and} \\ \langle f(x_o), x_o \rangle = 0 \end{array} \right.$$

These problems were defined in [C8].

We note that in [C8] we find also some existence theorems and some relations with the vector optimization.

C. The following complementarity problem is important since it is associated to some quasi-variational inequalities used in mechanics and optimal control theory etc.

Let $(E, \| \|)$ be a reflexive Banach space, E^* the topological dual of E, $A:E \to E^*$, $M:E \to E$ mappings and $K \subset E$ a closed convex cone.

Given $b \in E^*$ we consider the following system of Complementarity Equations:

$$\text{S.C.E.}(A,M,bK): \quad \left\| \begin{array}{l} \text{find } u \in (M(u) - K) \cap K \text{ and} \\ v \in (A(u) - b + K^*) \cap K^* \text{ such that} \\ \quad \langle u,v \rangle = 0 \\ \quad \langle M(u) - u, v - (A(u) - b) \rangle = 0 \end{array} \right.$$

Proposition 8.7.3

If (u,v) is a solution of the problem S.C.E. (A,M,b,K) then u is a solution of the following quasi-variational inequality:

$$\text{Q.V.I.}(A,M,b,K)\ \left\|\ \begin{array}{l} \text{find } u \in E \text{ such that} \\ 0 \le u \le M(u) \text{ and} \\ < w - u,\ A(u) - b > \ \ge 0; \text{ for all } w \text{ satisfying} \\ 0 \le w \le M(u). \end{array}\right.$$

Proof

Let (u,v) be a solution of the problem S.C.E.(A,M,b,K). If we denote $x_0 = v - A(u) + b$ then for every w satisfying $0 \le w \le M(u)$ we have

$$< w - u,\ A(u) - b > \ = \ < w - u,\ v - v + A(u) - b > \ = \ < w - u,\ v - x_0 > \ =$$

$$= \ < w - u,\ v > - < w - u,\ x_0 > \ = \ < w - u,\ v > + < - w + M(u) - M(u) + u,\ x_0 > \ =$$

$$= \ < w - u,\ v > + < - w + M(u),\ x_0 > + < - M(u) + u,\ x_0 > \ =$$

$$= \ < w,\ v > - < u,\ v > - < w - M(u),\ x_0 > - < M(u) - u,\ x_0 > \ \ge 0, \text{ since } < u,\ v > \ = 0$$

and $v,\ x_0 \in K^*$.

We say that E is directed (with respect to the ordering "≤") if for every $x,\ y \in E$ there is a $z \in E$ such that $x,y \le z$ and we say that E satisfies the Riesz condition if for every $x_1,\ x_2,\ y \in K$ with $y \le x_1 + x_2$ there exist $y_1,\ y_2 \in K$ such that, $y = y_1 + y_2$ and $y_i \le x_i$ $(i = 1,2)$.

Theorem 8.7.1

Let (E, ‖ ‖, K) be an ordered directed Banach space which satisfies the Riesz condition.

If u is a solution of the problem Q.V.I.(A,M,b,K) and $A(u) - b \in K^* - K^*$, then there is an element $v \in E^*$, mimimal with respect to the dual ordering, such that, (u,v) is a solution of the problem S.C.E.(A,M,b,K).

Proof

Since $A(u) - b \in K^* - K^*$ we have that for every order interval $[0,x]$ in K, $A(u) - b$ has a positive part defined by $< x,\ (A(u) - b)^+ > \ = \sup_{0 \le y \le x} < y, A(u) - b >$.

Since E satisfies the Riesz condition we have that $(A(u) - b)^+$ is additive [C33, chap. 2.6.1] and it can be uniquely extended to $E = K - K$ by a linear form [$E = K - K$ since E is directed].

Moreover, $(A(u) - b)^+$ is continuous [C33, chap.3.5]. Hence $(A(u) - b)^+ \in K^* \subset E^*$.

Since E is directed and satisfies the Riesz condition we have that E* is a lattice and $(A(u) - b)^+ = \sup \{A(u) - b, 0\}$.

Now, we show that $v = (A(u) - b)^+$ satisfies the problem S.C.E.(A,M,b,K).

Indeed, we put in Q.V.I. (A,M,b,K) $w = u - y$, for every y satisfying $0 \leq y \leq u$ and we obtain, $\langle u, (A(u) - b)^+ \rangle = 0$.

For every y with $0 \leq y \leq M(u) - u$ we put again $w = y + u$ in Q.V.I. (A,M,b,K) and we get, $\langle M(u) - u, (A(u) - b)^- \rangle = \langle M(u) - u, (A(u) - b)^+ - (A(u) - b) \rangle = 0$.

Since in practice we are interested to solve the problem Q.V.I.(A,M,b,K) it is interesting to study the problem S.C.E.(A.M,b,K) and to develop numerical methods to solve it.

In [A294] and [A295] we find som existence theorems. We note also that the problem C.S.E.(A,M,b,K) was defined by Vescan in [A294 - A295].

8.8 Some special problems

In this last section we inform the reader on some special problems on complementarity problems opened.

A. Global solvability

Let $\langle E,E^* \rangle$ be a duality of Banach spaces and $K \subset E$ a closed convex cone. Given a continuous mapping $f:K \to E^*$ and an element $q \in E^*$ we consider the following complementarity problems.

C.P. (f,K):
$$\left\| \begin{array}{l} \text{find } x_o \in K \text{ such that} \\ f(x_o) \in K^* \text{ and} \\ \langle x_o, f(x_o) \rangle = 0 \end{array} \right.$$

C.P. (f,q,K):
$$\left\| \begin{array}{l} \text{find } x_o \in K \text{ such that} \\ f(x_o) + q \in K^* \text{ and} \\ \langle x_o, f(x_o) + q \rangle = 0 \end{array} \right.$$

Definition 8.8.1

We say that the problem C.P. (f,K) is globally uniquely solvable (GUS) if for any vector $q \in E^*$ the problem C.P. (f,q,K) has a unique solution.

This problem was studied by Megiddo and Kojima in the particular case $E = R^n$ and $K = R^n_+$ [A211].

The principal result of the paper [H. Samelson, R.M. Thrall and O. Wesler: A partition theorem for euclidean n-space. Proc. Amer. Math. Soc. 9(1958), 805-807] implies the following result for linear complementarity problem in R^n.

Theorem 8.8.1

Let $A \in M_{nxn}(R)$ be an arbityrary matrix and $f(x) = Ax + b$, where $b \in R^n$.

Then the problem C.P. (f, R^n_+) is GUS if and only if the principal minors of A are positive. □

For the nonlinear case (in R^n) we remark also two classical existence results which can be considered as sufficient conditions to have the GUS property.

Theorem 8.8.2 [Karamardian]

Let $f:R^n_+ \to R$ be a continuous mapping.

If f is strongly monotone then the problem C.P. (f, R^n_+) is GUS. □

For the differentiable case we have the following result.

Theorem 8.8.3 [Cottle]

If $f:R^n_+ \to R^n$ is a differentiable mapping, such that all the principal minors of the Jacobian matrix of f are bounded between δ and δ^{-1}, for some $0 < \delta < 1$, then the problem C.P. (f, R^n_+) is GUS.

Proof

An interesting proof of this result is in [A211].

It is interesting to characterize the property GUS by necessary and sufficient conditions. Let $f:R^n_+ \to R^n$ be a mapping.

We denote by F the following extension of $f:F(x) = f(x^+) + x^-$

where $x \in R^n$, $x_i^+ = \begin{cases} x_i \text{ if } x_i \geq 0 \\ 0 \text{ otherwise} \end{cases}$

and $x_i^- = \begin{cases} x_i \text{ if } x_i \leq 0 \\ 0 \text{ otherwise.} \end{cases}$

We have F is from R^n into R^n.

Theorem 8.8.4 [Megiddo-Kojima]

Let $f: R^n_+ \to R^n$ be a continuous mapping. Then the problem C.P. (f, R^n_+) is GUS if and only if the extension F of f is a homeomorphism of R^n onto itself.

Proof

Since f is continuous and $x \to x^+$, $x \to x^-$ are continuous too, we have that F is continuous.

The problem C.P. (f, R^n) is GUS if and only if for every $q \in R^n$ there is a unique $x = x(q) \in R^n_+$ such that $f(x) + q \in R^n_+$ and $f_i(x) + q_i = 0$ for each i such that $x_i > 0$. The GUS property is thus equivalent to the existence of a unique $z = z(q) \in R^n$ such that $f(z^+) + q = - z^-$ or such that $F(z) = -q$. Hence the problem C.P. (f, R^n_+) has the GUS property if and only if F is a bijection of R^n. □

The remainder of the proof follows from the fact that the inverse of a continuous bijection of R^n is also continuous.

In [A211] we find other results in this sense but the problem is: what is the situation in an infinite dimensional space?

B. Unification of existence theorems

To unify the existence theorems in complementarity theory is another important problem.

In this sense, some interesting results were obtained by Kojima [A162], Karamardian [A158] etc. Important unification results were obtained byd Fisher and Tolle using the method of complementary pivoting on a triangulation of R^n [A9 7].

Given $f: R^n_+ \to R$ a continuous function, we consider the problem C.P. (f, R^n_+).

Definition 8.8.2

Suppose x_0, A and B are such that A is bounded and open in R^n and $x_0 \in A \cap R^n_+$. If $B = \partial A \cap R^n_+$ then we say that B separates x_0 from infinity.

Theorem 8.8.5 [Fisher and Tolle]

If there is a set $B = \partial A \cap R^n_+$ separating the origin from infinity such that for each $x \in B$ the following system is inconsistent:

$$(S_i): \begin{cases} f_i(x) + t = 0, \; x_i \geq 0 \\ f_i(x) + t \geq 0, \; x_i = 0 \\ t \geq 0 \end{cases}$$

then the problem C.P. (f,R_+) has a solution x_* with $x_* \in A$.

Proof

The proof is in [A97]. □

Corollary [Karamardian]

Let $G(x) = f(x) - f(0)$ be positively homogeneous of degree $d > 0$ and suppose the system

$$\begin{cases} G_i(x) + t = 0 \;,\; x_i > 0 \\ G_i(x) + t \geq 0 \;,\; x_i = 0 \\ t \geq 0 \end{cases}$$

is inconsistent for all $x \geq 0$, $x \neq 0$.

Then the problem C.P. (f,R^n_+) has a solution.

Proof

Consider the function $F(x) = G(x) + (1 - \sum_{i=1}^{n} x_i)\, f(0)$.

The function F satisfies the assumptions of Theorem 8.8.5 with

$B = \{x \geq 0 | \sum^{n} x = 1\}$. Hence the problem C.P.$(f,R^n_+)$ has a solution x_o with

$\sum_{i=1}^{n} x_{oi} < 1$.

Consider now the element $x^* = x_o/(1 - \sum_{i=1}^{n} x_{oi})^{1/d}$

Since G is homogeneous we have that $f(x^*) = F_i(x_o)/(1 - \sum_{i=1}^{n} x_{oi})$ and we can show that x^* is a solution of the problem C.P. (f,R^n_+). □

We denote by e the vector $e = (e_i)$ with $e_i = 1$; $\forall i = 1, 2 \ldots, n$.

Theorem 8.8.6 [Fisher and Tolle]

Suppose a set $B = \partial A \cap R^n_+$ separates $x_o \in R^n_+$ from infinity

If, for every $x \in B$ there is a $y \in R^n_+$ such that the system

$$(S_2): \begin{cases} \langle x - x_o - y, f(x) \rangle > 0 \\ \langle x - x_o - y, e \rangle \geq 0 \text{ when } x - x_o \in R^n_+ \end{cases}$$

has a solution, then the set $\bar{A}$ contains a solution of the problem C.P. (f, R^n_+)

Proof

The proof is in the paper [A97]. □

Corollary [Karamardian]

If there is a nonempty compact set C in R^n_+ such that for each $x \in R^n_+ \backslash C$ there is a $y \in C$ such that $\langle x-y, f(x) \rangle > 0$ then the problem C.P. (f, R^n_+) has a solution.

Proof

Let $x_o = 0$ and $r > 0$ be any scalar such that $B = \{x \in R^n_+ | \langle e, x \rangle = r\}$ separates each $y \in C$ from infinity.

Then the system (S_2) is Theorem 8.8.6 is satisfied for each $x \in B$.

Remark 8.8.1

We can show that for any $x \in B$ the system (S_2) has a solution with $y \in R^n_+$ if and only if the system (S_3) given below has a solution with $y \in R^n_+$:

$$(S_3): \begin{cases} \langle x - x_o - y, f(x) \rangle \geq 0, \\ \langle x - x_o, f(x) \rangle \neq 0 \text{ when } f(x) \in R^n_+, \\ \langle x - x_o - y, e \rangle > 0 \text{ when } x - x_o \in R^n_+. \end{cases}$$

From Theorem 8.8.6 and Remark 8.8.1 we obtain the following result.

Theorem 8.8.7

Suppose B separates $x_o \in R^n_+$ from infinity and there exists a positive vector d such that for every $x \in B$ there is a $y \in R^n_+$ satisfying one of the two equivalent systems:

$$(S_2'): \begin{cases} < x - x_o - y,\ f(x) >> 0, \\ < x - x_o - y,\ d >\geq 0 \text{ when } x - x_o \in R_+^n \end{cases}$$

$$(S_3'): \begin{cases} < x - x_o - y,\ f(x) >\geq 0 \\ < x - x_o,\ f(x) > \neq 0 \text{ when } f(x) \in R_+^n, \\ < x - x_o - y,\ d > > 0 \text{ when } x - x_o \in R_+^n \end{cases}$$

then the problem C.P. (f, R_+^n) has a solution. □

Corollary [Eaves]

The problem C.P. (f, R_+^n) has a solution if there exists a positive n-vector d, a positive scalar r and a set B separating $C = \{y \in R_+^n | <y,d> \leq r\}$ from infinity such that for each $x \in B$ there is a $y \in C$ for which $< x - y, f(x) >\geq 0$.

Proof

We put $x_o = 0$. Since B separates C from infinity, $< x,d >> r$ for each $x \in B$. Thus for each $y \in C$, $< x - y, d >> 0$. Now if $<x, f(x)> = 0$ when $f(x) \geq 0$, x is a solution of the problem C.P. (f, R_+^n). Otherwise the system (S_3') of Theorem 8.8.7 is satisfied for each $x \in B$. □

Can the Fisher and Tolle's results be extended to the infinite dimensional case?

C. The Parametric Complementarity Problems

Let $f: R_+^n \to R^n$ be a continuous function and $p \in R^n$.

The Parametric Complementarity Problem is the following family of Complementarity Problems:

$$\text{P.C.P. } (f, tp, R_+^n)_{t \geq 0} \quad \left\| \begin{array}{l} \text{find } x_o \in R_+^n \text{ such that} \\ f(x_o) + t\,p \in R_+^n \text{ and} \\ < x_o,\ f(x_o) + t\,p > = 0 \end{array} \right.$$

Supposing that P.C.P. (f, tp, R^n) has a solution for every $t \geq 0$ we denote by $x(t)$ the set of solutions of the problem P.C.P. (f, tp, R_+^n).

Generally, the mapping $t \to x(t)$ is a multivalued mapping from R_+ to R_+^n.

When $f(x) = Mx + q$, where $M \in \mathcal{M}_{nxn}(R)$ and $q \in R^n$ and M is a P-matrix we have that $x(t)$ has just one element for every $t \geq 0$, that is $x(t)$ is a function from R_+ to R^n_+.

Given a parametric complementarity problem, the problem is to study the properties of the function $x(t)$. In some problems, important in mechanics, we have a linear parametric complementarity problem and we are interested to know:when is $x(t)$ an increasing function?

A matrix $M \in \mathcal{M}_{nxn}(R)$ which is P- and Z-matrix simultaneously is called a <u>Minkowski matrix</u>.

In [A148] Kaneko proved the following result.

<u>Theorem 8.8.8</u>

<u>Let $M \in \mathcal{M}_{nxn}(R)$ be a P-matrix, $p,q \in R$ and $x(t; p,q)$ the solution of the problem L.C.P. $(M, q + tp, R^n_+)$.</u>

<u>Then the following statements are equivalent:</u>

1°) <u>M is Minkowski</u>

2°) <u>$x(t; p,q)$ is isotone in $t \in R_+$ for every $q > 0$ and every $p \in R^n$.</u> □

The monotonicity of the solution function $x(t)$ were also studied by Cottle [45), Megiddo [A208], [A207] and Kaneko [A144].

For the parametric Lineor Complementarity Problem another interesting problem is the following.

Given $M \in \mathcal{M}_{nxn}(R)$ and $p,q \in R^n$ we denote by $F(t)$ the feasible set of the problem L.C.P.$(M,q+tp, R^n_+)$, that is the set:

$$F(t) = \{x \in R^n_+ | q + tp + Mx \in R^n_+\}$$

We denote by $\Phi(t)$ the solution of the problem L.C.P.$(M,q + tp, R^n_+)$ which is the least element of the set $F(t)$, when this solution exists.

An interesting problem is to know if the function $\Phi(t)$ is monotone increasing in nonegative t.

For the problem P.C.P. $(f,tp,R^n_+)_{t\geq 0}$ we have an interesting result when f is differentiable in [[A207] or [<u>Math. Programming Study 7(1978), 142-150</u>]].

D. Stability

We consider the complementarity problem C.P.(f,R^n_+) where $f:R^n \to R$ is a continuous function.

Supposing that C.P.(f,R^n_+) has solutions, a very important problem is to study the behaviour of solutions near a particular solution x_* when the function f is perturbed.

For a continuous function $g:R^n \to R$ and $\bar{U}$ a set whose closure U is bounded, we define $\|g\|_U = \sup_{x \in \bar{U}} |g(x)|$, where $|\cdot|$ denotes the Euclidean norm on R^n. Given $\epsilon > 0$, $B(x,\epsilon)$ is the open ball of center x and radius .

Definition 8.8.3

The complementarity problem C.P.(f,R^n_+) is said to be stable at a solution point x_* if there is $\epsilon_o > 0$ such that for any ϵ satisfying $0 < \epsilon \le \epsilon_o$ there exists $\delta > 0$ such that for any continuous function $g:R^n \to R^n$ with $\|g-f\|_{B(x_*,\epsilon)} < \delta$ the problem C.P.(g,R^n_+) has a solution in $B(x_*,\epsilon)$.

So, to study the stability is to find conditions on f such that the problem C.P.(g,R^n_+) has solutions near x_* for g sufficiently close to f.

Given the problem C.P.(f,R^n_+) we introduce the following notations:

$$J: = \{i \mid x_{*i} > 0,\ f_i(x_*) = 0\},$$

$$K: = \{i \mid x_{*i} = 0,\ f_i(x_*) = 0\} \text{ and}$$

$$L: = \{i \mid x_{*i} = 0,\ f_i(x_*) > 0\}$$

where x_* is a solution.

We denote by x_J the vector in $R^{|J|}$ where components are components x_i of x for $i \in J$. Similar definitions hold for x_K, x_L, f_J, f_K and f_L.

For such g, if x is a solution of the problem C.P.(g,R^n_+) in $B(x_*,\epsilon')$ we must have (since $< x,g(x) > = 0$),

$$(1): \quad \begin{cases} x_i = 0, \text{ for } i \in L \text{ and} \\ g_i(x) = 0, \text{ for } i \in J, \end{cases}$$

Hence the problem C.P.(g,R^n_+) is reduced to the problem of finding $x = (x_J, x_K, 0)$ close to x_* satisfying the following system:

$$(2):\quad \begin{cases} g_J(x_J, x_K, 0) = 0 \\ g_K(x_J, x_K, 0) \geq 0,\ x_K \geq 0 \\ \text{and } < x_K, g_K(x_K, 0) > = 0 \end{cases}$$

Let $m = |J| + |K|$. Given a real number α we denote $\alpha^+ = \max(o,\alpha)$, $\alpha^- = \max(0, -\alpha)$ and for a vector $y \in R^m$ we put:

$$y^+ = (y_1^+, y_2^+, \ldots, y_m^+) \quad \text{and}$$

$$y^- = (y_1^-, y_2^-, \ldots, y_m^-).$$

Let $F:R^m \to R^m$ be the function defined by:

$$(4):\quad \begin{cases} F_J(z_J,z_K) = f_J(z_J,z_K^+,0) \quad \text{and} \\ F_K(z_J,z_K) = f_K(z_J,z_K^+,0) - z_K^- . \end{cases}$$

We can show that system (3) is equivalent to the equation.

$$(5):\qquad F(z) = 0$$

Indeed, if $z = (z_J,z_K)$ is a solution of equation (5) then $x = (x_J,x_K,0) = (z_J,z_K^+,0)$ is solution of system (3).

Conversely, suppose $(x_J,x_K,0)$ is a solution of system (3). We define $x_J = x_J$ and for $i \in K$,

$$z_i = \begin{cases} x_i, \text{ if } f_i(x) = 0 \\ -f_i(x), \text{ if } f_i(x) > 0 \end{cases}$$

and we can show that (z_J,z_K) is a solution of equation (5).

If $x_* = (x_{*J},0,0)$ is an isolated solution of the problem C.P.(f,R^n_+), then x_* is also an isolated solution of system (3) and so is $z_* = (x_{*_J},z_{*_K}) = (x_{*_J},0)$ of equation (5).

Hence, if x_* is an isolated solution of the problem C.P.(f,R^n_+) the index $i(F,z_*,0)$ is defined and equals $d(F,D,0)$ (the topological degree of F with respect to D and 0) for any open bounded neighborhood D of z_*, in which z_* is the unique solution of equation (5).

So, using the topological degree Ha proved in [Al13] the following very nice results.

Theorem 8.8.9

Let x_* be an isolated solution of the problem C.P.(f,R^n_+). If $i(F,z_*,0) \neq 0$ then the problem C.P.(f,R^n_+) is stable at x_*. □

Theorem 8.8.10

If the problem C.P.(f,R^n_+) is GUS then the problem C.P.(f,R^n_+) is stable at the unique solution x_*. □

The study of complementarity problems by the topological degree is a very interesting research direction.

BIBLIOGRAPHY

A Complementarity Problems

[1] Aashtiani H.Z. and Magnanti T.L.: *Equilibria on a congested transportation network.* SIAM J. Alg. Disc. Meth. 2, Nr 3, (1981), 213-226.

[2] Aganagic M.: *Contribution to complementarity theory.* Ph. D. Dissertation submitted to the Department of Operations Research School of Engineering, Stanford University (1978).

[3] -----------: *Variational inequalities and generalized complementarity problems.* Tech. Rept. SOL 78-11, Systems Optimization Laboratory, Department of Operations Research, Stanford University, Stanford CA (1978.

[4] -----------: *Iterative methods for the linear complementarity problems.* Tech. Rept. SOL 78-10, Department of Operations Research, Stanford University, (1978).

[5] -----------: *On diagonal dominance in linear complementarity.* Linear Algebra and its Appl. 39 (1981), 41-49.

[6] -----------: *Newton's method for linear complementarity problems.* Math. Programming 28,(1984), 349-362.

[7] Aganagic M. and Cottle R.W.: *A note on Q-matrices.* Math. Programming 16, (1979), 374-377.

[8] Ahn B.H.: *Solution of nonsymmetric linear complementarity problems by iterative methods.* J. Optimization Theory and Appl. 33 Nr. 2, (1981), 175-185.

[9] -----------: *Iterative methods for linear complementarity problems with upperbounds on primary variables.* Math. Programming 26, Nr. 3, (1983), 295-375.

[10] Al-Khayyal F.A.: *An implicit enumeration procedure for the general linear complementarity problem.* Tech. Rept. Georgia Institute of Technology (1984).

[11] -----------: *Linear, quadratic and bilinear programming approaches to the linear complementarity problem.* European J. Oper. Res., 24, (1986), 216-227.

[12] Allen G. *Variational inequalities, complementarity problems and duality theorems.* J. Math. Anal. Appl. 58 (1977), 1-10.

[13] Anderson E.J.: *A review of duality theory for linear programming over topological vector spaces.* J. Math. Anal. Appl. 97 (1983), 380-392.

[14] Balinski M.L. and Cottle R.W.: *Complementarity and fixed point problems.* North-Holland (1978).

[15] Bard J.F. and Falk J.E.: *A separable programming approach to the linear complementarity problem.* Comput. Operation Res., 9. Nr. 2 (1982), 153-159.

[16] ------------: *Computing quilibria via nonconvex programming.* Naval Res. Log. Quart., 27, Nr. 2 (1980), 233-255.

[17] Bazaraa M.S., Goode J.J. and Nashed M.Z.: *A nonlinear complementarity problem in mathematical programming in Banach space.* Proc. Amer. Math. Soc. 35, Nr. 1 (1972), 165-170.

[18] Bensoussan A, Gourset M. and Lions J.L.: *Contrôle impulsionnel et inéquations quasi-variationnelles stationnaires.* C.R. Acad. Sci. Paris 276 (1973), A. 1279-1284.

[19] Bensoussan A. and Lions J.L.: *Nouvelle formulation des problèmes de contrôle impulsionnel et applications.* C.R. Acad. Sci. Paris 276 (1973), A. 1189-1192.

[20] -----------: *Problèmes de temps d'arrêt optimal et inéquations variationnelles paraboliques.* Applicable Anal. (1973), 267-294).

[21] -----------: *Nouvelles méthodes en contrôle impulsionnel.* Applied Math. Optim. Nr. 1 (1974), 289-312.

[22] Bensoussan A.: *Variational inequalities and optimal stoping time problems.* D.L. Russel ed.: Calculus of variations and control theory. Academic Pres (1976), 219-244.

[23] Benveniste M.: *A mathematical model of a monopolistic world oil market.* The John Hopkins University, Ph.D. Dissertation (1977).

[24] -----------: *On the parametric linear complementarity problem: a generalized solution procedure.* J. Optimization Theory Appl., 37, Nr. 3 (1982), 297-314.

[25] Berman A.: *Complementarity problem and duality over convex cones.* Canad. Math. Bull., 17 (1) (1974), 19-25.

[26] Berschanskii Ya. M. and M.V. Meerov: *The complementarity problem: theory and methods of solution.* Automat. Remote Control, 44, Nr. 6 (1983), 687-710.

[27] Berschanskii Ya. M., Meerov M.V. and Litvok M.L.: *Solution of a class of optimal control problems for distributed multivariable systems I, II.* Avtomat. i Teleme. 4, 5-13 & 5, 5-15, (1976).

[28] Birge J.R. and Gana A.: *Computational complexity of Van der Heyden's variable dimension algorithm and Dantzig-Cottle's principal pivoting method for solving L.C.P.'s.* Math. Programming 26 (1983), 316-325.

[29] Bod P.: *On closed sets having a least element.* W. Oettli and K. Ritter eds.: Optimization and Operations Research. Lecture Notes in Economics and Math. Systems. Springer-Verlag, Berlin, Nr. 117 (1976), 23-34.

[30] -----------: *Sur un modèle non-linéaire des rapports interindustriels.* RAIRO Recherche Opér. 11, Nr. 4 (1977), 405-415.

[31] Borwein J.M.: *Generalized linear complementarity problems treated without fixed-point theory.* J. Optimization Theory and Appl. 43, Nr. 3 (1984), 343-356.

[32] -----------: *Alternative theorems for general complementarity problems.* Dalhousie Univ. Research Dept (1984).

[33] Borwein J.M. and Dempster M.A.H.: *The linear order complementarity problem.* Math. Oper. Research 14, Nr. 3 (1989), 534-558.

[34] Brandt A. and Cryer C.W.: *Multigrid algorithms for the solution of linear complementarity problems arising from free boundary problems.* M.R.C. Rept., Nr. 2131, Mathematics Research Center University of Wisconsin, Madison, WI.

[35] Capuzzo-Dolcetta I.: *Sistemi di complementarita a disequaglianze variazionali.* Ph.D. Thesis, Department of Mathematics, University of Rome (1972).

[36] Capuzzo-Dolcetta I., Lorenzani M. and Spizzichino F.: *A degenerate complementarity system and applications to the optimal stopping of Markov chains.* Boll. Un. Mat. Ital. (5) 17-B (1980), 692-703.

[37] Capuzzo-Dolcetta I. and Mosco U.: *Implicit complementarity problems and quasi-variational inequalities.* R.W. Cottle, F. Giannessi and J.L. Lions eds.: Variational inequalities and complementarity problems. Theory and Applications. John Wiley & Sons (1980), 75-87.

[38] Chan D. and Pang J.S.: *The generalized quasivariational inequality problem.* Math. Oper. Res., 7, Nr. 2 (1982), 211-222.

[39] Chandrasekaran R.: *A special case of the complementarity pivot problem.* Opsearch 7 (1970), 263-268.

[40] Cheng Y.C.: *Iterative methods for solving linear complementarity and linear programming problems.* University of Wisconsin, Madison, Wisconsin, Ph.D. Thesis (1981).

[41] -----------: *On the gradient-projection method for solving the nonsymmetric linear complementarity problem.* J. Optimization Theory Appl. 43, Nr. 4 (1984), 527-541.

[42] Cottle R.W.: *Note on a fundamental theorem in quadratic programming.* SIAM J. Appl. Math. 12 (1964), 663-665.

[43] -----------: *Nonlinear programs with positively bounded Jacobians.* SIAM J. Appl. Math 14, Nr. 1 (1966), 147-158.

[44] -----------: *On a problem in linear inequalities.* J. London Math. Soc. 43 (1968), 378-384.

[45] -----------: *Monotone solutions of the parametric linear complementarity problem.* Math. Programming 3 (1972), 210-224.

[46] -----------: *Solution rays for a class of complementarity problems.* Math. Programming Study 1 (1974), 59-70.

[47] -----------: *On Minkowski matrices and the linear complementarity problem.* R. Bulirsch, W. Oettli and J. Stoer eds.: Optimization, Springer-Verlag. Lecture Notes in Math., 477 (1975), 18-26.

[48] -----------: *Complementarity and variational problems.* Symposia Math. 19 (1976), 177-208.

[49] -----------: *Computational experience with large-scale linear complementarity problems.* S. Karamardian eds.: Fixed points algorithms and applications. Academic Press (1977), 281-313.

[50] -----------: *Numerical methods for complementarity problems in engineering and applied science.* Computing methods in Applied Sciences and Engineering (1977). Lecture Notes in Math. Springer-Verlag 704 (1979), 37-52.

[51] -----------: *Some recent developments in linear complementarity theory.* R.W. Cottle, F. Giannessi and J.L. Lions eds.: Variational inequalities and complementarity problems. Theory and Appl. John Wiley & Sons (1980), 97-104.

[52] -----------: *Solution on Problem 72-7. A parametric linear complementarity problem by G. Maier.* SIAM Rev. 15, Nr. 2 (1973), 381-384.

[53] Cottle R.W. and Dantzig G.B.: *Complementarity pivot theory of mathematical programming.* Linear Algebra and Appl. 1 (1968), 103-125.

[54] -----------: *A generalization of the linear complementarity problem.* J. Combinatorial Theory 8 (1970), 79-90.

[55] -----------: *Complementary pivot theory of mathematical programming.* Mathematics of decision sciences, Edited by G.B. Dantzig, A.F. Veinott Jr. (American Math. Soc. Providence (1968)), 115-135.

[56] Cottle R.W. and Goheen M.A.: *A special class of large quadratic programs.* "Nonlinear Programming 3". Edited by O.L. Mangasarian, R.R. Meyer, S.M. Robinson. Academic Press (1978), 361-390.

[57] Cottle R.W., Golub G.H. and Sacher R.S.: *On the solution of large structured linear complementarity problems: the block partitioned case.* Appl. Math. Optim. 4 (1978), 347-363.

[58] Cottle R.W. and Pang J.S.: *A least-element theory of solving linear complementarity problems as linear programs.* Math. Oper. Res., 3, Nr. 2 (1978), 155-170.

[59] -----------: *On solving linear complementarity problems as linear programs.* Math. Programming Study 7 (1978), 88-107.

[60] -----------: *On the convergence of a block successive overrelaxation method for a class of linear complementarity problems.* Math. Programming Study 17 (1982), 126-138.

[61] Cottle R.W., Von Randow R. and Stone R.E.: *On spherically convex sets and Q-matrices.* Linear Algebra and Appl. 41 (1981), 73-80.

[62] Cottle R.W. and Sacher R.S.: *On the solution of large structured linear complementarity problems: the tridiagonal case.* Appl. Math. Optim. 3, Nr. 4 (1976/1977), 321-340.

[63] Cottle R.W. and Stone R.E.: *On the uniqueness of solutions to linear complementarity problems.* Math. Programming 27 (1983), 191-213.

[64] Cottle R.W. and Veinott A.F. Jr.: *Polyhedral sets having a least element.* Math. Programming 3 (1972), 238-249.

[65] Cryer C.W.: *SOR for solving linear complementarity problems arising from free boundary problems.* Free boundary problems, Vol. I & II. Proceedings of a seminar, Pavia (1979). Instituto Nationale Di Alta Matematica, Franscesco Severi, Rome (1980.

[66] -----------: *The method of Christopherson for solving free boundary problems for infinite journal bearing by means of finite differences.* Math. Comput., 25, Nr. 115 (1971), 435-443.

[67] -----------: *The efficient solution of linear complementarity problems for tridiagonal Minkowski matrices.* ACM Trans. Math. Software, 9 (1983), 199-214.

[68] Cryer C.W. and Dempster M.A.H.: *Equivalence of linear complementarity problems and linear programs in vector lattice Hilbert spaces.* SIAM J. Control Optim., 18, Nr. 1 (1980), 76-90.

[69] Cryer C.W., Flanders P.M., Hunt D.J., Reddaway S.F. and Stransbury J.: *The solution of linear complementarity problems on a array processor.* Tech. Summary Rept., Nr. 2170, Mathematics Research Center. University of Wisconsin, Madison, January (1981).

[70] Dafermos S. and Nagurney A.: *Oligopolistic and competitive behavior spatially separated markets.* Preprint (1985). Lefschetz Center for Dynamical Systems. Brown University Providence, U.S.A.

[71] Dantzig G.B. and Cottle R.W.: *Positive (semi)-definite programming.* Nonlinear Programming. A course. J. Abadie ed., North-Holland, Amsterdam (1967), 55-73.

[72] Dantzig G.B. and Manne A.S.: *A complementarity algorithm for an optimal capital path with invariant proportions.* J. Econom. Theory, 9 (1974), 312-323.

[73] Dash A.T. and Nanda S.: *A complementarity problem in mathematical programming in Banach space.* J. Math. Anal. Appl., 98 (1984), 328-331.

[74] De Donato O. and Maier G.: *Mathematical programming methods for the inelastic analysis of reinforced concrete frames allowing for limited rotation capacity.* Internat. J. Numer. Methods Engrg, 4 (1972), 307-329.

[75] Dorn W.S.: *Self-dual quadratic programs.* SIAM J. Appl. Math., 9 (1961), 51-54.

[76] Doverspike R.D.: *A cone approach to the linear complementarity problem.* Ph.D. Thesis, Department of Mathematical Sciences. Rensselaer Polytechnic Institute Troy, N.Y. (1979).

[77] ------------: *Some perturbation results for the linear complementarity problem.* Math. Programming, 23 (1982), 181-192.

[78] Doverspike R.D. and Lemke C.E.: *A partial characterization of a class of matrices defined by solutions to the linear complementarity problem.* Math. Oper. Res., 7 (1982), 272-294.

[79] Du Val P.: *The unloading problem for plane curves.* Amer. J. Math., 62 (1940), 307-311.

[80] Eaves B.C.: *The linear complementarity problem in mathematical programming.* Ph.D. Thesis, Department of Operations Research, Stanford University, Stanford, CA (1969).

[81] ------------: *On the basic theorem of complementarity.* Math. Programming 1 (1971), 68-75.

[82] ------------: *The linear complementarity problem.* Management Sci., 17, Nr. 9,(1971), 612-634.

[83] ------------: *Computing stationary points.* Math. Programming Study, 7 (1978), 1-14.

[84] ------------: *A locally quadratically convergent algorithm for computing stationary points.* Tech. Rept. Department of Operations Research, Stanford University, Stanford, CA (1978).

[85] -----------: *Where solving for stationary points by LCPs is mixing Newton iterates.* B.C. Eaves, F.J. Gould, H.O. Peitgen and M.J. Todd eds.: Homatopy methods and global convergence. Plenum Press (1983), 63-77.

[86] -----------: *On quadratic programming.* Management Sci., 17 (1971), 698-711.

[87] -----------: *More with the Lemke complementarity algorithm.* Math. Programming, 15, Nr. 2 (1978), 214-219.

[88] Eaves B.C. and Lemke C.E.: *Equivalence of L.C.P. and P.L.S.* Math. Oper. Res., 6, Nr. 4 (1981), 475-484.

[89] -----------: *On the equivalence of the linear complementarity problem and system of piecewise linear equations.* B.C. Eaves, F.J. Gould, H.O. Peitgen and M.J. Todd eds.: Homotopy methods and global convergence. Plenum Press (1983), 79-90.

[90] Eckhardt U.: *Semidefinite liniare komplementar probleme.* Habilitationschrift RWTH Aachen (1978).

[91] Fang S.C.: *An iterative method for generalized complementarity problems.* IEEE Trans. Autom. Control A.C. - 25, Nr. 6 (1980), 1225-1227.

[92] -----------: *Generalized complementarity, variational inequality and fixed point problems: Theory and Applications.* Ph.D. Dissertation, Department of Industrial Engineering and Management Sciences. Northwestern University (1979).

[93] Fang S.C. and Peterson E.L.: *A fixed-point representation of the generalized complementarity problem.* J. Optimization Theory Appl., 45, Nr. 3 (1985), 375-381.

[94] Fathi Y.: *On the computational complexity of the linear complementarity problem.* Dissertation, the University of Michigan (Ann Arbor, MI, 1979).

[95] -----------: *Computational complexity of LCPs associated with positive definite symmetric matrices.* Math. Programming, 17 (1979), 335-344.

[96] Fisher M.L. and Gould F.J.: *A simplicial algorithm for the nonlinear complementarity problem.* Math. Programming, 6 (1974), 281-300.

[97] Fisher M.L. and Tolle J.W.: *The nonlinear complementarity problem: existence and determination of solutions.* SIAM J. Control Optim., 15, Nr. 4 (1977), 612-624.

[98] Freidenfelds J.: *Almost-complementary path in the generation of complementarity problems.* Fixed points. Algorithms and Applications. Academic Press, New York (1977), 225-247.

[99] Friesz T.L., Tobin R.L., Smith T.E. and Harker P.T.: *A nonlinear complementarity formulation and solution procedure for the general derived demand network equilibrium problem.* J. Region Sci., 23 (1983), 337-359.

[100] Fujimoto T .: *Nonlinear complementarity problems in a function space.* SIAM J. Control Optim., 18, Nr. 6 (1980), 621-623.

[101] -----------: *An extension of Tarski's fixed point theorem and its applications to isotone complementarity problems.* Math. Programming, 28 (1984), 116-118.

[102] Garcia C.B.: *The complementarity problem and its applications.* Ph.D. Thesis, Rensselaer Polytechnic Institute, Troy, N.Y. (1973).

[103] -----------: *Some classes of matrices in linear complementarity theory.* Math. Programming, 5 (1974), 299-310.

[104] ------------: *A note on a complementary variant of Lemke's method.* Math. Programming, 10, Nr. 1 (1976), 134-136.

[105] -----------: *A note on a complementarity problem.* J. Optimization Theory Appl., 21, Nr. 4 (1977), 529-530.

[106] Garcia C.B., Gould F.J. and Turnbull T.R.: *Relations between PL maps, complementarity cones and degree in linear complementarity problems.* B.C. Eaves, F.J. Gould, H.O. Peitgen and M.J. Todd eds.: Homotopy methods and global convergence. Plenum Press (1983), 91-144.

[107] Garcia C.B. and Lemke C.E.: *All solutions to linear complementarity problems by implicit search.* Paper presented at the 39th National Meeting of Oper. Res. Soc. of America (1971.

[108] Giannessi F.: *Theorems of alternative, quadratic programs and complementarity problems.* R.W. Cottle, F. Giannessi and J.L. Lions (eds): Variational inequalities and complementarity problems. Theory and Applications. John Wiley & sons (1980), 151-186.

[109] Gianessi F. and Tomasin E.: *Nonconvex quadratic programming, linear complementarity problems and integer linear programs.* Math. Programming in theory and practice (edited by: P.L. Hammer, Zontendijk), North-Holland (1974), 161-199.

[110] Glassey C.R.: *A quadratic network optimization model for equilibrium simple commodity trade flow.* Math. Programming, 14 (1978), 98-107.

[111] Gould F.J. and Tolle J.W.: *An unified approach to complementarity in optimization.* Discrete Math., 7 (1974), 225-271.

[112] Ha Cu D.: *Stability of the linear complementarity problem at a solution point.* Math. Programming, 31 (1985), 327-338.

[113] -----------: *Application of degree theory in stability of the complementarity problem.* Paper presented at the 12th International Symp. on Math. Programming M.I.T. Cambridge, Mass. U.S.A. (1985).

[114] Habetler G.J. and Price A.L.: *Existence theory for generalized nonlinear complementarity problems.* J. Optimization Theory Appl., 7 (1971), 223-239.

[115] -----------: *An iterative method for generalized nonlinear complementarity problems.* J. Optimization Theory Appl., 11 (1973), 36-48.

[116] Habetler G.J. and Kostreva M.M.: *Sets of generalized complementarity problems and P-matrice.* Math. Oper. Res., 5, Nr. 2 (1980), 280-284.

[117] -----------: *On a direct algorithm for nonlinear complementarity problems.* SIAM J. Control Optim., 16, Nr. 3 (1978), 504-511.

[118] Hallman W.P.: *Complementarity in mathematical programming.* Doctoral Dissertation. Department of Industrial Engineering, University of Wisconsin, Madison, WI (1979).

[119] Hallman W.P. and Kaneko I.: *On the connectedness of the set of almost complementarity paths of a linear complementarity problem.* Math. Programming, 16 (1979), Nr. 3, 384-385.

[120] Hansen T. and Manne A.S.: *Equilibrium and linear complementarity-an economy with institutional constraints prices.* G. Schwödiauer ed: Equilibrium and disequilibrium in economic theory. D.Reidel Publishing Company, Dordrecht (1977), 223-237.

[121] Hanson M.A.: *Duality and self-duality in mathematical programming.* SIAM J. Appl. Math., 12, Nr. 2 (1965, 446-449.

[122] Howe R.: *On a class of linear complementarity problems of variable degree.* B.C. Eaves, F.J. Gould, H.O. Peitgen and M.J. Todd eds: Homotopy methods and global convergence. Plenum Press (1983), 155-177.

[123] Howe R. and Stone R.: *Linear complementarity and the degree of mappings.* B.C. Eaves, F.J. Gould, H.O. Peitgen and M.J. Todd eds.: Homotopy methods and global convergence. Plenum Press (1983), 179-223.

[124] Ibaraki T.: *Complementary programming.* Operations Res., 19 (1971), 1523-1528.

[125] Ingleton A.W.: *A problem in linear inequalities.* Proc. London Math. Soc., 16 (1966), 519-536.

[126] Isac G.: *Un théorème de point fixe. Application au problème d'optimisation d'Ersov.* Sem. Inst. Mat. Appl. "Giovanni Sansone Univ. Firenze Rept. (1980), 1-23.

[127] -----------: *Complementarity problem and coincidence equations on convex cones.* K. Ritter, W. Oettli, R. Henn etc eds.: Methods of Operations Research, 51 (1984), 23-33. Verlagsgruppe Athenäum, Hain, Hanstein.

[128] -----------: *Nonlinear complementarity problem and Galerkin method.* J. Math. Anal. Appl., 108, Nr. 2 (1985), 563-574.

[129] -----------: *On implicit complementarity problem in Hilbert spaces.* Bull. Austral. Math. Soc., 32, Nr. 2 (1985), 251-260.

[130] ----------: *Complementarity problem and coincidence equations on convex cones.* Boll. Un. Mat. Ital. (6) 5-B (1986), 925-943

[131] -----------: *Problèmes de complémentarité. [En dimension infinie].* Mini-cours. Publications du Dép. de Math. et Informatique. Université de Limoges (France) (1985).

[132] Isac G. and Théra M.: *Complementarity problem and the existence of the post-critical equilibrium state of a thin elastic plate.* J. Optimization Theory Appl. (To appear).

[133] -----------: *A variational principle. Application to the nonlinear complementarity problem.* Dép. de Math. Collège Militaire Royal de St-Jean, St-Jean, Québec, Canada and Dép. de Math. Univ. Limoges, France. Rept. (1986).

[134] Jahanshahlou G.R. and Mitra G.: *Linear complementarity problem and tree search algorithm for its solution.* Survey Math. Programm. Proc. Ninth. Int. Math. Programm. Symp., 2, Acad. Kiado, Budapest (1979), 35-55.

[135] Jeroslow R.G.: *Cutting planes for complementarity constraints.* CORE Discussion Paper, 7707 - Univ. Cath. Louvain (1977).

[136] Jones Ph. C., Saigal R. and Schneider M.H.: *A variable dimension homotopy on Network for computing linear spatial equilibria.* Preprint, Dept. of Industrial Engineering and Management Sciences. Technological Inst., Northwestern University, Evanston, Illinois (1984).

[137] Judice J.J.: *A study of the linear complementarity problems.* Doctoral Thesis Brunel Univ. Uxbridge (1982).

[138] -----------: *Classes of matrices for the linear complementarity problem.* Linear Algebra and Appl. 54 (1983), 122-125.

[139] -----------: *On principal transforms of matrices.* Preprint (1985) Dept. de Matem. Universidade de Coimbra. Coimbra, Portugal.

[140] Judice J.J. and Mitra G.: *An enumerative method for the solution of linear complementarity problems.* Tech. Rept. TR/04/83, Dept. of Math. and Statistics, Brunel University (1983).

[141] -----------: *Reformulations of Mathematical Programming Problems as Linear Complementarity Problems and an investigation of their solution methods.* Preprint (1985), Dept. de Matem. Universidade de Coimbra, Coimbra, Portugal.

[142] Kaneko I.: *The nonlinear complementarity problem.* Term. Paper, OR340C, Department of Operations Research, Stanford University, Stanford, CA (1973).

[143] -----------: *The parametric linear complementarity problem in the De Donato-Maier analysis of reinforce concrete beams.* Tech. Rept. Sol. 75-13, Department of Operations Research, Stanford University, Stanford CA (1975).

[144] -----------: *Isotone solutions of parametric linear complementarity problems.* Math. Programming 12 (1977), 48-59.

[145] -----------: *A linear complementarity problem with an n by 2n "P"-matrix.* Math. Programming Study 7 (1978), 120-141.

[146] -----------: *The number of solutions of a class of linear complementarity problems.* Math. Programming, 17 (1979), 104-105.

[147] -----------: *A mathematical programming method for the inelastic analysis of reinforced concrete frams.* Internat. J. Numer. Methods Engrg., 11 (1977), 137-154.

[148] -----------: *Linear complementarity problems and characterization of Minkowski matrices.* Linear Algebra and Appl., 20 (1978), 111-129.

[149] -----------: *A maximization problem related to parametric linear complementarity.* Math. Programming, 15, Nr. 2 (1978), 146-154.

[150] -----------: *A maximization problem related to parametric linear complementarity.* SIAM J. Control Optim., 16, NR. 1 (1978), 41-55.

[151] -----------: *Complete solutions for a class of elastic-plastic structures.* Comput. Methods Appl. Mech. Engrg., 21 (1980), 193-209.

[152] -----------: *On some recent engineering applications of complementarity problems.* Math. Programming Study, 17 (1982), 111-125.

[153] ------------: *Piecewise linear elastic-plastic analysis.* Internat. J. Numer. Methods Engrg., 14 (1979) 757-767.

[154] Kaneko I. and Hallman W.P.: *An enumeration algorithm for a general linear complementarity problem.* Tech. Rept. WP-78-11 University of Wisconsin-Madison (1978).

[155] Karamardian S.: *The nonlinear complementarity problem with applications.* J. Optimization Theory Appl. 4 (1969) (I) 87-98; (II) 167-181.

[156] --------------: *Complementarity problems over cones with monotone and pseudomonotone maps.* J. Optimization Theory Appl. 18 (1976) 445-454.

[157] --------------: *Generalized complementarity problem.* J. Optimization Theory Appl. 8 (1971) 161-168.

[158] --------------: *The complementarity problem.* Math. Programming 2 (1972) 107-129.

[159] Kelly L.M. and Watson L.T.: *Q-matrices and spherical geometry.* Linear Algebra and Appl. 25 (1979) 175-190.

[160] Koehler G.J.: *A complementarity approach for solving Leontief substitution systems and (generalized) Markov decision processes.* R A I R O Recherche Opér. 13 Nr. 1 (1979) 75-80.

[161] Kojima M.: *Computational methods for solving the nonlinear complementarity problem.* Keio Engineering Repts. Vol. 27 Nr. 1 Department of Administrative Engineering Keio University, Yokohama, Japan (1974) 1-41.

[162] ---------: *A unification of the existence theorem of the nonlinear complementarity problem.* Math. Programming 9 (1975) 257-277.

[163] ---------: *A complementarity pivoting approach to parametric nonlinear programming.* Math. Oper. Res. 4 Nr. 4 (1979) 464-477.

[164] ---------: *Studies on piecewise - linear approximations of piecewise - C^1 mappings in fixed points and complementarity theory.* Math. Oper. Res. 3 Nr. 1 (1978) 17-36.

[165] Kojima M., Nishino H. and Sekine T.: *An extension of Lemke's method to the piecewise linear complementarity problem.* SIAM J. Appl. Math. Vol. 31 Nr. 4 (1976) 600-613.

[166] Kojima M. and Saigal R.: *On the number of solutions to a class of linear complementarity problems.* Math. Programming 17 (1979) 136-139.

[167] -----------: *On the number of solutions to a class of complementarity problems.* Math. Programming 21 (1981) 190-203.

[168] Kostreva N.M.: *Direct algorithms for complementarity problems.* Ph.D. Thesis, Rensselaer Polytechnic Institute, Troy, New York (1976).

[169] -----------: *Block pivot methods for solving the complementarity problem.* Linear Algebra and Appl. 21 (1978) 207-215.

[170] -----------: *Cycling in linear complementarity problems.* Math. Programming 16 (1979) 127-130.

[171] -----------: *Finite test sets and P-matrices.* Proc. Am. Math. Soc. 84 Nr. 1 (1982) 104-105.

[172] -----------: *Elasto-Hydrodynamic lubrication: a non-linear complementarity problem.* Internat J. Numer. Methods Fluids 4 (1984) 377-397.

[173] Lemke C.E.: *Bimatrix equilibrium points and mathematical programming.* Management Sci. 11 (1965) 681-689.

[174] ----------: *On complementary pivot theory.* G.B. Dantzig and A.F. Veinott Jr. eds Mathematics of the Decision Sciences. Am. Math. Soc. Providence R.I. (1968) 95-114.

[175] ----------: *Some pivot schemes for the linear complementarity problem.* Math Programming Study 7 (1978) 15-35.

[176] ----------: *Recent results on complementarity problem.* J.B. Rosen, O.L. Managasarian and K. Ritter eds.: Nonlinear Programming (1970).

[177] ----------: *A brief survey of complementarity theory.* C.V. Coffman and G.J. Fix eds.: Constructive approaches to Mathematical Models. Academic Press New York (1979).

[178] ----------: *A survey of complementarity theory.* R.W. Cottle, F. Giannessi and J.L. Lions eds.: Variational inequalities and complementarity problems. Theory and Applications. John Wiley & Sons (1980) 211-239.

[179] Lemke C.E. and Howson J.T.: *Equilibrium points of bimatrix.* SIAM J. Appl. Math. games. 12 Nr. 2 (1964) 413-413.

[180] Luk F.T. and Pagano M.: *Quadratic programming with M-matrices.* Linear Algebra and Appl. 33 (1980) 15-40.

[181] Luna G.: *A remark on the complementarity problem.* Proc. Amer. Math. Soc. 48 Nr. 1 (1975) 132-134.

[182] Lüthi H.J.: *A simplicial approximation of a solution for the nonlinear complementarity problem.* Math. Programming 9(1975) 278-293.

[183] ----------: *Komplementaritäts und fixpunktalgorithmen in der mathematischen programmierung spieltheory und ökonomie.* Lecture Notes in Economics and Mathematical Systems 129 (1976), Springer-Verlag.

[184] Lin Y. and Cryer C.W.: *An alternating direction implicit algorithm for the solution of linear complementarity problems arising from free boundary problems.* Appl. Math. Optim. 13, (1985) 1-17.

[185] Maier G. *A matrix structural theory of piecewise-linear elastoplasticity with interacting yield-planes.* Meccanica J. Italian Assoc. Theoret. Appl. Mech. 5 (1970), 54-66.

[186] -----------: *Problem on parametric linear complementarity problem.* SIAM Rev. 14, Nr. 2 (1972), 364-365.

[187] Maier G., Andreuzzi F., Giannessi F., Jurina L. and Taddei F.: *Unilateral contact, elastoplasticity and complementarity with reference to offshore pipeline design.* Comput. Methods Appl. Mech. Engrg. 17/18 (1979), 469-495.

[188] Maier G. Kaneko I.: *Optimum design of plastic structures under displacement constraints.* Comput. Methods Appl. Mech. Engrg. 27 (1981), 369-391.

[189] Mandel J.: *A multilevel iterative method for symetric positive definite linear complementarity problems.* Appl. Math. Optim 11 (1984), 77-95.

[190] Magnanti T.L.: *Model and algorithms for predicting urban trafic equilibria.* Working paper. M.I.T. Cambridge, Mass. U.S.A.

[191] Mangasarian O.L.: *Equivalence of the complementarity problem to a system of nonlinear equations.* SIAM J. Appl. Math. 31 (1976), 89-92.

[192] -----------: *Linear complementarity problems solvable by a single linear program.* Math. Programming 10 (1976), 263-270.

[193] -----------: *Solution of linear complementarity problem by linear programming.* G.W. Watson ed.: Numerical Analysis. Lecture Notes in Mathematics Nr. 506. Springer-Verlag, 166-175.

[194] -----------: *Solution of symmetric complementarity problems by iterative methods.* J. Optimization Theory Appl. 22, Nr. 4 (1977), 465-485.

[195] -----------: *Characterization of linear complementarity problems as linear programs.* Math. Programming Study 7 (1978), 74-87.

[196] -----------: *Characterization of bounded solutions of linear complementarity problems.* Math. Programming Study 19 (1982), 153-166.

[197] -----------: *Locally unique solutions of quadratic programs, linear and nonlinear complementarity problems.* Math. Programming 19 (1980), 200-212.

[198] -----------: *Simplified characterization of linear complementarity problems as linear programs.* Math. Oper. Res., 4 (1979), 268-273.

[199] -----------: *Iterative solution of linear programs.* SIAM J. Numer. Anal., 18, Nr. 4 (1981), 606-614.

[200] -----------: *Sparsity-preserving SOR algorithms for separable quadratic and linear programming.* Comput. Operations Res., 11, Nr. 2 (1984), 105-112.

[201] Mangasarian O.L. and McLinden L.: *Simple bounds for solutions of monotone complementarity problems and convex programs.* Math. Programming, 32 (1985), 32-40.

[202] Mathiesen L.: *Computational experience in solving equilibrium models by a sequence of linear complementarity problems.* Opérations Res., 33, Nr. 6 (1985), 1225-1250.

[203] McLinden L.: *The complementarity problem for maximal monotone multifunction.* R.W. Cottle, F. Giannessi and J.L. Lions eds.: Variational inequalities and complementarity problems. Theory and Appl. John Wiley & Sons (1980), 251-270.

[204] -----------: *An analogue of Moreau's proximation theorem, with application to the nonlinear complementarity problem.* Pacific J. of Math, 88, Nr. 1 (1980), 101-161.

[205] -----------: *Stable monotone variational inequalities.* MRC Technical Summary Report, Nr. 2734 (1984), Math. Research Center, Univ. of Wisconsin-Madison.

[206] Meerov M.V., Bershchanskiy Ya. M. and Litvak M.L.: *Optimization methods for one class of multivariable systems.* Internat. J. Control, 3, Nr. 2 (1980), 239-270.

[207] Megiddo N.: *On the parametric nonlinear complementarity problem.* Department of Statistics, Tel Aviv University, Tel Aviv (August 1975, Revised).

[208] -----------: *On monotonicity in parametric linear complementarity problems.* Math. Programming, 12, Nr. 1 (1977), 60-66.

[209] -----------: *A monotone complementarity problem with feasible solutions but no complementary solutions.* Math. Programming Study, 12, Nr. 1 (1977), 131-132.

[210] -----------: *On the parametric nonlinear complementarity problem.* Math. Programming Study, 7 (1978), 142-150.

[211] Megiddo N. and Kojima M.: *On the existence and uniqueness of solutions in nonlinear complementarity theory.* Math. Programming, 12 (1977), 110-130.

[212] Meister H.: *A parametric approach to complementarity theory.* J. Math. Anal. Appl., 101 (1984), 64-77.

[213] Mitra G.: *An exposition of the (linear) complementarity problem.* Internat. J. Math. Ed. Sci. Tech., 10, Nr. 3 (1979), 401-416.

[214] Mitra G. and Jahanshalou G.R.: *Linear complementarity problem and a tree search algorith for its solution.* A. Prekopa ed.: Survey of Math. Programming, Budapest (1979), 35-56.

[215] Mohan S.R.: *Parity of solutions for linear complementarity problems with Z-matrices.* Opsearch, 13 (1976), 19-28.

[216] -----------: *Existence of solution rays for linear complementarity problems with Z-matrices.* Math. Programming Study, 7 (1978), 108-119.

[217] -----------: *Degenerate complementarity cones induces by a K_0-matrix.* Math. Programming, 20 (1981), 103-109.

[218] -----------: *On the simplex method and a class of linear complementarity problems.* Linear Algebra and Appl., 14, Nr. 1 (1976), 1-9.

[219] Moré J.J.: *Class of functions and feasibility conditions in nonlinear complementarity problems.* Math. Programming 6 (1974) 327-338.

[220] ---------: *Coercivity conditions in nonlinear complementarity problems.* SIAM Rev. 16 Nr. 1 (1974) 1-16.

[221] Moré J.J. and Rheinboldt W.: *On P- and S- functions and related classes of n-dimensional nonlinear mappings.* Linear Algebra and Appl. 6 (1973) 45-68.

[222] Mosco U.: *On some non-linear quasi-variational inequalities and implicit complementarity problems in stochastic control theory.* R.W. Cottle, F. Giannessi and J.L. Lions eds. Variational inequalities and complementarity problems. Theory and Appl. John Wiley & Sons (1980) 271-283.

[223] Mosco U. and Scarpini F.: *Complementarity systems and approximation of variational inequalities.* RAIRO Recherche Oper. (Analyse Numer.) (1975) 83-104.

[224] Murty K.G.: *On the number of solutions to the complementarity quadratic programming problem.* Doctoral Dissertation, Engineering Science, University of California, Berklay (1968).

[225] ----------: *On the characterization of P-matrices.* SIAM J. Appl. Math. 20 (3) (1971) 378-384.

[226] ----------: *On the number of solutions to the complementarity problem and the spanning properties of complementarity cones.* Linear Algebra and Appl. 5 (1972) 65-108.

[227] ----------: *On the Bard-type scheme for solving the complementarity problem.* Opsearch 11 (1974) 123-130.

[228] ----------: *An algorithm for finding all the feasible complementary bases for a linear complementarity problem.* Tech. Rept. 72-2 Dept. of Industrial Engineering, Univ. of Michigan (1972).

[229] ----------: *Computational complexity of complementarity pivot methods.* Math. Programming Study 7 (1978) 61-73.

[230] ----------: *On the linear complementarity problem.* Operations Research Verfahren 31 (1978) 425-439.

[231] Murty K.G. and Fathi Y.: *A critical index algorithm for nearest point problem on simplicial cones.* Math. Programming 23 (1983) 206-215.

[232] Nanda S. and Nanda S.: *On the stationary points and the complementarity problem.* Bull. Austral. Math. Soc. 20 (1979) 77-86.

[233] ----------: *A nonlinear complementarity problem in mathematical programming in Hilbert space.* Bull. Austral. Math. Soc. 20 (1979) 233-236.

[234] Nanda S. and Patel U.: *A nonlinear complementarity problem for monotone functions.* Bull. Austral. Math. Soc. 20 (1979) 227-231.

[235] Niccolucci F.: *On the equivalence among integer programs, nonconvex real programs, special complementarity problems, variational inequalities and fixed-point problems.* Pubblicazioni del Dipartimento di Ricerca Operativa e Scienze Statistiche. Seria A nr. 8, Universita Di Pisa (1973).

[236] Pang J.S.: *Least element complementarity theory.* Ph. D. Dissertation, Department of Operations Research, Stanford University (1976).

[237] ----------: *A note on an open problem in linear complementarity.* Math. Programming 13 (1977) 360-363.

[238] ----------: *One cone orderings and the linear complementarity problem.* Linear Algebra and Appl. 22 (1978) 267-287.

[239] ----------: *On a class of least-element complementarity problems.* Math. Programming 16 (1979) 111-126.

[240] ----------: *A column generation technique for the computation of stationary points.* Math. oper. Res. 6 (1981) 213-224.

[241] ----------: *The implicit complementarity problem.* O.L. Mangasarian, R.R. Meyer and S.M. Robinson eds.: Nonlinear Programming 4 (Academic Press - 1981) 487-518.

[242] ----------: *On the convergence of a basic iterative method for the implicit complementarity problem.* J. Optimization Theory Appl. 37 (1982) 149-162.

[243] -----------: *A parametric linear complementarity technique for optimal portfolio selection with a risk-free asset.* Operations Res., 28, Nr. 4 (1980), 927-941.

[244] -----------: *Necessary and sufficient conditions for the convergence of iterative methods for the linear complementarity problem.* J. Optimization Theory Appl., 42 (1984), 1-17.

[245] -----------: *More results on the convergence of iterative methods for the symmetric linear complementarity problem.* J. Optimization Theory Appl. 49 Nr 1 (1986), 107-134.

[246] -----------: *On Q-matrices.* Math. Programming, 17 (1979), 243-247.

[247] -----------: *A new and efficient algorithm for a class of portfolio selection problems.* Operations Res., 28 (1980), 754-767.

[248] Pang J.S. and Chan D.: *Iterative methods for variational and complementarity problems.* Math. Programming, 24 (1982), 284-313.

[249] Pang J.S., Kaneko I. and Hallman W.P.: *On the solution of some (parametric) linear complementarity problems with applications to portfolio selection structural engineering and actuarial graduation.* Math. Programming, 16 (1979), 325-347.

[250] Pang J.S. and Lee P.S.C.: *A parametric linear complementarity technique for the computation of equilibrium prices in a single commodity spatial model.* Math. Programming, 20 (1981), 81-102.

[251] Pardalos P.M. and Rosen J.B.: *Global optimization approach to the linear complementarity problem.* Tech. Rept., Computer Science Department, University of Minnesota (1985).

[252] Parida J. and Sahoo B.: *A note on generalized linear complementarity problems.* Bull. Austral. Math. Soc., 18, Nr. 2(1978), 161-168.

[253] -----------: *An existence theorem for the generalized complementarity problem.* Bull. Austral. Math. Soc., 19, Nr. 1 (1978), 51-58.

[254] Patrizi G.: *The equivalence of an LCP to a parametric linear program in a scalar variable.* Preprint, Instituto di Economia, Univ. degli Studi di Siena, Italy (1985).

[255] Peterson E.L.: *The conical duality and complementarity of price and quality for multicomodity spatial and temporal network allocation problems.* Discussion paper 207. Center for Mathematical Studies in Economics and Management Science, Northwestern University (1976).

[256] Ramarao B. and Shetty C.M.: *Application of disjunctive programming to the linear complementarity problem.* Naval Res. Log. Quart., 31 (1984), 589-600.

[257] Ravindran A.: *Computational aspects of Lemke's complementary algorithm applied to linear programs.* Opsearch, 7, Nr. 4 (1970), 241-262.

[258] Reiser P.M.: *A modified integer labeling for complementarity algorithms.* Math. Oper. Res., 6, Nr. 1 (1981), 129-139.

[259] Riddell R.C.: *Equivalence of nonlinear complementarity problems and least element problems in Banach lattices.* Math. Oper. Res., 6, Nr. 3 (1981), 462-474.

[260] Rockafellar R.T.: *Lagrange multipliers and variational inequalities.* R.W. Cottle, F. Giannessi and J.L. Lions eds: Variational inequalities and complementarity problems. John Wiley & Sons (1980), 303-322.

[261] Sacher R.S.: *On the solution of large, structured complementarity problems.* Ph. D. Dissertation, Department of Operations Research, Stanford University (1974).

[262] Saigal R.: *A note on a class of linear complementarity problem.* Opsearch, 7 (1970), 175-183.

[263] ----------: *Lemke's algorithm and a special linear complementarity problem.* Opsearch, 8 (1971), 201-208.

[264] ----------: *On the class of complementarity cones and Lemke's algorithm.* SIAM J. Appl. Math., 23, Nr. 1 (1972), 46-60.

[265] ----------: *A characterization of the constant parity property of the number of solutions to the linear complementarity problem.* SIAM J. Appl. Math. 23 (1972), 40-45.

[266] ----------: *Extension of the generalized complementarity problem.* Math. Oper. Res., 1, Nr. 3 (1976), 260-266.

[267] Saigal R. and Simon C.B.: *Generic properties of the complementarity problem.* Math. Programming, 4 (1973), 324-335.

[268] Saigal R. and Stone R.E.: *Proper, reflecting and absorbing facets of complementary cones.* Math. Programming, 31 (1985), 106-117.

[269] Sargent R.W.H.: *An efficient implementation of the Lemke algorithm and its extension to deal with upper and lower bounds.* Math. Programming Study, 7 (1978), 36-54.

[270] Schneider M.H.: *A complementary pivoting algorithm for linear network problems.* Preprint. School of Engineering and Applied Science, Princeton University (1985).

[271] ----------: *Single - Commodity Spatial - Equilibria: A network complementarity approach.* Ph. Dissertation, Dept. of Industrial Engineering and Management Sciences. Northwestern Univ. Evanston, Illinois (1984).

[272] Shiau T.H.: *Itertive linear programming for linear complementarity and related problems.* Computer Sciences Technical Report #507, University of Wisconsin, Madison (1983).

[273] Smale S.: *On the average number of steps of the simplex method of linear programming.* Math. Programming, 27 (1983), 241-262.

[274] Smith T.E.: *A solution condition for complementarity problems with an application to spatial price equilibrium.* Appl. Math. Comput., 15 (1984), 61-69.

[275] Smith M.J.: *A descent algorithm for solving monotone variational inequalities and monotone complementarity problems.* J. Optimization Theory Appl., 44, Nr. 3 (1984), 485-496.

[276] Stickney A. and Watson L.: *Digraph models of bard-type algorithms for the linear complementarity problem.* Math. Oper. Res., 3, Nr. 4 (1978), 322-333.

[277] Stone R.E.: *Geometric aspects of the linear complementarity problem.* Tech. Rept., SOL 81-6, Department of Perations Research, Stanford University. Stanford, CA (1981).

[278] -----------: *Linear complementarity problem with an invariant number of solutions.* Math. Programming, 34, Nr. 3 (1986), 265-291.

[279] Talman D. and Van der Heyden L.: *Algorithms for the linear complementarity problem which allow an arbitrary starting point.* B.C. Eaves, F.J. Gould, H.O. Peitgen and M.J. Todd eds: Homotopy methods and global convergence. Plenum Press (1983), 267-285.

[280] Tamir A.: *On a characterization of P-matrices.* Math. Programming, 4 (1973), 110-112.

[281] -----------: *The complementarity problem of mathematical programming.* Ph.D. Dissertation. Department of Operations Research, Case Western Reserve University (1973).

[282] -----------: *Minimality and complementarity properties associated with Z-functions and M-functions.* Math. Programming, 7 (1974), 17-31.

[283] -----------: *On the number of solutions to the linear complementarity problem.* Math. Programming, 10 (1976), 347-353.

[284] Thoai N.V. and Tuy H.: *Solving the linear complementarity problem through concave programmming.* Zh. Vychisl. Mat. J. Mat. Fiz., 23 (3) (1983), 602-608.

[285] Todd M.J.: *Abstract complementary pivot theory.* Ph.D. Thesis, Department of Administrative Sciences, Yale University (1972).

[286] -----------: *Orientations in complementary pivot algorithm.* Math. Oper. Res., 1, Nr. 1 (1976), 54-66.

[287] -----------: *Extension of Lemke's algorithm for the linear complementarity problem.* J. Optimization Theory Appl, 20, Nr. 4 (1976), 397-416.

[288] Tomlin J.A.: *User guide for LC.PL. a program for solving linear complementarity problems by Lemke's method.* Tech. Rept. SOL, 76-16, Systems Optimization Laboratory Dept. of Operations Research, Stanford University (1976).

[289] -----------: *Robust implementation of Lemke's method for the linear complementarity problem.* Math. Programming Study, 7 (1978), 55-60.

[290] Tuy H. and Thoai Ng. V.: *Solving the linear complementarity problem via concave programming.* Meth. of Operations Research Proc. V. Sympos. Oper. Res. Köln. R.E. Burkard, T. Ellinger eds, 175-178.

[291] Tuy H, Thieu T.V. and Thai Ng.Q.: *A conical algorithm for globally minimizing a concave function over a closed convex set.* Math. Oper. Res., 10, Nr. 3 (1985), 498-514.

[292] Van der Heyden L.: *A variable dimension algorithm for the linear complementarity problem.* Math. Programming, 19, Nr. 3 (1980), 328-346.

[293] Van der Panne C.: *A complementary variant of Lemke's method for the linear complementarity problem.* Math. Programming, 7 (3) (1974), 283-310.

[294] Vescan R.T.: *Un problème variationnel implicite faible.* C.R. Acad. Sc. Paris, t. 299, série A, Nr. 14 (1984), 655-658.

[295] -----------: *A weak implicit variational inequality.* Preprint, AL. I. Cuza, University of Iasi, (Romania) (1984).

[296] Watson L.T.: *A variational approach to the linear complementarity problem.* Doctoral Dissertation, Dept. of Mathematics University of Michigan, Ann Arbor (1974).

[297] -----------: *An algorithm for the linear complementarity problem.* Internat J. Comput Math., 6, Nr. 4 (1978), 319-325.

[298] -----------: *Solving the nonlinear complementarity problem by a homotopy method.* SIAM J. Control Optim., 17, Nr. 1 (1979), 36-46.

[299] -----------: *Some perturbation theorems for Q-matrices.* SIAM J. Appl. Math., 31, Nr. 2 (1976), 379-384.

[300] Werner R. and Wetzel R.: *Complementary pivoting algorithms involving extreme rays.* Math. Oper. Res., 10, Nr. 2 (1985), 195-206.

[301] Wierzbicki A.P.: *Note on the equivalence; of Kuhn-Tucker complementarity conditions to an equation.* J. Optimization Theory Appl., 37, Nr. 3 (1982), 401-405.

[302] Wilmuth R.: *A computational comparison of fixed point algorithms which use complementary pivoting.* S. Karamardian ed.: Fixed points algorithms and applications. Academic Press (1977), 250-279.

[303] Wintgen G.: *Indifferente optimierungs probleme.* Beitrag zur Internationalen Tagung. Mathematik und Kybernetik in der Okonomie. Berlin 1964, Konferenzprotokoll, Teil II. (Akademie Verlag, Berlin), 3-6.

[304] Yoo Y.: *Iterative methods for linear complementarity problems and large scale quadratic programming problems.* M.S. Thesis, Korea Adavanced Institute of Science and Technology (Seul, Korea 1981).

B General references

[1] Browder F.E.: *On the unification of the calculus of variations and the theory of monotone nonlinear operators in Banach spaces.* Proc. Natl. Acad. Sci. U.S.A., 56 (1966), 419-425.

[2] ------------: *Existemce and approximation of solutions of nonlinear variational inequalities.* Proc. Natl. Acad. Sci. U.S.A., 56 (1966),1080-1086.

[3] -----------: *Existence theorems for nonlinear partial differential equations.* Global Analysis, Proc. Symp. Pure Math., Nr. 16, A.M.S. Providence, 1-60.

[4] Eilenberg S. and Montgomery D.: *Fixed point theorems for multivalued transformations.* Amer. J. Math., 68 (1946), 214-222.

[5] Fidler M. and Ptak V.: *On matrices with nonpositive off-diagonal elements and positive principal minors.* Czechoslovak Math. Journal, 12 (1962), 382-400.

[6] Hartman P. and Stampacchia G.: *On some nonlinear elliptic functional differential equations.* Acta Math., 115 (1966), 271-310.

[7] Kinderlehrer D. and Stampacchia G.: *An introduction to variational inequalities and their applications.* Academic Press (1980).

[8] Köthe G.: *Topological vector spaces I.* Grundlehren der Math. Wiss, 159, Springer-Verlag, Berlin-Heidelberg-New York, (1969), Reivised in 1983.

[9] -----------: *Topological vector spaces II.* Grundlehren der Math. Wiss, 237, Springer-Verlag, Berlin-Heidelberg - New York (1979).

[10] Mosco U.: *Convergence of convex sets and solutions of variational inequalities.* Adv. in Math., 3 (1969), 520-585.

[11] Ortega J.M. and Rheinboldt W.C.: *Iterative solution of nonlinear equations in several variables.* Academic Press, New York, London (1970).

[12] Peressini A.: *Ordered topological vector space.* Harper & Row, New York (1967).

[13] Rheinboldt W.C.: *On M-functions and their applications to nonlinear Gauss-Seidel iterations and to network flows.* J. Math. Anal. Appl. 32 (1970), 274-307.

[14] Rockafellar R.T.: *On the maximality of sums of nonlinear monotone operators.* Trans. Amer. Math. Soc., 149 (1970), 75-88.

[15] Samelson H., Thrall R.M. and Wesler O.: *A partition theorem for euclidian n-spaces.* Proc. Amer. Math. Soc., 9 (1959), 805-807.

[16] Schaefer H.: *Topological vector spaces.* Springer-Verlag, Graduate texts in Maths. 3, Berlin-Heidelberg - New York (1971).

[17] Yun K.K.: *On the existence of a unique and stable market equilibrium.* J. Econom. Theory, 20 (1979), 118-123.

C Supplementary references

[1] Baronti M.: *The nonlinear complementarity problem and related questions.* Rend. Mat. Apl. 6 (1986), 313-319.

[2] Berman A.: *Matrices and the linear complementarity problem.* Linear Algebra and Appl. 40 (1981), 249-256.

[3] Bodo E.P. and Hanson M.A.: *Class of continuous non-linear complementarity problems.* J. Optim Theory and Appl. 24 (1978), 243-262.

[4] Borwein J.M.: *Alternative theorems and general complementarity problems.* Lecture Notes in Economics and Math. Systems. Springer-Verlag Nr. 259 (1985), 194-203.

[5] Brézis H.: *Équations et inéquations non-linéaires dans les espaces vectoriels en dualité.* Ann. Inst. Fourier 10 (1968), 115-176.

[6] Brézis H. and Hess P.: *Nonlinear mappings of monotone type in Banach space.* J. Funct. Anal. 11 (1972), 251-294.

[7] Browder F.E.: *On the fixed point index for continuous mappings of locally connected spaces.* Summa Brasiliensis Mathematicae 4 Nr. 7 (1960), 253-293.

[8] Chen Guang-Ya. and Yang Xiao-Qi: *The vector complementarity problem and its equivalences with the weak minimal element in ordered spaces.* J. Math. Anal. Appl. 153 (1990), 136-158.

[9] Cottle R.W.: *Completely Q-matrices.* Math. Programming 19 (1980), 347-351.

[10] -----------: *Observations on a class of nasty linear complementarity problems.* Discrete Appl. Math. 2 (1980), 89-111.

[11] Cottle R.W., Pang J.S. and Venkateswaran V.: *Sufficient matrices and the linear complementarity problem.* Linear Algebra Appl. 114/115 (1989), 231-249

[12] Cottle R.W. and Sacher R.S.: *Solution of large, structural linear complementarity problems - Tridiagonal case.* Appl. Math. Optim. 3 (1977), 321-340.

[13] Eagambaram N. and Mohan S.R.: *On some classes of linear complementarity problems with matrices of order n and rank (n-1).* Math. Operations Res. 15 Nr. 2 (1990), 243-257.

[14] Eaves B.C.: *Complementarity pivot theory and Markovian decision chains.* Fixed Points: Algorithms and Applications, S. Karamardian (ed) Academic Press (1977), 59-85.

[15] Fan K.: *A minimax inequality and applications.* Inequalities III (Ed. O. Shisha) Academic Press (1972).

[16] Fang S.C.: *A linearization method for generalized complementarity problems.* IEEE Trans. Automatic Control 29 (1984), 930-933.

[17] Friesz T.L., Tobin R.L., Smith T.E. and Harker P.T.: *A nonlinear complementarity formulation and solution procedure for the general derived demand network equilibrium problem.* J. Regional Science 23 (1983), 337-359.

[18] Gowda M.S.: *Complementarity problems over locally compact cones.* SIAM J. Control Opt. 27 Nr-4 (1989), 836-841.

[19] ----------: *On Q-matrices.* Math. Programming 49 (1990), 139-141.

[20] Gowda M.S. and Seidman T.I.: *Generalized linear complementarity problems.* Math. Programming 46 (1990), 329-340.

[21] Harker T. and Pang J.S.: *Finite-dimensional variational inequality and nonlinear complementarity problems: a survey of theory, algorithms and applications.* Math Programming 48 (1990), 161-220.

[22] Harker T. and Xiao B.: *Newton's method for the nonlinear complementarity problem: A B-differentiable equation approach.* Math. Programming 48 (1990), 339-357.

[23] Ingleton A.W.: *Linear complementarity problem.* J. London Math. Soc. 2 (1970), 330-336.

[24] Joffe A.D. and Tihomirov V.M.: *Theory of extremal problems.* North-Holland, New York (1979).

[25] Isac G.: *Fixed point theory and complementarity problems in Hilbert space.* Bull. Austr. Math. Soc. 36 Nr. 2 (1987), 295-310.

[26] -------: *Fixed point theory, coincidence equations on convex cones and complementarity problem.* Contemporary Math. 72 (1988), 139-155.

[27] -------: *On some generalization of Karamardian's theorem on the complementarity problem.* Bull. U.M.I (7) 2-B (1988), 323-332.

[28] -------: *The numerical range theory and boundedness of solutions of the complementarity problem.* J. Math. Anal. Appl. 143 Nr. 1 (1989), 235-251.

[29] -------: *A special variational inequality and the implicit complementarity problem.* F. Fac. Science, Univ. Tokyo Sec. 1A, 37 Nr. 1 (1990), 109-127.

[30] Isac G. and Théra M.: *A variational principle application to the nonlinear complementarity problem.* Nonlinear and convex Analysis (B.L. Lin and S. Simons (eds)) Marcel Derkker Inc. (1987), 127-145.

[31] Isac G. and Németh A.B.: *Projection methods, isotone sprojection cones and the complementarity problem.* J. Math. Anal. Appl. 153 Nr. 1 (1990), 258-275.

[32] Jansen M.J. and Tijs S.H.: *Robustness and nondegenerateness for linear complementarity problems.* Math. Programming 37 (1987), 293-308.

[33] Jameson G.: *Ordered linear spaces.* Lecture Notes in Math. 141 (1970), Springer-Verlag.

[34] Jeroslow R.G.: *Cutting-planes for complementarity constraints.* SIAM J. Control Opt. 16 (1978), 56-62.

[35] Jeter M.W. and Pye W.C.: *Some properties of Q-matrices.* Linear Alg. Appl. 57 (1987), 169-180.

[36] ---------------------: *The linear complementarity problem and a subclass of fully semimonotone matrices.* Linear Alg. Appl. 87 (1987), 243-256.

[37] Jones P.C.: *Even more with the Lemke complementarity algorithm.* Math. Programming 34 (1986), 239-242.

[38] Kaneko I.: *The number of solutions of a class of linear complementarity problems.* W.P. 76-10, Department of Industrial Engineering, Univ. Wisconsin-Madison (Aug. 1976).

[39] ---------: *Reduction theorem for linear complementarity problem with a certain patterned matrix.* Linear Alg. Appl. 21 (1978), 13-34.

[40] Kelly L.M. and Watson L.T.: *Erratum: some perturbation theorems for Q-matrices.* SIAM J. Appl. Math. 34 (1978), 320-321.

[41] Kojims M., Mizuno S. and Noma T.: *A new continuation method for complementarity problems with uniform P-functions.* Math Programming 43 (1989), 107-113.

[42] Kyparisis J.: *Uniqueness and differentiability of solutions of parametric nonlinear complementarity problems.* Math. Programming 36 (1986), 105-113.

[43] Larsen R.: *Functional Analysis.* Dekker, New York (1973).

[44] Mangasarian O.L.: *Simple computable bounds for solutions of linear complementarity and linear programs.* Math. Programming Study 25 (1986), 1-12.

[45] Mangasarian O.L. and Shiau T.H.: *Error bounds for monotone linear complementarity problems.* Math. Programming 36 (1986), 81-89.

[46] ------------------------------: *Lipschitz continuity of solutions of linear inequalities, programs and complementarity problems.* SIAM J. Control and Opt. 25 (1987), 583-595.

[47] Mangasarian O.L. and De Leone R.: *Parallel successive overrelaxation methods for symmetric linear complementarity problems and linear programs.* J Optim Theory Appl. 54 Nr. 3 (1987), 437-446.

[48] Mas-Collel A.: *A note on a theorem of F. Browder.* Math. Programming 6 (1974), 229-233.

[49] Mathiesen L.: *An algorithm based on a sequence of linear complementarity problems applied to a Walrasian equilibrium model: an example.* Math Programming 37 (1987), 1-18.

[50] -----------: *Computation of economic equilibria by a sequence of linear complementarity problems.* Math. Programming Study 23 (1985), 144-162.

[51] Mc Callum C.J.: *Existence theory for the complex linear complementarity problem.* J. Math. Anal. Appl. 40 (1972), 738-762.

[52] -------------: *Solution of the complex linear complementarity problem.* J. Math. Anal. Appl. 44 (1973), 643-660.

[53] Megiddo N.: *On the expected number of linear complementarity cones intersected by random and semi-random rays.* Math. Programming 35 (1986), 225-235.

[54] Murty K.G.: *Linear complementarity, linear and nonlinear programming.* Heldermann Verlag, Berlin (1988).

[55] Oh K.P.: *The formulation of the mixed lubrication problem as a generalized nonlinear complementarity problem.* Trans. ASME, Journal of Tribology 108, (1986), 598-604.

[56] Pang J.S.: *A unification of two classes of Q-matrices.* Math. Programming 20 (1981), 348-352.

[57] ---------: *Inexact Newton methods for the nonlinear complementarity problem.* Math. Programming 36 (1986), 54-71.

[58] Pang J.S. and Chandrasekaren R.: *Linear complementarity problems solvable by a polynomially bounded pivoting algorithm.* Math. Programming Study 25 (1985), 13-27.

[59] Parida J., Sen A. and Kumar A.: *A linear complementarity problem involving a subgradient.* Bull. Austral. Math. Soc. 37 (1988), 345-351.

[60] Parida J. and Sen A.: *A class of nonlinear complementarity problem for multifunctions.* J. Optim Theory Appl. 53 (1987), 105-113.

[61] Pardalos P.M. and Rosen J.B.: *Bounds for the solution set of linear complementarity problems.* Discrete Appl. Math. 17 (1987), 255-261.

[62] Paris Q.: *Perfect aggregation and disaggregation of complementarity problems.* Amer. J. Agricultural Economics 62 (1980), 681-688.

[63] --------: *Revenue and cost uncertainty, generalized mean-variance and the linear complementarity problem.* Amer. J. Agricultural Economics 61 (1979), 268-275.

[64] Pshenichnyi B.N. and Sasnovskii A.A.: *The complementarity problem.* Translated from Kibernetika 4 (1988), 42-46.

[65] Ravindran A.: *Comparison of primal-simplex and complementarity pivot methods for linear programming.* Naval Research Logistics Quarterly 20 (1973), 95-100.

[66] Saviozzi G.: *On dependency in linear complementarity problems.* Discrete Applied Math. 11 (1985), 311-314.

[67] Schmidt J.W.: *On tridiagonal linear complementarity problems.* Num. Math 51 (1987), 11-21.

[68] Stevens S.M. and Lin P.M.: *Analysis of piecewise linear resistive newtworks using complementarity pivot theory.* IEEE Transactions on Circuits and Systems 28 (1981), 429-441.

[69] Todd M.J.: *Complementarity in oriented matroids.* SIAM J. on Algebraic and Discr. Methods 5 (1984), 467-485.

[70] ---------: *Polynomial expected behavior of a pivoting algorithm for linear complementarity and linear programming problems.* Math. Programming 35 (1986), 173-192.

[71] Van der Laan G., Talman J.J. and Van der Heyden L.: *Simplicial variable dimension algorithms for solving the nonlinear complementarity problem on a product of unit simplices using a general labelling.* Math. Oper. Res 12 Nr. 3 (1987), 377-397.

[72] Van der Laan G. and Talman A.J.J.: *Simplicial approximation of solutions to the nonlinear complementarity problem with lower and upper bounds.* Math. Programming 38 (1987), 1-15.

[73] Van Eijndhoven J.T.J.: *Solving the linear complementarity problem in circuit simulation.* SIAM J. Control Opt. vol. 24 Nr. 5 (1986), 1050-1062.

[74] Wilson R.: *Bilinear complementarity problem and competitive equilibria of piecewise linear economic models.* Econometrica 46 (1978), 87-103.

[75] Boyd D.W. and Wong J.S.: *On nonlinear contraction.* Proc. Amer. Math. Soc. 20 (1969), 458-464.

[76] Fan K.: *A generalization of Tychonoff's fixed point theorem.* Math. Annalen 142 (1961), 305-310.

[77] Noor M.A.: *Generalized complementarity problems.* J. Math. Anal. Appl. 120 Nr. 1 (1986), 321-327.

[78] Noor M.A. and Zarae S.: *Linear quasi complementarity problems.* Utilitas Math. 27 (1985), 249-260.

[79] Browder F.E.: *Non-expansive non-linear operators in Banach spaces.* Proc. Nat. Acad. Sci. U.S.A. 54 (1965), 1041-1044.

[80] Hilbert D.: *Neue begründung der Bolya-Lobatshefskyschen geometrie.* Math Ann. 57 (1903), 137-150.

[81] Isac G. and Németh A.B.: *Isotone projection cones in Hilbert spaces and the complementarity problem.* Boll. Un. Nat. Ital. (In press).

[82] ------------------------: *Every generating isotone projection cone is latticial and correct.* J. Math. Anal. Appl. 147 Nr. 1 (1990), 53-62.

[83] ------------------------: *Isotone projection cones in euclidean spaces. Preprint.*

[84] ------------------------: Monotonicity of metric projections onto positive cones of ordered euclidean spaces. *Arch. Math. (Basel) 46 (1986), 568-576. Corrigendum.*

[85] *Ishikawa S.:* Fixed points and iteration of nonexpansive mapping in a Banach space. *Proc. Amer. Math. Soc. 59 Nr. 1 (1976), 65-71.*

[86] *Krasnoselskii M.A.:* Positive solutions of operator equations. *Noordhoff Groningen, (1964).*

[87] *Moreau J.:* Décomposition orthogonale d'un espace hilbertien selon deux cônes mutuellement polaires. *C.R. Acad. Sci. Paris 225 (1962), A238-240.*

[88] *Potter A.J.B.:* Applications of Hilbert's projective metric to certain classes of non-homogeneous operators. *Quart. J. Math. Oxford (2), 28 (1977), 93-99.*

[89] *Sadovskii B.N.:* A fixed point principle. *Funct. Analiz. i Prilog 1 (1967), 74-76.*

[90] *Thompson A.C.:* On certain contraction mappings in a partially ordered vector space. *proc. Amer. Math. Soc. 14 Nr. 3 (1963), 438-443.*

D Recent papers

[1] Cottle R.W., Pang J.S. and Stone R.E.: *The linear complementarity problem*, Academic Press (1992).

[2] Chang S.S. and Huang N.J.: *Generalized multivalued implicit complementarity problems in Hilbert spaces*. Math. Japonica 36, Nr. 6 (1991), 1093-1100.

[3] De Wit G.: *Investigations of a certain class of linear complementarity problems*. J. Opt. Theory Appl. 72 Nr. 1 (1992), 65-90.

[4] Gowda M.S.: *A degree formula of Stewart*, Preprint, Dept. Math. Univ. Maryland, Baltimore County (1992).

[5] Gowda M.S.: *Application of degree theory to linear complementarity problem*, Preprint, Dept. Math. Univ. Maryland, Baltimore County (1992).

[6] Isac G.: *Iterative methods for the general order complementarity problem*, Approx. Theory, Spline Funct. Appl. (Ed. S.P. Singh) Kluwer Academic Publ. (1992), 365-380.

[7] Isac G. and Kostreva M.: *Kneser's theorem and the multivalued generalized order complementarity problem*. Appl. Math. Lett. 4 Nr. 6 (1991), 81-85.

[8] Isac G. and Kostreva M.: *The generalized order complementarity problem*, J. Opt. Theory Appl. 71, Nr. 3 (1991), 517-534.

[9] Kojima M., Megido N. and Noma T.: *Homotopy continuation methods for nonlinear complementarity problem*, Math. Oper. Res. 16 Nr. 4 (1991), 754-774.

[10] Morris W.D. Jr.: *On the maximum degree of an LCP map*, Math. Oper. Res. 15, Nr. 3 (1990), 423-429.

[11] Noor M.A., Noor K.I. and Rassias Th. M.: *Some aspects of variational inequalities*, Preprint, Math. Dept. College of Science King Sand Univ., Riyad (1991).

INDEX

Printing: Druckhaus Beltz, Hemsbach
Binding: Buchbinderei Schäffer, Grünstadt